ENGINEERING PRINCIPLES OF PHYSIOLOGIC FUNCTION

Daniel J. Schneck
Virginia Polytechnic Institute and State University

NEW YORK UNIVERSITY PRESS
New York *and* London

Library of Congress Cataloging-in-Publication Data
Schneck, Daniel J., 1941–
    Engineering principles of physiologic function / Daniel Schneck.
        p.   cm.—(Biomedical engineering series : 5)
    Includes bibliographical references and index.
    ISBN 0-8147-7905-0 (cloth)
    1. Biomedical engineering.   2. Biophysics.   3. Physiology.
I. Title.   II. Series: New York University biomedical engineering
series ; 5.
R856.S36   1990
612'.014—dc20   90-40992
                                                            CIP

New York University Press books are printed on acid-free paper,
and their binding materials are chosen for strength and durability.

To Judi, Patti, and Cyndi,

Because that's what it's all about ...

Table of Contents

List of Figures

List of Tables

List of Symbols

a Characteristic Transport Acceleration, $\dfrac{L}{t^2}$, $f(t)$.

a^* Characteristic Mean Transport Acceleration,
$f(t^*) = f_1(\nabla e)$

$\vec{a}$ Linear Acceleration Vector $= a_x \hat{i} + a_y \hat{j}$

$\vec{a}_{CG}$ Linear Acceleration Vector Of Center Of Gravity Of Limb $= \vec{a}_{CG/O} + \vec{a}_O$

$\vec{a}_{CG/O}$ Linear Acceleration Vector Of Center Of Gravity Of A Limb Relative To The Proximal Joint, O, $-\bar{r}(\dot{\Theta})^2 \, \hat{i}_n + \bar{r}(\ddot{\Theta}) \, \hat{i}_t$

a_i Coefficients Of Cosine Terms In The Fourier Series Expansion Of A Complex Periodic Function.

a_h Horizontal Component Of The Linear Acceleration Vector Of The Proximal Joint Of A Limb.

a_m $\dfrac{\varepsilon_3 [Y_m][\rho_S \delta]}{(MW)A}$, $\dfrac{moles}{L^3 - t}$, c.f., equation [3-74]

$\dfrac{a_o}{2}$ Leading Term Coefficient In The Fourier Series Expansion Of A Complex Periodic Function

$\vec{a}_O$ Linear Acceleration Vector Of A Proximal Limb Joint (elbow) $= a_h \hat{i} + a_v \hat{j} = \vec{a}_{O/O'} + \vec{a}_{O'}$

$\vec{a}_{O'}$ Linear Acceleration Vector Of Shoulder Joint

$\vec{a}_{O/O'}$ Linear Acceleration Vector Of Elbow Joint Relative To Shoulder Joint, $-L_u(\dot{\Theta}_1)^2 \, \hat{i}'_n + L_u(\ddot{\Theta}_1) \, \hat{i}'_t$

a_v Vertical Component Of The Linear Acceleration Vector Of The Proximal Joint Of A Limb.

b^2 $\dfrac{\alpha_S A_S}{2}$, having The Same Dimensions as α_S.

$b*^2$ $\qquad$ $\dfrac{\overline{F}Zr^2}{\kappa*MW}\sqrt{2\pi\rho k}$, $\dfrac{L^2}{t}$.

b' $\qquad$ Arbitrary Exponent (see equation [5-64])

b'_0, b'_1, b'_2 $\qquad$ Arbitrary Coefficients (see equation [5-65])

b_i $\qquad$ Coefficients Of Sine Terms In The Fourier Series Expansion Of A Complex Periodic Function.

b_m $\qquad$ $\dfrac{a_m}{K_m}$, dimensions $\dfrac{1}{t}$.

c $\qquad$ Speed Of Light, $\dfrac{L}{t}$.

$c*$ $\qquad$ Any Real Number

c_A $\qquad$ Latent Heat Of Adsorption, $\dfrac{L^2}{t^2}$.

c_L $\qquad$ Latent Heat Of Fusion Or Of Vaporization, $\dfrac{L^2}{t^2}$

c_p $\qquad$ Specific Heat At Constant Pressure, $\dfrac{L^2}{t^2 T}$.

c_v $\qquad$ Specific Heat At Constant Volume, $\dfrac{L^2}{t^2 T}$.

d $\qquad$ Pore Diameter, L.

d_a $\qquad$ Mean Alveolar Diameter Of Lung, L.

d_1 $\qquad$ Pore diameter At One End Of A Pore, L.

d_2 $\qquad$ Pore Diameter At The Other End Of A Pore, L.

e $\qquad$ Energy Per Unit Mass ("Effort"), $\dfrac{L^2}{t^2}$. Also, Napierien Base For Natural Logarithms, 2.718.

e^- $\qquad$ Electron

$e_0(t)$; $e_1(t)$; $e_2(t)$ $\qquad$ Exponential Time Functions (see equations [5-66] through [5-69]).

f $\qquad$ Frequency, $\dfrac{1}{t}$.

f^*	"Flow", $\dfrac{M}{t}$.
$f(t)$	Acceleration, in general (see a above).
$f(t^*)$	Acceleration at time t^* (see a^* above).
$f_1(\nabla e)$	Force Per Unit Mass Propelling The Transport Process (see a^* above).
$f_2\left(\dfrac{1}{e}\right)$	$\dfrac{\tau}{\alpha}$
f_{I_a}	Actual Firing Frequency Of Muscle Spindle Type I_a Afferent Sensory Nerve Fibers, $\dfrac{1}{t}$.
f_B	Firing Frequency Of Baroreceptors, $\dfrac{1}{t}$.
f_H	Helmholtz Free Energy, $\dfrac{L^2}{t^2}$; (f_{H_i} = initial, f_{H_f} = final).
$f_j{}^*$	$j = 0, 1, 2, \ldots$; Coefficients Of A Hamiltonian Expansion Of The Form $f_o{}^*\psi_o{}^* + f_1{}^*\psi_1{}^* + \cdots$
f_n	Natural Frequency Of A System, $\dfrac{1}{t}$.
f_o	Fundamental Frequency Of A Complex Periodic Function, $\dfrac{1}{t}$.
f_t	Firing Frequency Of An α-motoneuron Necessary To Bring A Muscle Fiber To Threshold So It Will Contract (Fire).
f_α	Actual Firing Frequency Of The α-Motoneuron Fibers, $\dfrac{1}{t}$.
f_γ	Actual Firing Frequency Of The γ-Motoneuron Fibers.
g	Gibbs Free Energy $= RT \ell n\,[S]$, $\dfrac{ML^2}{\text{moles} - t^2}$
g_i	Gibbs Free Energy For Species i (Total Energy Per Mole).
g^*	Total Gibbs Free Energy $= gN$, $\dfrac{ML^2}{t^2}$
g^o	Gibbs Free Energy, Per Mole Of S, In A One-Molar Solution Of S, $\dfrac{ML^2}{\text{moles} - t^2}$
Δg^{*o}	"Standard" Gibbs Free Energy

g_o — Gravitational Constant, 32.2 ft/sec^2 = 9.81 m/sec^2, $\dfrac{L}{t^2}$.

$g_{1,2}$ — Initial (1) and Final (2) Gibbs Free Energy

h — Convection Coefficient, $\dfrac{M}{t^3 T}$.

h^* — Human Body Dynamics Constraint Equation, $F^*(x, y, z)$; Also, $h^*{}_j$, $j = 1, 2, 3, \dots n$.

h_E — Enthalpy, $\dfrac{L^2}{t^2}$

h_o — Planck's Constant, 6.625×10^{-27} erg-sec, $\dfrac{ML^2}{t}$.

h_P — Magnetic Pole Strength, $\dfrac{ML^2}{tq}$.

i — Integer Index, 1, 2, 3, … ; Or, $2 - j$ ($j = 0, 1, 2, \dots$); Also, $\sqrt{-1}$

$\hat{i}$ — Unit Vector, With $\hat{j}$ and $\hat{k}$ Forms An Orthogonal Coordinate System.

$\hat{i}_n$ — Unit Vector Normal To A Path (unprimed = with respect to forearm; primed = with respect to upper arm).

$\hat{i}^*{}_n$ — Unit Vector Defining An Arbitrary Orientation In Space Of Any Limb Centerline.

$\hat{i}_r$ — Unit Vector Defining The Spatial Direction Of An Arbitrary Axis Of Rotation.

$\hat{i}_t$ — Unit Vector Tangent To A Path (unprimed = with respect to forearm; Primed = with respect to upper arm).

j — Integer Index, 1, 2, 3, …

j^* — Number Of Kinematic And Kinetic "Unknowns" In Any Physical System.

$\hat{j}$ — Unit Vector, with $\hat{i}$ and $\hat{k}$ Forms An Orthogonal Coordinate System.

k — Electric Coulomb Constant, 9×10^9 Joule-meter/coulomb^2, $\dfrac{ML^3}{t^2 q^2}$.

k^* — Electrical Specific Resistivity, $\dfrac{ML^3}{tq^2}$.

$\hat{k}$ — Unit Vector, With $\hat{i}$ and $\hat{j}$ Forms An Orthogonal Coordinate System.

k_a — Dimensionless Proportionality Constant Between Thermodynamic Pressure And Mean Translational Molecular Kinetic Energy Per Unit Volume.

k_b — Boltzmann's Constant, 1.38×10^{-16} Erg/Deg; $\dfrac{ML^2}{Tt^2}$.

k_h — Dimensionless Henry's Law Coefficient

k_m — Magnetic Coulomb Constant, $\dfrac{q^2}{ML}$.

k_p — Forward Rate Constant For Enzyme-Catalyzed Formation Of Products From Reactants, $[(k_p)_1 =$ Uncatalyzed; $(k_p)_2 =$ Catalyzed]; $\left(\dfrac{L^3}{\text{moles} - t} \right)^2$.

k_r — Radius Of Gyration, L.

k_s — Stefan-Boltzmann Constant, 0.1714×10^{-8} Btu/hr-ft^2-$^o R^4$, $\dfrac{M}{t^3 T^4}$

k_t — Thermal Conductivity, $\dfrac{ML}{Tt^3}$.

k_x — Reverse Rate Constant For Enzyme-Catalyzed Dissociation Of Products Back To Reactants (Dimensions Same As For k_p).

ℓ — Some Characteristic Reference Length Scale, L; Also, Liters.

ℓ^* — Height Of An Individual, L

$\bar{\ell}$ — Distance Between Elbow Joint And Insertion Of Biceps Musculature Into The Lower Arm (Or Forearm).

$\ell^*{}_p$ — Characteristic Perspiration Length Scale, L.

m — Mass of one molecule of species S, M; subscripts $1, 2, 3, \ldots, N^*$ Identify *specific* molecules of S; Also used As An Integer Subscript or Superscript $(1, 2, 3, \ldots)$, and As An Abbreviation For Meter.

$\vec{m}$ — Vector Part Of A Quarternion.

m_a — Mass Of An atomic Particle, M.

m_B — Total Blood Mass, $\rho V^*{}_B$.

m_f — Mass Of Forearm Including The Hand.

$m^*{}_f$ — Mass Of Forearm Excluding The Hand.

m_w Mass of Fist (Hand)

n Number Of Constraints Limiting Musculoskeletal Function; Also, As A Subscript, Indicates A Direction *Normal* To A Path; Also Used To Represent Integers $1, 2, 3, \ldots$, Or The Numbers 0, $\frac{1}{2}$, and 1.

n^* Number Of Pores Per Unit Area Of membrane Surface, $\frac{1}{L^2}$.

$\bar{n}$ Dimensionless Non-Newtonian Index For Blood.

$\hat{n}$ Unit Vector Normal To dA.

p Pressure, $\frac{M}{Lt^2}$; Also, as a Subscript, means Products of a Biochemical Reaction.

p_a Ambient Partial Pressure Of Water Vapor, mm Hg.

p_s Vapor Pressure Of Water At Skin Temperature, mm Hg.

p_t Threshold Pressure Required To Fire A Baroreceptor.

$p[u_{cj}(t)]$ Probability Density Function.

q, q_1, q_2 Electric Charge (Fundamental Dimension).

$\bar{q}$ Quaternion, $\vec{m} + s^*$, (c.f., equation [5-61]).

$\bar{q}^*$ Quaternion (c.f., equation [5-62]).

r Radial Coordinate, L.

$\bar{r}$ Location Of Limb Center Of Gravity (e.g., forearm, or wrist) Relative To Proximal Joint (e.g., O) Of That Limb.

$\vec{r}$ Differential Spatial Displacement, $dx\,\hat{i} + dy\,\hat{j} + dz\,\hat{k}$.

$\bar{r}_f$ Location Of Forerm Center Of Gravity (*not* including separate wrist joint) Relative To Elbow Joint.

r_i Inner Tube Radius, L.

r_o Outer Radius Of Spherically-Shaped Particle, L.

$\bar{r}_w$ — Location Of The Center Of Gravity Of The Fist Relative To The Wrist Joint.

s — Specific Entropy, $\dfrac{L^2}{Tt^2}$.

s^* — Scalar Part Of A Quaternion

t — Time (Fundamental Dimension); Also used as a subscript to indicate "With Respect To Time," or, "With Respect To A Direction *Tangent* To A Particle Path."

t^* — Some Characteristic Value Of t During the Elapsed Period τ.

t' — Dummy Variable For Integration With Respect To Time.

t_p — Arbitrary Sampling Time (Range In t) Over Which A Physiologic Signal Is Measured.

u — Internal Energy Per Unit Mass, $\dfrac{L^2}{t^2}$.

u^* — Generalized Function, $\phi^* + vF^*$.

u_a — Maximum (Initial) Value Of The Exponentially Decreasing Amplitude Of The Oscillations Of u_o Around u_r In An Underdamped Feedback Control System.

u_c — Controlling Signal Of A Feedback Control System.

u_d — Disturbance To A Feedback Control System.

u_e — Steady-State Error In Feedback Control System.

u_o — Controlled Signal Of Feedback Control System; Feedback Signal.

u_p — Primary Feedback Signal Of Feedback Control System.

u_r — Reference Quantity Of Feedback Control System.

u_w — Physiologic "Warning" Signals.

v — Characteristic Transport Speed, $\dfrac{L}{t}$; (Subscripts $r, \Theta =$ In Radial and Azimuthal Directions, Respectively; Subscripts $i, j =$ In all Directions i and j in space, $i, j = 1, 2, 3$, Respectively; Subscripts $x, y, z =$ In Cartesian Directions, x, y, z, Respectively).

$\vec{v}$ Characteristic transport velocity, $\dfrac{L}{t}$; Subscripts as above for v; also used to identify specific particle having characteristic velocity $\vec{v}$.

$\overline{v_x{}^2}$ Mean-Square Value Of v_x; Same As $\overline{v_y{}^2}$, $\overline{v_z{}^2}$, and/or $\dfrac{1}{3}\overline{v^2}$ For Particle Undergoing Purely Random Brownian Motion.

$\overline{v^2}$ Mean-Square Value of $\vec{v}$; $\overline{v_x{}^2} + \overline{v_y{}^2} + \overline{v_z{}^2}$.

v_{Θ} Azimuthal Velocity, $\bar{r}\dot{\Theta}$.

w Specific Weight, $\dfrac{M}{L^2 t^2}$; Also Used As A Subscript To Designate Conditions At A "Wall".

x Cartesian Coordinate; Also Used As A Subscript To Designate The Substrates Of A Biochemical Reaction.

$\dot{x}$ $\dot{x}(x, t) = \dfrac{dx}{dt}$.

y Cartesian Coordinate; Also Used As A Subscript To Designate The Undetermined Exponents, $y_i (i = 1, 2, 3, \dots)$ In The Buckingham Pi-Formulation Of Dimensionless Parameters.

z Cartesian Coordinate; Also Designates Height Above the Center Of The Earth, L.

A Area, L^2

A_a Total area of contact (down to the level of the alveoli) between the air inspired and the surface lining the respiratory tract.

A_i Dimensionless Scaling Coefficient In Equation [2-2] and [2-4]; $(i = H_2O$ for Osmosis, c.f., Equation [3-34]; $i = S$ for Ordinary Diffusion Of Species S, c.f., Equation [3-20]).

A_j Dimensionless Scaling Coefficient In Equation [2-3].

A^*_j Dimensionless scaling coefficient for $v^2 = A^*_j \phi_j$; $(j = H_2O$ for water flux due to a hydrostatic pressure gradient, c.f., equation [3-34a]).

A_N Total Body Surface Area, L^2.

A_o Dimensionless constant in Van't Hoff and Arrhenius equation [4-26].

A^*_o $A_o + \ell n\, k_x$.

A_P — Dimensionless Fractional Free-Cross-Section Of Porous Membrane.

A_w — Perspiratory Wetted Skin Surface Area, L^2.

B — Voltage, $\dfrac{ML^2}{qt^2}$; (subscript p = products; Subscript x = substrates); Also Used As A Subscript To Designate The Body, or Blood, As Defined In The Text.

B^* — Generalized Redox Potential.

B_o — Generalized Redox Potential At A Concentration Of One Mole Per Liter And An Arbitrary pH.

$B_o{}'$ — Generalized Redox Potential At A Concentration Of One Mole Per Liter And A pH Equal To 7 (i.e., Neutral).

ΔB — Transmembrane Voltage, $\dfrac{ML^2}{qt^2}$.

ΔB^* — Redox Potential, $B_p - B_x$.

ΔB_m — Membrane Equilibrium Resting Potential, $\dfrac{ML^2}{qt^2}$.

$\Delta B'_o$ — Standard Reduction Potential At A pH Of 7.00.

ΔB_r — $B^* - B_o{}'$.

ΔB_s — $|\Delta B_t - \Delta B_m|$; Membrane Depolarization To Threshold.

ΔB_t — Membrane Threshold Potential, $\dfrac{ML^2}{qt^2}$.

C — Capacitance, $\dfrac{q^2 t^2}{ML^2}$.

C^*_m — Membrane Capacitance.

C_o — Feedback Control Error Signal Response Constant

C_T — Total Thermal Capacity Of Blood, $\dfrac{ML^2}{Tt^2}$.

D — A Transfer Group, Such As Carboxyl ($-COOH$), Amine ($-NH_2$), And Phosphate ($-PO_3H$)

D_i — Any Dimensional Quantity Affecting A Physical Process (see Section On Buckingham ¶-Theorem); Dimensions Depend On Specific Quantity; $i = 1, 2, 3, \ldots$

D_S Binary Effective Diffusion Coefficient, or, Fick's Mass Diffusivity, $A_s \alpha_s \dfrac{MW}{2}$; $\dfrac{M}{M^*} \times \dfrac{L^2}{t}$.

D^*_S Facilitated Transport Binary Effective Diffusion Coefficient, Or, Facilitated Diffusivity (Same Dimensions As D_S).

D_{sm} Effective Diffusion Coefficient For A Membrane, $P_S \dfrac{A}{\delta}$, $\dfrac{L^2}{t}$.

E Energy, $\dfrac{ML^2}{t^2}$; Also, Abbreviation For Erythropoietin; Also, $E, E_1,$ and E_2 Represent Enzymes In A Biochemical Reaction.

E^* Potential Energy (Same Dimensions As E).

$\overline{E}^*$ $E^* - vh^*$

E_a Energy Of Activation, $\dfrac{ML^2}{moles - t^2}$; (Added subscript 1 = For Uncatalyzed biochemical reaction; Added subscript 2 = For Catalyzed Biochemical Reaction).

E_d Dynamic Elastic Modulus, $\dfrac{M}{Lt^2}$.

E_e Electric Field Intensity, $\dfrac{ML}{qt^2}$

E_s Conventional Static Young's Modulus Of Elasticity, $\dfrac{M}{Lt^2}$.

F Transport Driving Force, $\dfrac{ML}{t^2}$; Subscripts x, y, z indicate respective $x - y - z$ components along the three mutually-perpendicular directions Of A Cartesian Coordinate System.

$F(t)$ Quantity of material infused into the extracellular fluid space.

$F(0)$ Value of $F(t)$ at $t = 0$.

$\overline{F}$ Faraday Constant, $\dfrac{q}{mole}$ (Monovalent Ions).

$\vec{F}$ Force Vector, in General; Biceps Force, In Particular $= - F_x \hat{i} + F_y \hat{j}$, Generating A Moment Around The Elbow Joint.

$F^*_j(x, y, z)$ Arbitrary Constraint Equations $(j = 1, 2, 3, \dots, n)$ In Human Body Dynamics; When $j = 1$ *only*, $F^*_1 = F^*(x, y, z) = h^*$.

$\Sigma \vec{F}_{external}$ Resultant Of All Of the External Forces Acting On A Human Body Segment $= \Sigma F_x \hat{i} + \Sigma F_y \hat{j}$.

G Linear Momentum, $\dfrac{ML}{t}$.

G_1, G_2, G_3 — Sensitivity Coefficients For Frequency-Response Characteristics Of Carotid Artery Baroreceptor Reflex Mechanisms $(G_1, G_2 = \frac{Lt^2}{M}, G_3 = \frac{Lt}{M})$.

H — Magnetic Field Intensity, $\frac{q}{Lt}$.

H^* — Hamiltonian Function, $\psi_o^* + \psi_1^* + \cdots$

I — Electric Current, $\frac{q}{t}$.

I^* — Stimulus Current Required To Generate An Action Potential.

I_{CG} — Moment Of Inertia Of A Limb With Respect To Its Corresponding Center Of Gravity, ML^2.

I_O — Moment of Inertia Of A Limb With Respect To Its Proximal Joint (e.g., Forearm With Respect To Elbow Joint O).

I_r^* — Rheobase Current; Minimum Stimulus Intensity Necessary To Elicit A Response From An Excitable Membrane.

J — Current Density, $\frac{q}{L^2 t}$.

J^* — Number of joints involved in a prescribed musculoskeletal movement.

K — Dimensionless Distribution Coefficient For The Partitioning Of Solute, S, Between Aqueous Phase And The Membrane Material.

K_1 — Overall Transfer Function Of Controlled Element Of Feedback Control System.

K_2 — Overall Transfer Function Of Controlling Element Of Feedback Control System; Also, Overall Transfer Function Of The Comparator And Controller Element Of A Feedback Control System.

K_1' — Transfer Function Defining The Controlled Element Response To A Control Signal In A Feedback Control System.

K_2' — Proportional Control Transfer Function Of The Comparator Element In a Feedback Control Loop.

K_1'' — Transfer Function Defining The Controlled Element Response To A Disturbing Signal In A Feedback Control System.

K_2'' — Differential Control Transfer Function Of The Comparator Element In A Feedback Control Loop.

K_2''' — Integral Control Transfer Function Of The Comparator Element In A Feedback Control Loop.

K_e — Equilibrium Constant For Enzyme-Catalyzed Biochemical Reaction (Dimensionless).

K_m — Active Transport Dissociation Constant, "Affinity Factor," Or Michaelis Constant

K^*_m — Facilitated Diffusion Constant

K_α — Muscle Spindle Transfer Function For Active Contraction

K_γ — Muscle Spindle Transfer Function For Active γ-Innervation (Biasing)

$K_{\xi*}$ — Muscle Spindle Transfer Function For Passive Stretch.

L — Length (Fundamental Dimension)

L_f — Length Of Forearm From Elbow Joint To Wrist Joint

L_w — Length Of "Fist" From Wrist Joint To Second Set Of Knuckles In Clenched Hand.

L_u — Length Of Upper Arm From Shoulder Joint To Elbow Joint.

M — Mass, expressed in Kilograms (A Fundamental Dimension), $M^* \times (MW)$.

M^* — Mass, expressed in Moles, $\dfrac{M}{MW}$.

$\overline{M}$ — Mean Value Of A Function

$\vec{M}$ — Moment (Subscript Designates With Respect To Which Point Or Axis, e.g., O = Elbow Joint: CG = Center Of Gravity Of Limb; $j, i = 1, 2, 3, \ldots, J^*$ = Joint i or j); If arrow is omitted From the symbol, It Indicates Only the *Magnitude* Of The Corresponding Moment.

MW — Molecular Weight.

N — Number Of Moles Of A Species; Also, An Abbreviation For Newtons.

N^* — Total Number Of Molecules Contained In A Sample Control Volume Of A Continuum.

N — Respiration Rate, $\dfrac{1}{t}$.

N_i — Number Of Moles Of Species i.

$\vec{O}$ — Elbow Joint Force, $O_x \hat{i} + O_y \hat{j}$

$\vec{O}'$	Shoulder Joint Force, $O_x'\hat{i} + O_y'\hat{j}$
P	Pressure, expressed in Atmospheres ($P = 1, 2, 3, ...$)
P_f	Facilitated Transport Permeability Coefficient For Membrane, $\dfrac{L}{t}$
P_i	Inorganic Phosphate.
P_m	Active Transport Permeability Coefficient For Membrane m, $\dfrac{L}{t}$
P_S	Permeability Coefficient Of Membrane For Species S, $\dfrac{L}{t}$ ($S = H_2O$ For Water).
$P_m{}^*$	Maximum Value Of P_m.
$P_f{}^*$	Maximum Value Of P_f.
$P_S{}^*$	Electrical Permeability ($S = H_2O$ For Water).
$P[u_c(t)]$	Probability Distribution Function.
Q	Heat Energy Per Unit Mass, $\dfrac{L^2}{t^2}$.
Q^*	*Total* Heat Energy, $\dfrac{ML^2}{t^2}$.
R	Universal gas constant, $\dfrac{ML^2}{M^*Tt^2}$; also, kidney erythropoietic factor.
$\mathscr{R}$	Magnitude Of The Complex Potential $\Omega = \phi + i\psi$, $\dfrac{L^2}{t^2}$.
R^*	Electrical Resistance, $\dfrac{ML^2}{tq^2}$.
$R_m{}^*$	Membrane Resistance.
$R_T{}^*$	Thermal Resistance, $\dfrac{Tt^3}{ML^2}$.
$R_t(\tau)$	Autocorrelation Coefficient.
$R_t{}^*(\tau)$	Normalized Autocorrelation Coefficient Whose Value Is Zero For $\tau = 0$.
S	Any quantity of material to be moved (mass, energy, momentum).

$[S_i]_j$ Concentration of species i (expressed in moles per liter) in spatial region j ($i = 1, 2, 3, \ldots, n$; j = extracellular region 1, intracellular region 2, or, 1, 2, 3, $\ldots$, total number of compartments; see also Table 2-1).

$[\bar{S}]$ Mean Concentration Of S, $\frac{1}{2}\left([S]_{\text{inside}} + [S]_{\text{outside}}\right)$

$\Delta[S]$ $\left|[S]_{\text{inside}} - [S]_{\text{outside}}\right|$.

$S_{nm}(t)$ Total Amount Of Material $S_n(n = 1, 2, 3, \ldots)$ Located In Compartment m ($m = 1$ or j, $i, j = 1, 2, 3, \ldots$) At Time t.

S_{nm}^o $S_{nm}(0)$, $m = i$ or j as above.

$S^*{}_{ni}(t)$ External Input To Compartment i At Time t (Rate Of Infusion Of S_n to Compartment i At Time t, $\frac{M}{t}$).

$S^*{}_{ni}{}^o$ Flow Of S_{ni} Out To The Environment.

$S_{nji}(t)$ Total Amount, Per Unit Time, Of Material S_n Flowing From Compartment j to Compartment i At Time t.

T Temperature, T (A Fundamental Dimension)

T^* Kinetic Energy, $\dfrac{ML^2}{t^2}$.

T_c Body Core Temperature.

T_1 Temperature of Heat Sink.

T_2 Temperature Of Heat Source.

ΔT $T_2 - T_1$.

T_a Ambient Temperature Of The Environment, °C.

T_B Absolute Skin Temperature, °K.

T_E Absolute Ambient Temperature, °K.

T_s Skin Temperature, °C.

V Volume, L^3.

$\dot{V}$ Volumetric Diffusion Rate $\left(\dfrac{\text{liters}}{\text{hr-m}^2}, \dfrac{L}{t}\right)$.

V_a Tidal Volume (L^3).

V_B Volume Of Displaced Fluid (Buoyancy), L^3.

V^*_B Total Blood Volume, L^3.

$\overline{V}_w$ Partial Molar Volume Of The Solvent (Water), $\dfrac{L^3}{\text{mole}}$.

$\dot{V}_P$ Volumetric Perspiration Rate, $\dfrac{L}{t}$.

$\overline{V}_S$ Partial Molar Volume Of The Solute, S, $\dfrac{L^3}{\text{mole}}$.

W Work, $\dfrac{ML^2}{t^2}$; Also, Gibbs Free Energy.

$\dot{W}$ Power, Rate Of Energy Expenditure, Metabolic Rate, $\dfrac{ML^2}{t^3}$.

W_E Nondimensional Water Content Of Expired Air (Kg H_2O Per Kg Dry Air).

W_I Nondimensional Water Content Of Inspired Air (See Above).

X Number Of Governing Laws Of Nature That Must Be Satisfied By A Physical System.

X_i Amplitude Of i'th Component Of The Frequency Spectrum Of A Complex Periodic Function.

Y, Y^* Represent Carrier Molecules

$[Y]$ Actual concentration of carrier molecules in cell membrane, $\dfrac{\text{moles}}{L^3}$

$[Y_m]$ Total concentration of carrier molecules in cell membrane, $\dfrac{\text{moles}}{L^3}$

Z Valence

α_n Kinematic Transport Coefficient, $\dfrac{L^2}{t}$, (Generalized Momentum); $n = d$, Insensible Perspiration Coefficient; $n = e$, "Effective Diffusion Coefficient" For Transudation Through "Tortuous" Pores; $n = h$, Convection Thermal Diffusivity; $n = k$, Conduction Thermal diffusivity; $n = \ell$, Insensible Respiratory Evaporation Coefficient; $n = L$, Sensible Respiratory Convection Coefficient; $n = p$, Sensible Perspiration Coefficient; $n = r$, Radiation Thermal Diffusivity; $n = S$,

Ordinary Diffusion Coefficient; $n = \mu$, Conduction Momentum Diffusivity; $n = v$, Convection Momentum Diffusivity.

α^* Nondimensional "Activity Factor."

$\vec{\alpha}$ Angular Acceleration Vector, $\dfrac{1}{t^2}$.

∇ Gradient Operator.

β Staverman Factor, Dimensionless Reflection Coefficient.

β^*_i "Sensitivity Coefficients" Involved In Musculoskeletal Mechanics.

∞ Infinity Sign

ψ Scalar Intensive Energy-Dissipation Function (per unit mass) For Nonconservative Force Systems; $\dfrac{L^2}{t^2}$.

ψ^*_j Individual Elements Of An Extensive Hamiltonian Function, see Equation [5-48], $j = 0, 1, 2, \ldots$

Ψ_w The Volume Fraction Of Water In The Membrane (Dimensionless).

ϕ General Scalar Intensive (per unit mass) Potential Energy Function For Conservative Force Systems, $\dfrac{L^2}{t^2}$.

ϕ_j Specific Intensive Potential Energy Function Per Unit Mass For Species j.

$\phi^*(x, y, z)$ or $\phi^*(M_1, M_2, M_3, \ldots, M_n, t)$ Scalar *Extensive* Energy Functions, $\dfrac{ML^2}{t^2}$.

Φ_n Flux Of Quantity n (See Table 2.1), $\dfrac{M}{t^3}$, $n = d$, Diffusional Heat Flux By Evaporation; $n = h$, Heat Convection; $n = \ell$, Insensible Respiratory Heat Flux By Evaporation; $n = L$, Insensible Respiratory Heat Flux By Convection; $n = r_t$ Heat Radiation; $n = s$, Perspiration Sensible Heat Flux By Evaporation.

ε Nondimensional Strain In A Material

$\varepsilon_1, \varepsilon_4$ Association Rate Constants, $\dfrac{(L^3/\text{mole})}{t}$

$\varepsilon_1^*, \varepsilon_4^*$ Association Rate Constants $\dfrac{(L^3/\text{mole})^2}{t}$

$\varepsilon_2, \varepsilon_3, \varepsilon_5, \varepsilon_6,$ $\varepsilon^*_2, \varepsilon^*_3$ Dissociation Rate Constants, $\dfrac{1}{t}$.

$\varepsilon_{nji}(t)$ Fractional Rate Constant For The Flow Of S_n From Compartment j To Compartment i, $\dfrac{S_{nji}(t)}{S_{nj}(t)}$, $\dfrac{1}{t}$; Flow Is In The Reverse Direction If Indices i and j Are Reversed.

$<$ Indicates That Quantity To The Left Of The Sign Is Less Than The Quantity To The Right Of The Sign.

λ Molecular Mean Free Path, L.

η Viscous Modulus, $\dfrac{M}{Lt^2}$.

ι Number Of Electrons Transferred Per Mole In An Oxidation-Reduction Reaction.

$>$ Indicates That The Quantity To The Left Of The Sign Is Greater Than The Quantity To The Right Of The Sign.

κ Radiation Emissivity, Dimensionless.

κ^* Dimensionless Dielectric Constant.

κ_E Emmissivity Of The Environment.

κ_B Emmissivity Of Human Skin.

ω Angular Velocity (Radians Per Second), $\dfrac{1}{t}$.

Ω Complex Potential, $\phi + i\psi$.

$\vec{\Omega}$ Vector Representation Of The Complex Potential For Nonconservative Force Systems (see Above), $\dfrac{L^2}{t^2}$; $\phi\hat{i} + \psi\hat{j}$.

Ω_{ij} Second-Order Tensor $= \nabla\vec{\Omega}$.

μ Dynamic Viscosity, $\dfrac{M}{Lt}$.

μ_e Magnetic Permeability, $\dfrac{ML}{q^2}$.

μ_c Electrochemical Potential, $\dfrac{ML^2}{M^*t^2}$.

μ_f Dynamic Viscosity Of Suspending Medium Or Solvent.

∂ Partial Differentiation Symbol.

ν_j LaGrange Multipliers $(j = 1, 2, 3, \ldots, n)$.

ρ Mass Density, $\dfrac{M}{L^3}$.

ρ_a Air Density At Ambient Temperature And Pressure, $\dfrac{M}{L^3}$.

ρ_f Mass Density Of Suspending Medium, solvent f.

ρ_S Mass Density Of Species S.

$\dot{\gamma}$ Spatial Gradients In Velocity, $\dfrac{1}{t}$.

θ Argument Of The Complex Potential, $\Omega = \phi + i\psi$; Arc Tan $\dfrac{\psi}{\phi}$.

θ_i Phase Angle Of The i'th Component Of The Frequency Spectrum Of A Complex Periodic Function.

θ_o Feedback Control System Phase Angle.

$\Theta(t)$ Sagittal Plane Angular Displacement Of The Forearm During A Discrete Flexion.

$\dot{\Theta}(t)$ Sagittal Plane Angular Velocity Of The Forearm During A Discrete Flexion.

$\ddot{\Theta}(t)$ Sagittal Plane Angular Acceleration Of The Forearm During A Discrete Flexion.

Θ_o Final Angular Displacement Of The Forearm At $t = \tau$.

$\dot{\Theta}*$ Maximum Value Of $\dot{\Theta}(t)$, Same As $\dot{\Theta}_{max}$.

Θ_1 Angular Displacement Of The Upper Arm During A Saggital Movement.

$\dot{\Theta}_1$ Angular Velocity Of The Upper Arm During A Saggital Movement.

$\ddot{\Theta}_1$ Angular Acceleration Of The Upper Arm During A Saggital Movement.

Θ_2 Angle Of Forearm Relative To Position of Upper Arm In The Saggital Plane, $\Theta - \Theta_1$.

$\Theta_i, \dot{\Theta}_i$ Generalized Angular Coordinates (Displacement and Velocity, Respectively) Which Specify The Physical Orientation Of A System At Any Given Instant In Time.

σ	Shear Stress, $\dfrac{M}{Lt^2}$.
σ_e	Electrical Conductivity, $\dfrac{q^2 t}{ML^3}$.
Σ	Summation Symbol.
τ	Characteristic Transport Time Scale, t; Also used To Represent An Arbitrary Time Delay, or The Period Of A Discrete Process (Such As A Forearm Flexion, c.f., Chapter 5).
τ^*	Time Period Necessary For I^* To Change ΔB_m To ΔB_t, i.e., For Stimulus I^* To Generate An Action Potential.
τ_b	Build-up, or rise-time of an underdamped feedback control system.
τ^*_c	Chronaxy Time, $0.693\ \tau^*_m$.
τ_d	Decay Time-Response Constant Of A Feedback Control System.
τ^*_d	Decay-Time Of An Underdamped Feedback Control System.
τ_{ii}	Normal Components Of The Generalized Material Stress Tensor.
τ_{ij}	Stress Tensor Components, $\dfrac{M}{Lt^2}$; $i, j = 1, 2, 3$, Respectively.
τ^*_m	Characteristic Physiologic Membrane Time-Constant, t; $R^*_m C^*_m$.
τ_n	Natural Period Of A Feedback Control System.
τ_p	Time Period Of A Complex Periodic Function.
τ_s	Frictional Coefficient Which Quantifies The Interaction Between Solute And Boundary Material, $\dfrac{M}{L^2 t\text{-moles}}$.
τ_t	"Period" Of The Damped Response Characteristics Of An Underdamped Feedback Control System.
τ_{sw}	Frictional coefficient Which Quantifies The Mutual Drag Of Solvent And Solute On Each Other, $\dfrac{M}{L^2 t\text{-moles}}$.
τ_w	Frictional Coefficient Which Quantifies The Drag Between Solvent And Bounding Surface, $\dfrac{M}{L^2 t\text{-moles}} = \dfrac{\text{Force Per Unit Volume}}{\text{Moles Per Unit Time}}$.
$\rightarrow$	Arrow Over Any Quantity Indicates That The Quantity Is A Vector.
ξ	Dimensionless Tortuosity Factor For A Pore.

$\xi*$ A Passive Stretch-Disturbance Applied To A Muscle Spindle.

Ξ Bulk Modulus Of A Liquid, $\dfrac{M}{Lt^2}$; $V\dfrac{dp}{dV}$.

$\propto$ Symbol Meaning, "Is Proportional To".

δ Membrane Thickness, L; Also Used To Signify A Small Differential Change In Any Quantity That Follows This Symbol.

$\delta(t)$ Dirac Delta Function

δ_ℓ Logarithmic Decrement Of Exponentially Decaying Oscillations.

Δ Symbol Indicating A Finite Change In Any Quantity Which Follows It.

$\equiv$ Symbol Indicating, "Identically Equal To," Implying A Definition Or Exact Identity.

v Specific Volume, $\dfrac{L^3}{M}$; $\dfrac{1}{\rho}$.

ζ Surface Tension, $\dfrac{M}{t^2}$.

$\zeta*$ Nondimensionalized Damping Ratio Of Underdamped, Second-Order Feedback Control System.

$\simeq$ Symbol Meaning, "Is Approximately Equal To".

π Pi = 3.14159; The Ratio Of The Circumference Of A Circle, To Its Respective Diameter.

Π Osmotic Pressure.

$\P_n$ Dimensionless Buckingham Pi-Parameter: $n = 1$, Thermal Time Constant (Thermal Capacitance, C_T, Times Thermal Resistance, h/k_f^2) Divided By Kinematic Time Constant, $\dfrac{1}{\omega}$; $n = 2$, Cube Of The Nusselt Number; $n = 3$, Specific Heat Ratio, c_p/c_v; $n = 4$, Prandtl Number; $n = 5$, Reynolds Number; $n = 6$, Peclet Number, $n = 7$, Stanton Number; $n = 8$, pH (Alkalinity Or Acidity Of A Solution); $n = 9$, Hematocrit; $n = 10$, Ambient Relative Humidity Ratio; $n = 11$, Radiation Emmissivity; $n = 12$, Dielectric Constant; $n = 13$, Non-Newtonian Index, $\bar{n}$; $n = 14$, Solution Activity Factors; $n = 15$, Stefan Number; $n = 16$, Radiation Number, $n = 17$, Boltzmann Number, $n = 18$, α_v/α_h.

$\underset{=}{m}$ Symbol meaning, "Is A Measure Of," or, "Is Measured By The Dimensions That Follow The Symbol."

Foreword

The *New York University Biomedical Engineering* series is a series of books
and monographs which will attempt to present a number of the more
important contributions which engineering has recently made in the fields
of medicine and biology. For many years the contributions of the physical sciences to medicine were primarily restricted to the field of radiology.
However, with the explosive growth of engineering theory and technology
in the last 40 years, the applications of engineering to medicine have become important in fields as diverse as cardiology, cardiac surgery,
neurology, neurosurgery, orthopedics, ophthalmology, urology, respiratory medicine, and indeed in general medical diagnosis.

In these applications it has been desirable to draw upon many classical engineering disciplines, such as electrical engineering, electronics,
computer engineering, computer science, mechanical engineering, mechanics, hydraulics, material science, and chemical engineering, as well as
such recent disciplines as systems theory and information theory.

There has been a widespread impression that work in such interdisciplinary fields as biomedical engineering is impeded by the difficulties in
communication among the physicians, life scientists, physical scientists,
and engineers. This series of books is evidence that such a blockage in
communications no longer exists and that the scientists involved in exploring this interdisciplinary field have learned each other's language.

The successes of engineering methodologies in medicine have been
substantial in recent years. In cardiology we see patient monitoring,
cardiac output measurement, noninvasive assessment, implantable
cardiac pacemakers, intra-aortic balloon pumps, and artificial hearts. In
imaging we see computer-aided tomography, ultrasonic analysis, digital
fluorography, magnetic resonance imaging, and positron emission

tomography. In biomaterials and biomechanical engineering we see artificial joints, artificial limbs, artificial skin, gait analysis, connective tissue studies, and myoelectrically controlled prostheses. Finally, in the neurological sciences we see electroencephalography, electromyography, electrooculography, evoked potentials, radionuclide brain imaging, vision analysis, and the study of visual fields and thresholds.

It is these successes and many others that are described in the books in this series, and in each case a careful engineering analysis is presented with appropriate mathematics to elucidate the various physiological and anatomical mechanisms discussed.

Walter Welkowitz
Rutgers University

Preface and Acknowledgments

A colleague of mine once said, "We do not know anything about the human body, and I can prove it: If we knew anything about the human body, physiology textbooks wouldn't be so thick!" His point, of course, was that there must be some finite set of fundamental principles that underlie all physiologic function; and if we understood these, then such function could be summarized quite concisely in a relatively brief text. It is with this philosophy in mind that we developed a course at VPI&SU entitled, "Engineering Analysis Of Physiologic Systems;" and this text-book is the culmination of 15-years' worth of accumulated class notes that were prepared for this course.

Obviously, this is *not* a thin book, so I do not profess to have lived up to my colleague's contention -- nor to know everything there is to know about physiologic function. However, the objective of the material presented is to move in the direction of examining not so much the *details* associated with the structure and function of *specific* organ systems -- as is the case in "traditional" courses that deal with Anatomy and Physiology -- but, rather to identify physical features and engineering principles that govern the behavior of *all* organ systems uniformly, i.e., that are the common denominators of *all* physiologic function. Toward this end, the six chapters that follow approach the study of physiology from the point of view that it involves basically: the *Transport* -- of mass, energy, momentum, and information; the *Utilization* of same by the various systems and subsystems of the organism; and the careful *Control* of all physiologic mechanisms concerned with the transport and utilization of mass, energy, momentum, and information.

Chapter 1 takes you on a whirl-wind trip through the entire human body, introducing the reader to some basic anatomical vocabulary and physiologic concepts. Physiologic function is defined within the frame-work of considering the living organism to be a fine-tuned electrochemical engine that functions in space and time according to basic principles of conservation of mass, energy, and momentum. The engine interacts with its environment; its output is essentially mechanical; and it has the unique ability to reproduce itself. The various Tables, and other "words" in Chapter 1 constitute a vocabulary list, if you will, for the material that follows in subsequent chapters. It is recommended that the reader have at his or her disposal a handy medical dictionary to assist him or her in becoming familiar with this anatomical terminology.

Chapter 2 addresses basic principles of mass, energy, and momentum *Transport* in physiologic systems -- more or less in general; and Chapter 3 gets more specific in examining the quantitative aspects of passive, facilitated and active mass transport mechanisms in such systems. Chapter 4, then, introduces the reader to some basic principles of bioenergetics, metabolism and thermoregulation in physiologic systems, which is followed in Chapter 5 by a discussion of basic principles of mass, energy, and momentum *Utilization* in physiologic systems.

Although all five chapters to this point include various aspects of *Control* as they relate to the physiologic processes being examined, Chapter 6 summarizes the basic principles involved by addressing information transport, feedback control, and the concept of homeostasis in physiologic systems. Here, the reader is introduced to general principles of feedback control theory, and to how these are manifest in the day-to-day operation of the living organism.

The material in this book is presented at a level that should be comfortable for a fourth-year engineering student or a first-year graduate student. In a way, it defines the unique body of knowledge that distinguishes *Biomedical* engineering from the other more "traditional" engineering disciplines. That is to say, one of the aspects of a profession that gives it a sense of identity is the existence of a body of knowledge that is *unique* to that profession. And biomedical engineering certainly fits this criterion. Consider, for example, the indeterminateness that arises when one tries to analyze human body function in the presence of "will" as an undefined variable. In no other field of engineering is one confronted with the idea that something moves because it "wants to", or that behavior is motivated by pain, fatigue or discomfort as well as the natural laws of mechanics. Obviously, this adds a new dimension to what engineers traditionally think of when they sit down to analyze a system; and Chapter 5 illustrates some of the ways in which this variable, "will", may be taken into account from an analytical point of view.

As illustrated in Chapter 1, if one agrees with nothing else, one must certainly concede that the biomedical engineer is the only engineer specifically trained to speak and understand the language of medicine as well as his or her own. And as illustrated in Chapter 6, the sophistication of the concept of homeostasis, "floating" set points, adaptation, and the nonlinearities inherent in physiologic feedback control mechanisms, puts these in a realm well beyond traditional engineering dogma. Indeed, the biomedical engineering graduate is among the few individuals who do not leave school believing that all fluids are Newtonian, all solids are Hookean Elastic and all systems are at least quasi-linear!

And the list goes on, to include the concept of design in a self-healing environment. The knowledge required to include such constraints as "bio-compatible," "host rejection," "bio-degradable," or even "hypoallergenic" is surely new and distinct from that which is traditionally covered in engineering design coursework. This additional degree of freedom (or lack thereof, if you will) is certainly unconventional in terms of traditional design criteria, and it establishes an entirely different point of view towards the design process. So, too, do considerations related to physiological monitoring and the development of instrumentation for diagnostic and therapeutic purposes. The biomedical engineer, unlike other engineers, must be sensitive to concepts related to noninvasive testing, and those that involve powering implantable devices within human tolerances. Furthermore, (s)he must learn that design in and for living systems must be within strict safety standards, regulations, and reliability parameters that far exceed the constraints imposed on other types of designs. So, think of this book as an introduction to the body of knowledge that distinguishes biomedical engineering from other fields of study.

For those readers who are not engineers, anyone working in the life-science, or related fields, should also find this book useful as a reference text, since it contains a great deal of meaningful anatomical and physiological information that can be assimilated without necessarily having a highly technical background.

There is no way that a work like this could come to fruition without the help and encouragement of many individuals. Writing a book is not an endeavor -- it is an obsession! The writer becomes "Possessed", and loved ones who must live with this individual need to be endowed with limitless patience, love, understanding, and tolerance. So, at the very top of the list goes a great big hug, kisses, and sincere "Thank You" to my dear wife, Judi, and to the best pair of 18-year-old twin daughters -- Cyndi and Patti -- that a father could ever hope to have. You guys are great, and I couldn't have done it without you! Thanks, also, to Dr. Walter Welkowitz, editor of this series, for giving me the opportunity to formalize the class notes developed for the afore-mentioned course into this textbook; and to the folks at the New York University Press (in particular, consulting editor Kenneth I. Werner, editor Jason Renker, and director Colin Jones) for backing the series. Ms. Sandy Jordan did a wonderful job in preparing all of the text Figures (she *loved* putting in all of the dots in Figures 2-1, 3-1, and 5-2, ...); Ms. Connie Callison gets superlatives for putting the manuscript in type-set format; and Virginia Tech stood behind me all the way. Finally, thanks to all of the great minds whose brains I was able to pick in formulating the material for this book. Many of them are listed in the bibliography at the end of the text, and others I have had the opportunity to "shoot the breeze with" as we sat around philosophizing on how the human body works. Indeed, one will find almost as much *philosophy* of physiologic function in this book, as *engineering principles*, per se, of physiologic function. I hope this will serve to stimulate some meaningful dialogue that will ultimately allow us to improve the health, comfort, and understanding of humankind.

Blacksburg, Virginia
Daniel J. Schneck
June 1, 1990.

Chapter 1

Basic Anatomical Vocabulary and Physiologic Concepts

Introduction to the Physiologic System

One of the main characters in a television series of the late 1970s and early 1980s that depicted life in the 21st century was a small, lovable robot named "Tweeky". Tweeky existed for the purpose of transporting around a small, very sophisticated, extremely intelligent, high-speed supercomputer. This computer not only spoke clearly (like a certain automobile in a later series called, "Knight Rider"), but it also had an almost infinite memory capacity, total recall ability, and incredible deductive powers that allowed it to reason its way logically to unmistakably correct and rational conclusions. The small robot could often be seen scampering about with the portable computer hung around its neck -- the latter doing all of the thinking and the former dutifully obeying.

There are those who would argue in an analogous sense that the only reason our own bodies exist is to provide mobility to the brain and to keep it alive and functioning. While we carry our brain inside of us, rather than around our neck, the principle is essentially the same, i.e., it does the thinking, and the rest of us obeys. We are mobile creatures so that we can roam about finding food to sustain the brain. We are also expected to transport it from place to place and to protect it from harm or injury. And all the rest of our physiologic function is merely designed to keep the brain working, much like all of the peripheral devices that are required simply to interact with the main central processing units (CPUs) of a computer.

This is a very powerful theory of life, which is at the root of such concepts as: powers of positive (or constraints of negative) thinking (see Chapter 6), transcendental meditation (ibid.), biofeedback (see Chapter 5), psychosomatic ailments, the relaxation response (Benson, 1975), mind control over matter, brainwashing, sensori-motor bases of physiologic response mechanisms (ibid.), behavioral therapy (teaching the body to heal itself), and other approaches to the treatment of physiologic conditions which are based on the power of the brain (mind) to control all that goes on (good or bad) within the human system.

But the above is certainly not the only theory of life. Indeed, others would argue the exact opposite. That is, they would assert that our very body (or perhaps its "soul") is the center of life, not the brain. Although they might concede its ability to do all of the above, they would point out that these abilities serve merely to *control* and *coordinate* physiologic processes -- not for the brain's *own* sake, but to keep the *entire* organism alive for some higher (perhaps not yet understood) purpose. Theologians, of course, propose that this purpose centers around the will of God and his ultimate plan for humankind.

Atheists and/or Agnostics are not so sure that such is the case. They might propose that the purpose of life is *life*, per se, with the ultimate cause and essential nature of things being unknown and/or unknowable. Leaning more towards Darwinian doctrine, these individuals profess survival of the fittest as being at the root of life, itself, and they view the theory's premise of ultimate perfection as being an end in and of itself. If that be the case, then we exist solely for the purpose of reproducing, and, hopefully, physiologically improving our structure and function within the framework of some optimization scheme. The purpose of life is then simply to perpetuate the species. Man lives to make more men (and women), and reproductive processes become the central focus of our existence.

The point of all of this is that we, as a civilization, have been pre-occupied for centuries with attempts to put life in perspective -- to define it (i.e., to give it some meaning), to justify our existence, and to understand where we fit into some overall scheme of things. In short, without even realizing it, we have attempted to take a *systems* approach to understanding the *purpose* of life. That is to say, we have studied life as a complex assemblage or combination of things that are inter-related to the extent that they represent a cohesive set of orderly events concerned with the same ultimate purpose (whatever that purpose might be). This is precisely what systems analysis is all about.

But, ironically, we have not taken the same approach to studying the physiological *organism itself.* Rarely is physiologic function examined from a *systems* point of view, the word, "systems", being used here in the engineering context as described above, rather than in the physiologic context as defined below. Researchers tend to settle into their own little niche when it comes to defining the processes of life, and, in the proverbial sense, they lose sight of the forest from the trees.

Because of the complexity of physiological processes, this may, indeed, be unavoidable -- perhaps even justifiable. However, let us not forget that our body *does* function as a *whole*, and its various organs are not self-contained and independent of one another. Thus, the systems approach to the examination of biological processes should actually lead to a more significant understanding of physiologic function, from which the *details* could start to make more sense.

Such an understanding of physiologic processes from a systems point of view might help us to control disease and malfunction much more effectively. This is the basis of so-called "Whole-istic Medicine", which recognizes the importance in managing disease of the unified treatment of the *whole* patient -- both the *somatic* (whole physical body) aspects of the individual, as well as the *psychic* (concerning the state of mind) aspects of the individual. It might also help us to understand exactly what we are doing as we develop a technology to replace human body parts with man-made synthetic devices (prosthetics). And, it might further help us to comprehend the envelope of human performance capabilities, together with the effects on this envelope of environmental stress.

If nothing else, the systems overview of physiologic function can shed further light on the kinds of *questions* that we, as engineers, should be asking about the body if we are to contribute to the health, comfort and understanding of humankind.　With this in mind, the material that follows in this and subsequent chapters is formulated under the premise that, despite its seeming complexity, the human physiologic organism may be analyzed from an *engineering* point of view -- in the sense that it functions in space and time in accordance with some very fundamental common scientific denominators that govern *all* physiologic processes uniformly.　These involve principles of mass, energy, and momentum transport and utilization (including associated control mechanisms) as they apply to isobaric, isothermal, electrochemical engines (see Figure 1-1).

Basic Engineering Concepts

The human body shall be treated in this text from an *engineering* point of view as a sophisticated "system", known as a *physiologic* system. Within this system are various *subsystems* that perform specific functions vital to the overall welfare of the entire organism.　These functions all have certain common denominators that include:

(1)　Highly selective mass, energy, and momentum *transport* mechanisms that operate within the framework of *Compartmental Kinetics.* Compartmental kinetics may be defined to be the collection of processes that are responsible for and control all transport across biological membranes -- in such a way that desirable substances are allowed access to the organism and undesirable substances are effectively excluded or excreted. These, and processes associated with getting mass, energy and momentum more generally from one point in the organism to another are discussed in Chapters 2 and 3 which follow.

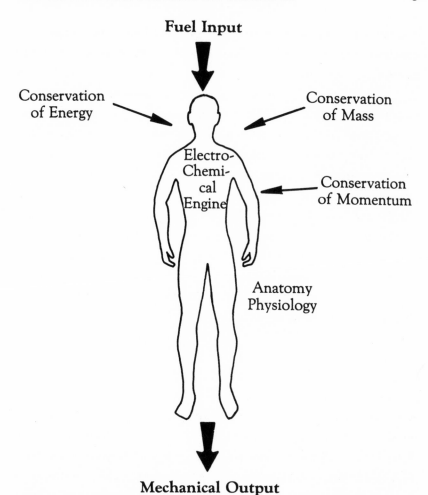

Figure 1-1 The biomedical engineer views the human body as a fine-tuned electro-chemical engine which functions in space and time in accordance with certain fundamental laws of nature. The output of the engine is mechanical. It runs on fuel and has a sophisticated exhaust system.

(2)   Bioenergetic mechanisms concerned with *Metabolism* and *Thermoregulation*. Metabolism may be defined to be all of the energy and material transformations that occur within living organisms. It includes the thermodynamics of anabolic (building up) and catabolic (breaking down) processes, enzyme (catalytic) kinetics control of biochemical reactions, and the chemical kinetics of the reactions, themselves. Thermoregulation may be defined to be all of the processes concerned with maintaining body temperature at 37°C, since we are, ultimately, an isobaric (atmospheric pressure), isothermal (98.6°F), approximately isovolumetric electrochemical engine. These considerations, within the framework of certain optimization schemes that prevail for the *utilization* of mass, energy, and momentum by physiologic systems, are addressed in Chapters 4 and 5 which follow.

(3) *Action Potential* mechanisms concerned with the digital transfer of bits of information, especially as they relate to the feedback and feedforward *control* of processes concerned with *Homeostasis*. The latter may be defined to be the steady-state internal environment that is required for the maintenance of life -- including those processes by which physiologic variables are maintained within well-defined and carefully prescribed limits. Action potentials may be defined to be the biological signals that constitute the "language" of the human body. They are transported by a nerve (communications) network that receives (sensory), processes, stores, retrieves, and transmits (motor control) information that is coded both *spatially* (i.e., *which* nerve fires at any given time) and *temporally* (involving the *rate* of conducted action potentials, a process known as pulse frequency modulation, FM). This, as considered more fully in the material of Chapter 6, makes the organism a digital machine, not an analog one.

Within the context of the above common denominators, all of the systems and subsystems of the physiologic organism operate according to certain fundamental laws of physics -- these being conservation of mass, energy and momentum; and the bottom line in all cases appears to be an attempt on the part of the body to economize on the energy expenditure associated with all physiologic function (see later). The engineering approach to studying and understanding physiologic function thus exploits the above principles through the use of both deductive and inductive techniques of analysis, employing basically three methods: purely *mathematical* ones (closed-form analytic solutions, computerized numerical and finite-element procedures, perturbation schemes, and so on); *experimental* ones (empirical laboratory observations); and *order-of-magnitude* (dimensional and/or nondimensional analysis) ones. We shall have cause to say a few words about all of these methods as we explore the basic principles involved in this engineering approach to the study of physiologic function, beginning with a very short introduction to the human "engine".

Basic Physiologic Concepts

Cells, Tissues, and Organs: Composition of the Human Body

CELLS, ICF AND ECF FLUIDS

The basic functional unit of all systems within the physiologic organism is the *cell*, which might be thought of as one of the "cylinders" of our engine (see Figure 5-2). In that sense, we are a one-hundred-trillion-cylinder-engine, since various estimates suggest that we have at least 10^{14} or more total cells in our body. About 0.003% of these (some 3 billion) die and are replenished (mostly in the skin and blood) every minute of the day, suggesting that there is a complete cellular turnover each three-plus weeks or so. This, of course, is somewhat misleading because not all cells live the same length of time -- some survive for just a few seconds, others for the entire life span of the individual.

The entire amount of fluid contained *within* (i.e., inside of) the sum total of all the cells of the body is called the *Intracellular Fluid* (ICF), or the intracellular fluid compartment of the body. It may amount to 21 to 32 liters of total liquid, and represent some two-thirds of the total fluid content of the body. The latter, amounting to anywhere from 38 to 49 liters, as shown in Table 1-I, may account for 45 to 75% of an individual's total body weight, averaging 60%. As also shown in the table, more than half (23 liters) of the total body water (TBW) exists in the musculature of the organism, with just under another third of it (13.6 liters) divided roughly two-to-one between the skin (9 ℓ) and the blood (4.6 ℓ). Thus, 84% of all the fluid in the body is confined to three physiologic subsystems.

Since water accounts for 60% of an individual's body weight, and since H_2O is $\frac{8}{9}$'ths oxygen, it is not surprising to learn that oxygen is *the most* prevalent chemical constituent of the organism, accounting, together with its occurrence in the organic material of the physiologic system, for some 65% of the average adult weight (see Table 1-I). In fact, Carbon, Hydrogen, and Oxygen (the ingredients comprising the group of chemical substances called *carbohydrates*) together comprise about 93% of an individual's body weight (see Table 1-I). Add Nitrogen (a primary ingredient in proteins), Calcium (85% of the mineral matter of bones and 75% of the body ash left behind when all else has been completely incinerated), and Phosphorus (a key ingredient in bones and the compounds that provide sources of energy for the system), and you have accounted for nearly 99% of the weight of the body. The remainder is primarily in the form of electrolytes (Sodium, Chlorine, Magnesium, Potassium) and other so-called "trace elements" (Iodine, Iron, Zinc, Cobalt, Sulphur, Manganese, Copper, and virtually every other member of the periodic table of the elements).

All of the fluid in the body that lies *outside* of the cells, themselves, is called the *Extracellular Fluid* (ECF), or the Extracellular Fluid compartment. Just about half of the ECF (some 44% representing 7.5 or so liters) includes Interstitial fluid that lies between and bathes the cells themselves, and lymph (see later). The remainder includes blood plasma (about 2.8 liters or 16.5% of the ECF), supporting tissue (collagen, connective tissue, bone, cartilage, and tendon) fluid (about 5 liters, or 29.5% of the ECF), and Trans-cellular fluids (about 1.8 liters, or 10% of

TABLE 1-1

BODY COMPOSITION OF THE AVERAGE ADULT OF WEIGHT Mg_o

Total Body Water (38 – 49 Liters, Average = 43.6) (45 – 75% Body Weight, Average = 60%)						Organic Materials (35 to 40% of Mg_o)				Minerals 3 – 6% Of Total Body Weight, largely in the Skeleton
Muscle (23 ℓ)	Skin (9 ℓ)	Bone (2 ℓ)	Fat (1 ℓ)	Blood (4.6 ℓ)	Other (4 ℓ)	Fat 16% Mg_o	Carbohydrate 0.4 – 0.5% of Mg_o		Protein 16.0 % of total body weight	
							Glycogen 300 grams mostly in Liver and Muscle	Glucose 10 – 20 Grams in ECF		
Intra-Cellular ICF Fluid 21 – 32 ℓ 67% TBW 40% Mg_o						Range 10 – 40% Depending On Many Factors >25% is Considered to be "Obese"	Daily Caloric Requirement: 900 Dietary Calories of Fat 300 Dietary Calories of Protein 1400 Dietary Calories of Carbohydrate			Physiologic Chemical Elements
Extra-Cellular ECF Fluid 17.1 ℓ, 33% of Total Body Water (TBW) 20% of Total Body Weight (Mg_o)										
Extravascular ECF, 14.3 ℓ					Intra-Vascular ECF 2.8 ℓ Blood Plasma 16.5% of ECF					
Inter-stitial Fluid	Trans-Cellular Fluids 10% of ECF 1.8 ℓ		Support-ing Tissue ECF 29.5% of ECF 5.0 ℓ							
Between Blood Vessels and Cells, Approximately 1.8 ℓ is contained within the Red Blood Cells	Pleural Intra-Ocular Cochlear Endolymph Cerebro-Spinal Peri-Cardial Synovial Peri-toneal Endocrine Hormones Exocrine Secret-ions		Collagen Digestive Connec-tive Tissue Bone 3 – 3.5 ℓ Cartilage 2 – 2.1 ℓ Tendon							
44% of ECF 7.5 ℓ										

Physiologic Chemical Elements:

Oxygen = 0.65Mg_o Carbon = 0.18Mg_o Iodine = 0.0000003Mg_o

Hydrogen = 0.10Mg_o Iron = 0.00004Mg_o Nitrogen = 0.03Mg_o

Calcium = 0.0175Mg_o Sodium = 0.0012Mg_o Zinc = Trace

Magnesium = 0.0004Mg_o Cobalt = Trace Sulphur = 0.002Mg_o

Potassium = 0.0029Mg_o Phosphorus = 0.01Mg_o

Manganese = 0.000003Mg_o Chlorine = 0.0015Mg_o

Copper = 0.0000015Mg_o Other Trace Elements = 0.004355Mg_o

Note: All values in this table are approximations based on averages accumulated from a variety of sources. Specific values for a given individual may vary, but they should be within a reasonable range of the means quoted above.

the ECF). As shown in Table 1-I, included in the latter are digestive fluids, cerebrospinal fluid, intraocular fluid, pleural fluid, pericardial fluid, peritoneal fluid, synovial (joint, lubricating) fluid, cochlear fluid (endolymph), sweat and other secretions from the exocrine glands, and hormones and other secretions from the endocrine glands (see later in this chapter).

The part of the body that is not fluid consists of organic and inorganic materials that include: (1) Fats (Lipids) -- *should be* about 14 to 18% (average 16%) or so of total body weight, but can vary from as little as 10% to as much as 40% (anything over 25% usually bordering on "obesity"); (2) Proteins -- about 16% of total body weight; (3) Carbohydrates (pure) -- 300 grams contained as stored glycogen in liver and muscle, 10 to 20 grams as glucose in the ECF and ICF, representing a negligible percentage of total body weight in this absolute form; and, (4) Minerals -- mostly in the skeleton, representing 3 to 6% of total body weight as listed in Table 1-I. Collectively, these organic and inorganic materials represent some 30 to 40% of the extremes of total body weight.

TISSUES AND CELL FUNCTION

Hypothetically, *every* cell in the body is *coded* genetically to manufacture one complete physiologic organism. Indeed, studies have confirmed that cells transplanted from one organ in the body to another soon begin to function and learn to reproduce cells of *that* organ. In actuality, however, relatively early in the reproductive process, by mechanisms that are not yet totally understood, cells become differentiated into *tissues.* These are organized collections of cells that are placed in one of six different categories, depending on their similarity of anatomic structure, metabolic characteristics, intercellular substance, and the specific functions they perform. The six basic tissue classifications are: (1) Epithelial; (2) Connective; (3) Muscular; (4) Nervous; (5) Fluid; and (6) Parenchymal.

Epithelial Membranes are cells arranged in a continuous sheet, consisting of one or more layers, and responsible for covering, lining, *sheathing*, or encasing things. They form the epidermis of the skin, cover the surfaces of organs, line cavities and canals, and form tubes and ducts and secreting portions of glands. Epithelial tissue may be *smooth* (as in endothelial vascular linings and dermal skin coverings); *ciliated* (with hair-like projections, as in the lining of the small intestine); *glandular* (secretory, as in the glands); and/or *phagocytic* (capable of engulfing and digesting, as in the reticuloendothelial system of the spleen, see later).

Connective Supporting tissue, as the name implies, consists of cells designed to support and connect other tissues and parts of the body. They include, therefore, *dense* material such as cartilage and bone (osseous tissue), and muscle tendons; and, more *delicate* fibers forming interlacing networks, such as is found in muscle tissue, bone marrow, the spleen, liver, lungs, kidneys, and mucous membranes of the gastrointestinal tract. Connective tissue may be: (i) *fibrous* (made mostly of collagenous and elastic fibers, such as *areolar* or loose connective, *white* fibrous, and *yellow* fibrous or elastic); (ii) *adipose* (made mostly of fat, indeed, *synonymous* with *fat* tissue, which comprises around 16% of total body weight); (iii) *mucous* (jelly-like, consisting mostly of water); or (iv) *reticular* (combinations of all three, fibrous, adipose, and mucous). Of these, areolar is rather widely distributed, forming the interstitial tissue (connective "substance") of most organs, the adventitial membranes surrounding blood vessels and nerves, and constituting the principal portion of fascia (band-like coverings). White fibrous tissue forms tendons, ligaments, and resistant membranes. Yellow fibrous tissue is found in certain ligaments and in the intimal and medial wall layers of larger arteries.

Muscle and *Nerve* tissues are characterized as being *excitable*, i.e., capable of generating an *action potential* in response to the proper stimulus. In the case of muscles, the action potential incites to a working state excitable cells that are concerned with generating movement, both internal and external. Internal movement includes peristalsis (a "wavelike milking along" of material in hollow tubes of the body, such as the small intestine), breathing, gastric churning, cardiovascular pumping and tone, and the type of mobility encountered in arteries and veins. External movement involves posture and the locomotion of parts or all of the body (see Chapter 5). Muscle may be of the *vascular* and *visceral smooth* type, associated mostly with internal motion; of the *cardiac* type, specifically concerned with the pumping action of the heart; or, of the *striated skeletal* type, which is responsible for external locomotion. Cardiac muscle is put in a class by itself because, although it has many of the anatomical and physiological characteristics of striated skeletal muscle, it possesses as well many of the functional attributes of smooth muscle tissue.

Nerve tissue consists of nerve cells (neurons, dendrites, axons, nerve endings), glial cells, and synapses, all of which are specifically concerned with the transmission of information and the establishment of a communications network within the human organism. This is accomplished through the generation and propagation of action potentials (signals) that are discussed in Chapters 5, 6, and elsewhere in this text.

Parenchymal tissue includes those cells that represent the essential parts of an organ as related to its actual *function*, as opposed, or in contradistinction to its anatomical *framework*. Thus, the functional cells of the liver, the kidneys, the lungs, and other *viscera* (internal organs of the torso) are called parenchymal tissue.

Fluid "tissue" includes blood, lymph, water, ECF and ICF, as described earlier and summarized in Table 1-I. Although not strictly a conglomeration of cells, per se, in the sense of the definition of physiologic tissue (hence the quotation marks around the word), body fluid satisfies the definition in spirit, if not to the letter of the "law". That is, it does constitute material organized for a specific purpose -- mostly to affect and control physiologic transport as discussed in the chapters to follow.

ORGANS AND THE TASKS OF ORGANIZED TISSUES

Just as cells are arranged according to function into tissues, tissues, in turn, are *organized* according to their collective performance of a specific, well-defined task, into *organs*. These are identifiable parts of the body that have a specialized architecture designed to perform that specific task, which may include, for example, manufacturing, processing, transducing, protecting, pumping, and so on -- all of the processes concerned with the maintenance of life and the procreation of the species.

Among the organs concerned with *manufacturing* are *Glands*. These produce a substance called a *secretion*, which is discharged from the gland and either used in some other part of the body, or excreted, or both. On the basis of their secretion, glands are classified as being *mucous* -- producing a viscous, slimy substance; *serous* -- producing a clear, watery substance; or *mixed* -- producing both. On the basis of the anatomical presence or absence of ducts, and whether or not the glandular secretion eventually reaches the external surface of the organism, glands are, respectively, *Exocrine* or *Endocrine* ("ductless organs of internal secretion"). And finally, on the basis of the manner by which secretion is accomplished, glands are classified as being *Merocrine* -- wherein secretions form within cells and are subsequently passed through cell membranes into excretory ducts; *Apocrine* -- where the secretion forms in the apical end of the cell, which subsequently breaks off to form an actual part of the secretion (as is the case in mammary glands); or *Holocrine* -- within which the entire cell, together with its contents, is extruded in the secretion process and leaves the gland entirely, such as occurs in sebaceous glands. All of these are discussed further later in this chapter.

Among the organs concerned with *processing* are the lungs -- within which blood is oxygenated while giving off carbon dioxide; the kidneys -- within which blood is cleansed of the undesirable by-products of metabolism; and the spleen -- within which worn-out red blood cells are systematically removed from the circulation at the incredible rate of 2 to 10 million per second (around a half-billion per minute, representing a daily erythrocyte turnover rate of about 1%). In other words, processing organs *do* something to the material that courses through them.

Among the organs concerned with *transducing* are the various sense organs of the body (see later in this and subsequent chapters). These are designed to allow us to *perceive* the attributes (inertia, three-dimensional space, time, and so on) of the environment within which we exist, including specific ranges of a significant portion of the electromagnetic spectrum. The sense organs convert (transduce) the form of energy to which they are receptive into afferent electrochemical digital impulses (action potentials) that carry information to the central nervous system (CPU) of the organism. Muscles -- to the extent that they, too, convert chemical potential energy into work -- are in some sense chemico-mechanical transducers.

Among the organs concerned with *protecting* the body are the skin and the axial skeleton. In fact, weighing about six (to as much as 18 or more) pounds, skin is the largest single organ of the body. It protects against injuries and parasitic invasion; it houses sense organs concerned with the perception of pain, pressure, heat, cold, and "touch"; it acts as a thermal insulator; and it takes an integral part in the overall regulation of body temperature. None of these have anything to do with its color, which, unfortunately, is what gets most of the attention.

Among the organs concerned with *pumping* is the heart, which drives blood through the entire vascular system. More on this later.

Systems, and the Collective Mission of Organs

A *collection* of organs working together towards the ultimate accomplishment of a given generic mission is referred to as a subsystem, or, simply, *system* of the body. This concept will become clearer in the material which follows, wherein our intent is simple: we shall make no effort to define the purpose of life, or to seek to justify our existence, but we shall take a candid look at the overall living system purely for what it is -- a fine-tuned, physico-electrochemical thermodynamic engine, which functions in space and time in accordance with certain fundamental laws of nature (i.e., conservation of mass, energy, and momentum) that govern the behavior of an isothermal, isobaric, isovolumetric, open thermodynamic system.

Like most engines, this one takes in fuel. In fact, it can move about to find, procure, and ingest its fuel, often with great pleasure and satisfaction. Having done so, the engine "burns" (not literally), or metabolizes the fuel for energy (not heat energy, since the system is isothermal), and, it has a sophisticated exhaust system. The engine has the unique ability to interact with its environment in order to seek sustenance, to escape from predators and other life-threatening forms of life or adverse ambient conditions, and to communicate with other creatures. The output of the engine is mechanical, i.e., it does *work*. The fine-tuning is accomplished through a variety of complex feedback control mechanisms that maintain variables critical to life within prescribed limits (homeostasis). And, finally, this engine has the remarkable ability to make other engines just like it, i.e., to procreate. Indeed, the human organism is truly an "incredible machine".

But, then again, it *should* be! After all, it is the perfected product of an estimated two million (some say as much as two *billion*) years of evolution, or, if you will, research and development. I maintain that if you were to give *me* that long to design a well-functioning piece of machinery -- efficient, effective, micro-miniaturized, economical, optimized, and all that good stuff -- I would do a pretty good job, myself. The machine that is the 20th-century-model Homo Sapiens has survived attack from the environment, from hungry predators, and from pathological processes by which it could conceivably have been annihilated. At least *today*, *this* 20th century model *is* the fittest. It has developed the ability to selectively absorb and concentrate (by appropriate mass transport processes) from the environment desirable chemical nutrients and gases; and to excrete the undesired end-products of metabolism. It has learned to optimize its performance (through, for example, appropriate reflexes and specific enzyme systems), and to economize on the energy expenditures associated with the activities of daily living. It has brought into delicate balance homeostatic control and immune mechanisms that can ward off environmental threats and disturbances. It even has a built-in ability to modify itself in response to persistent environmental changes that might otherwise make it obsolete. In short, the human machine has learned to do what it needs to do to survive, and we shall allude to all of these attributes throughout the text that follows. In fact, from an *engineering* point of view, Darwinism may be interpreted as the manifestation of physiologic processes that concern themselves with *Transport, Optimization,* and *Control.*

Bearing all of this in mind, the engineering study of physiologic function can be approached with the subsystems features described below as a basis for analysis. Each feature (there are basically four general categories) heads a list of the physiologic systems involved, and these are broken down further into their essential anatomical elements. The elements constitute a "vocabulary", if you will, of important anatomical terms with which the biomedical engineer should be familiar. Thus, consider the following features of our engine:

Feature 1: The Human Machine Takes in Fuel, Metabolizes the Fuel for Energy, and Has a Sophisticated Exhaust System

In fact, no less than five major systems of the body are concerned with this one single function, which, coupled with the work output, classifies the body in a general sense as an engine. The five organ systems occupy virtually the entire trunk -- i.e., the Thoracic, or chest cavity plus the Abdominal, or belly cavity, separated from one another by the muscular Diaphragm -- much of the extremities (arms and legs), and significant portions of the head (face, skull and brain) and neck regions. Included in this "engine function" are the: Alimentary System (Digestive), Respiratory System, Circulatory System, Renal (Urological) System, and Excretory System.

Like an internal combustion engine, the fuel (nutrients) is mixed with air (oxygen) and ultimately oxidized to release chemical (bond) energy, which is the source of power for all physiologic processes. But, *unlike* an internal combustion engine, the energy utilized by the physiologic system to do work is chemical *bond* energy, not heat. Thus, exergonic (bond-energy-releasing) chemical reactions drive endergonic (bond-energy-absorbing) chemical reactions to power all physiologic processes, while attempting to generate as little heat as possible. This aspect of human-engine function is addressed in Chapter 4.

The primary fuel for our engine is the simple, six-carbon (hexose), monosaccharide (sugar) called d-Glucose, or Dextrose, designated by the chemical formula $C_6H_{12}O_6$. The chemical energy extracted from this carbohydrate is stored in the high-energy phosphate bonds of the Adenosine Triphosphate (ATP) molecular complex (see Figure 5-4). ATP is to physiologic function what a battery is to an automobile engine, and Glucose corresponds to the gasoline, the energy from which is used to recharge the battery. Many glucose molecules strung together and packed for storage purposes comprise the Glycogen granules that can be seen floating around in the cytoplasm (fluid matrix) of muscle and liver tissue cells (see Figure 5-2). The breakdown of glycogen to release glucose is called *Glycogenolysis*. The manufacture of glucose from raw materials *other than* simple carbohydrates -- for example, fat and proteins -- is called *Gluconeogenesis*. The cellular "organelles" (literally, "small organs") that are responsible for extracting energy from nutrients and trapping the energy released by exergonic oxidative processes in the high-energy chemical bonds of the sugar derivative ATP are the *Mitochondria* (see Figure 5-2). In other words, the mitochondria are the power plants of the cell, and their role in the energy-conversion processes that drive all aspects of metabolism is discussed in Chapter 5. Let us now take a little bit of a closer look at the five organ systems that comprise the engine-function of the body.

The Alimentary System, or Digestive Tract

Akin to an "Oil Refinery", this organ system is responsible for taking in the fuel (food) and converting it into a form suitable for use by the body. The latter is accomplished primarily in the *Gastrointestinal* (stomach plus small and large intestinal portions of the digestive tract), or, simply "GI" segments of this system. The Alimentary Canal (as it is sometimes called) also disposes of (exhausts) fuel that it either does not want, or cannot use (digest), as well as the remnants of red-blood-cell destruction (hence the typical brown color of stool or feces) and other physiologic wastes. The large intestine is also the site of major fluid absorption from the intestinal contents -- which can result in diarrhea if the absorption is deficient, or constipation if the absorption is excessive. Table 1-II lists some of the major anatomical features of this system.

The Respiratory System

Akin to a "carburetor", this organ system is responsible for taking in (inspiring) air, extracting from it oxygen, which is carried by red blood cells (forming the oxyhemoglobin complex as the fluid courses through the lungs), and expiring water vapor and carbon dioxide as the products (wastes) of metabolism (see Chapters 2 and 4). The respiratory system also plays a significant role in the regulation of body temperature, since, from as little as a negligible amount (perhaps 5% or less) to as much as 20% or more of the heat generated by physiologic metabolic processes leaves the body through the lungs. Most of this heat loss is by evaporation into water vapor at the level of the 300 million tiny air sacs, called *Alveoli* (*latent*, or *insensible* heat loss). Some of it is by convection resulting from the controlled processes of inhaling and exhaling (*sensible* heat loss), and, in extreme cases, panting. Control of body temperature is also handled by the cardiovascular system, the integumentary system (skin), the musculoskeletal system, and other homeostatic mechanisms (see Chapter 4 for more details of thermoregulation).

Listed in Table 1-III are some of the major anatomical features of the Respiratory System. Starting at the *trachea*, the bronchial tree progresses through 16 generations of bifurcations (the *conductive zone* including *bronchi*, *bronchioles*, and *terminal bronchioles*); and then through a *transitional zone* consisting of 3 bi-and-trifurcating generations of *respiratory bronchioles*, before reaching the *respiratory zone*. The latter in-

TABLE 1-II

MAJOR ANATOMICAL FEATURES OF THE ALIMENTARY SYSTEM

Lips and Mouth (System Inlet)	Abdominal Cavity
32 Permanent Teeth:	Four Lobes of Liver (the largest
8 incisors	organ <u>inside</u> the body)
4 canines in two upper	Bile (formed in the liver)
8 premolars and two	Gall Bladder (Bile storage)
12 molars lower sets	Common Bile Duct (Bile transport)
Tongue and Glottis	Pancreas
Six Salivary Glands:	Trypsin enzyme (Proteolytic)
2 Parotid Glands	Maltase enzyme (Carbohydralytic)
2 Submaxillary Glands	Steapsin (Fat Lipase; Lipolytic)
2 Sublingual Glands	Amylopsin (Starch enzyme)
Auxiliary Buccal Glands (Ptyalin)	Pancreatic Duct
Pharynx (Throat)	Islets of Langerhans (Insulin
The Fauces, opens into the	and Glucagon)
Esophagus	Splanchnic (Visceral)
Esophagus (Food Pipe) and Uvula	Circulation, including:
Cardiac Orifice opens into the	4 Esophageal Arteries
Stomach	2 Mediastinal Arteries
The Stomach ("Mixer")	Diaphragmatic Artery
The Pylorus, opens into the	Inferior Mesenteric Artery
Small Intestine	Superior Mesenteric Artery
20 feet of coiled Small Intestine	Anterior Cecal Artery
including:	Posterior Cecal Artery
Duodenum (First Branch)	Celiac Axis Trunk
Jejunum (Second Branch)	Left Gastric Artery
Ileum (Final Three-Fifths)	Right Gastric Artery
Peritoneum (Abdominal Lining)	Short Gastric Artery
Mesentary (blood supply)	Gastroduodenal Artery
Vermiform Appendix	Cystic Artery
The Cecum, opens into the	Common Hepatic Artery
Large Intestine	2 Suprarenal Arteries
5 feet of Large Intestine	Splenic (Lienal) Artery
including:	Superior Pyloric Artery
The Ascending Colon	Middle Sacral Artery
The Transverse Colon	Hepatic Vein
The Descending Colon	Portal Vein
The Sigmoid Colon	Pelvic Vein
5 inches of Rectum	Perineal Vein
Anus (System Outlet)	Epigastric Veins

TABLE 1-III

MAJOR ANATOMICAL FEATURES OF THE RESPIRATORY SYSTEM

Nose (Intake and Outlet)	Bronchial Circulation: Feeding
Nasal Septum (Wall)	Main Bronchial Artery
Sinuses: Paranasal Groups	Left Bronchial Artery
2 Frontal	Right Bronchial Artery
2 Maxillary	Left Bronchial Vein
Sphenoidal (Generally Paired)	Right Bronchial Vein
Ethmoidal (Generally Paired)	Pulmonary Circulation: Processing
Epiglottis	Pulmonary (Pulmonic) Trunk
Larynx (Voice Box)	Left Pulmonary Artery
Pharynx (Throat)	Right Pulmonary Artery
Thyroid Cartilage	Pulmonary Arterioles
Circoid Cartilage	Pulmonary Metacapillaries
The Trachea (Wind Pipe)	Pulmonary Capillaries
Two Bronchi (Left and Right)	Precapillary Sphincters
The Carina Ridge Between Bronchi	Pulmonary Venules
Bronchiole Branches	Superior Left Pulmonary Vein
14 Million Alveolar Ducts	Inferior Left Pulmonary Vein
300 Million Alveoli (Air Sacs)	Superior Right Pulmonary Vein
The Diaphragm	Inferior Right Pulmonary Vein
The Thoracic (Chest Cavity)	The Lobes of The Lungs:
Pleural Lining of the Chest	Two Oblique Fissures
The Cupula (above the Clavicle)	One Horizontal Fissure
Hilum (slits in the lungs)	Two Superior Lobes
The Mediastinum (Wall)	One Right Middle Lobe
The Surfactant Coating	Two Inferior Lobes

cludes 3 generations of irregularly branching *alveolar ducts*, terminating in a final branching network of *alveolar sacs*, which are the 23rd generation off of the original trachea.

The Circulatory System (Transport)

In general, this system forms the major "highway network" of our "engine", providing miles and miles of tubing for the transport of mass, energy, and momentum to and from a differential neighborhood surrounding each and every one of the one-hundred-trillion or so cells (or "cylinders") that comprise the microscopic functional units of the human machine. The Circulatory System actually includes two subsystems -- namely, the Hemato(Blood)-Cardio(Heart)-Vascular(Blood Vessels) sub-. system, and the Lymphatic subsystem.

THE CARDIOVASCULAR SYSTEM, INCLUDING BLOOD

This subsystem is responsible for transporting nutrients (fuel) and oxygen to the cells of the body, and for removing waste products from these cells -- the so-called *metabolic* function of circulation. However, it also acts as a counter-current heat exchanger, which gives it a *thermoregulatory* function, as well; and it distributes hormones, enzymes and other biochemical constituents throughout the body, giving it a *control* function. The Cardiovascular system is composed of three subsystems, namely, the *Hematologic* subsystem (Blood), the *Cardiac* subsystem (Heart), and the *Vascular* subsystem (blood vessels and circulatory pathways).

THE HEMATOLOGIC SUBSYSTEM: THE FLUID, BLOOD

Blood, accounting for about 8% of total body weight, consists of a complex, heterogeneous suspension of formed elements in a continuous, straw-colored suspending medium. The formed elements comprise the blood cells, collectively termed *hematocytes*, from the Greek, "haima", meaning, "blood", and "kytos", meaning, "cell". As shown in Table 1-IV, hematocytes include three basic types of cells: Red Blood Cells (Erythrocytes, from the Greek, "erythros", meaning, "red"), White Blood Cells (Leukocytes, from the Greek, "leukos", meaning, "white"), and Platelets (Thrombocytes, from the Greek, "thrombos", meaning, "lump"). The formed elements in blood are manufactured in the bone marrow, the spleen, and some specialized lymphatic tissue (hence the designation, "Lymphocytes" for some types of white blood cells). In the course of the average life span of any given individual, bone marrow will manufacture more than half-a-ton of red blood cells, each of which takes six days to complete, and survives about 120 days in the circulation. These oxygen-carrying cells (see Chapter 2) are removed from the circulation by the Spleen, the largest collection of reticuloendothelial cells (page 29) in the body. The phagocytic cells of this large lymphatic organ can also store and discharge blood cells into the vascular system as necessary. The term, *Hematocrit* refers to the percentage by volume of red blood cells in blood, which can normally range from 30 to 54% for adults (average, 45%).

The fluid contained within hematocytes accounts for something on the order of 10% of the body's total intracellular fluid space, even though the total number of *cells*, themselves, account for some 25% of the total number of cells in the body. Removal of all hematocytes (blood cells or corpuscles) from blood by centrifugation or other separating techniques leaves behind the aqueous (90% water by volume, 91% water by weight), saline (0.15 Normal Salt) suspending medium called *plasma*. Plasma is the fluid portion of the blood, in which all the cellular elements are suspended, and it accounts for some 15 to 18% of the body's total extracellular fluid space (see Table 1-I).

Seven to 8% by weight of total plasma consists of the plasma proteins. As listed in Table 1-IV, these are of four major types: Fibrinogen, Albumin, the Globulins, and Prothrombin. The remainder of this straw-colored fluid, although accounting for only 2%, or so, of its total weight, consists of a rather impressive long list (which seems to be almost endless)

TABLE 1-IV

MAJOR ANATOMICAL AND BIOCHEMICAL FEATURES OF BLOOD

Formed Elements: Hematocytes
    Red Blood Cells (25 trillion
        Erythrocytes)
    Reticulocytes (250 billion)
    Hemoglobin
    White Blood Cells (20 - 55
        billion Leukocytes)
Differential White Blood Cells:
    Agranulocytes (23 - 58%),
        including:
        Lymphocytes (20 - 50%)WBC
        Monocytes (3 - 8%)WBC
    Granulocytes (Polymorphonu-
        clear Phagocytes, PMN's,
        42 to 77% of all WBC's)
        including:
        Neutrophils (40 - 72%)WBC
        Eosinophils, or Acidophils
            (1 - 4%)WBC
        Basophils (0.5 - 1%)WBC
    Sub-classes (Variable Numbers)
        Segmented Granulocytes:
            Phagocytes
                Macrophages
                Microphages
            Myelocytes
            Band Cells
        Agranulocytes:
            B-Lymphocytes
            T-Lymphocytes
    Platelets (700 billion to
        2 trillion Thrombocytes)
Suspending Medium: Plasma
    Water (91% Aqueous Solution)
    Plasma Proteins (7 - 8%) = TP
        Fibrinogen (Clotting Fac-
            tor I; 4% of TP)
        Prothrombin (Clotting Fac-
            tor II; 0.25% of TP)
        Albumin (45 - 60% of TP)
        Globulins (35 - 50 % of TP)
            including:
            Lipoproteins (6% of TP):
                Very-Low-Density
                    (VLDL, α - 2, 22%)

        Low-Density (LDL, β-
            Globulins, 42%)
        High-Density (HDL, α-1,
            HDL-2, 9%, HDL-3, 27%)
        Immunoglobulins, IgA, 6%;
            IgD, 2%; IgG = γ-Glo-
            bulin, 88%; IgM, 4%
            (14% of Total Protein)
        Transport Carrier Free
            Globulins (23% TP)
    Plasma Electrolytes:
        Primary Salt Cations: mg/dℓ
            Na (317 mg/dℓ); K (15 mg/dℓ)
            Ca (10 mg/dℓ); Mg (2.28)
        Primary Anions: mg/dℓ
            Cl (360); HCO_3^- (27 meq/ℓ)
    Buffering Ions: mg/100 mℓ serum:
        H_2PO_4 (3.75); $C_6(H_2O)_4O_3$ (2.5);
        NH_4^+ (0.16); $SO_4^=$ (1.00)
    Dissolved Nutrients: Free Fatty
        Acids; Amino Acids; Glucose;
        Chylomicrons; Carbohydrates
    Clotting Factors: Thromboplastin
        (III); Proaccelerin (V); Pre-
        Convertin (VII); 8 Others.
    Dissolved Gases (O_2; CO_2)
    Additional Electrolytes:
        Organic Acids; Acetic Acid;
        Formic Acid; Lactic Acid
    Emulsified Fats: mg/100 mℓ plasma
        Cholesterol (LDL, 180)
        Triglycerides (VLDL, 160)
        Phospholipids (HDL, 160)
        Lipoprotein Protein (200)
    Hormones (Endocrine Secretions)
    Enzymes
    Minerals (Fe, PBI, Pb, Cu, S, Bo)
    Vitamins (A, B-Complex, C, D, E,
        K, Biotin, PABA, Folic Acid)
    Antibodies (Antigens)
    Trace Elements (Virtually every
        element in the Periodic Table)
    Other Serum Constituents

of additional constituents -- emphasizing the importance of this "river of life" as a mass transport medium. Without even attempting to be exhaustive, the list (and Table 1-IV) includes, by weight: electrolytes (0.9 to 1.0%); carbohydrates (0.7 to 0.8%); organic acids (0.12%); nonprotein nitrogenous compounds (0.10%); free lipids (0.04%); vitamins (0.03%); hormones (0.00014%); enzymes (trace); antibodies (trace); and less than 0.01% total of other inorganic and organic molecules that encompass a wide variety of species. Removal of the protein fibrinogen from plasma leaves behind a clear, slightly yellowish, low-viscosity fluid called *serum*, the study of which is called *Serology*.

THE CARDIAC SUBSYSTEM: THE PUMP, HEART

Table 1-V lists the major anatomical features of this four-chambered organ, which is actually two pumps arranged anatomically in parallel (i.e., side-by-side separated by a thick muscular wall, the interventricular septum), but physiologically in series. The right heart (pump) receives blood from the systemic circulation and sends it through the pulmonary circulation (see Table 1-III) to be oxygenated. The left heart (pump) receives blood from the pulmonary circulation and sends it through the systemic circulation (see Table 1-VI) to feed all of the cells of the organism, from whence the fluid returns to the right heart to start all over again. The complete journey out the left ventricle, through the systemic circulation into the right atrium, and then out the right ventricle, through the pulmonary circulation and back into the left atrium, takes from as little as 18 to 24 seconds to as much as a minute under normal circumstances. This is called the *circulation time*.

Weighing some three-quarters of a pound, the heart is about the size of a 3-by-5-inch index card (the approximate dimensions of the "fist" of the individual involved), and it rests on the diaphragm, between the lower part of the two lungs, leaning mostly to the left side of the body between the 3rd and 6th ribs. It holds a total of about one-half quart of fluid, and it pumps some 70 mℓ of blood (the *stroke volume*) out of each ventricle

TABLE 1-V

MAJOR ANATOMICAL FEATURES OF THE HEART

The Base of The Heart (Top)	The Coronary Circulation:
The Apex of The Heart (Bottom)	Right Aortic Sinus of Valsalva
The Chambers of The Heart:	Right Coronary Artery
Left and Right Atria (Auricles)	Conus Artery
Left and Right Ventricles	Superior Vena Caval Branch, or
Left and Right Atrial Appendage	Sinus Node Artery
Interventricular Septum (Wall)	Anterior Right Atrial Branch
Excitatory Conduction Pathways:	Right Intermediate Atrial
Sino-Atrial (SA) Node	Artery (Right Atrial Artery)
(Cardiac Pacemaker)	Right, or Dorsal Interventri-
Anterior Internodal Pathway	cular Artery
(Bachmann's Bundle)	Right Marginal Artery
Posterior Internodal Tract	Posterior Descending Artery
Middle Internodal Tract	Posterior Septal Arteries
Torus Loweri to AV Node	Anterior Septal Arteries
Atrio-Ventricular (AV) Node	Left Aortic Sinus of Valsalva
Common Bundle Musculature	Main Left Coronary Artery
The Right Bundle Branch	Left Anterior Descending Branch
The Left Bundle Branch	Left Circumflex Branch
Bundles of His	Left Atrial Circumflex Branch
Sub-Endocardial Branches	Diagonal Branch of the Anterior
Terminal System Purkinje Fibers	Interventricular Artery
The Layers Of The Heart:	First Main Septal Artery
Outer Fibrous Pericardium	Left Marginal Artery
Pericardial Fluid	Deep Myocardial Branches
Serous Parietal Pericardium	Left Marginal Vein
Visceral Pericardium =	Great Cardiac Vein
Epicardium	Posterior Vein, Left Ventricle
Myocardium (Muscle Layer)	Oblique Vein of Left Atrium
Inner Endocardium	Middle Cardiac Vein
Contractile Musculature:	Small Cardiac Vein
15 Atrial Muscle Bundles	Right Marginal Vein
Pectinate Musculature of Atria	Anterior Cardiac Veins
Superficial Bulbospiral Bundle	Coronary Sinus Vein
Deep Bulbospiral Bundle	The Valves Of The Heart:
Deep Sinuspiral Bundle	Pulmonary Semilunar Valve
Interventricular Longitudinal	Aortic Semilunar Valve
Bundles	Tricuspid Valve
Septal (Partition) Band	Bicuspid (Mitral) Valve
Moderator Band	Eustachean Valve
Papillary Muscles (Five)	Coronary Sinus Valve of
Chordae Tendineae (Twenty)	Thebesius (Thebesian Valve)
Intercalated Discs	Right (Acute, or Inferior) Margin
Parietal (Wall) Band	Left (Obtuse, or Superior) Margin
Septo-Marginal Trabecula	Atrio-Ventricular Sulcus
Syncytium	Longitudinal Sulcus or Groove

at the average cyclic rate of 70 times per minute (the *heart rate*), yielding a total *cardiac output* of some five (plus or minus 0.2) liters per minute under routine conditions related to the activities of daily living (see later chapters).

THE VASCULAR SUBSYSTEM: CIRCULATORY PATHWAYS

The systemic circulation arises from the left ventricle via the largest artery of the body -- the Aorta, which has four major divisions: the Ascending Aorta, the Aortic Arch, the Thoracic Aorta, and the Abdominal Aorta. These give off a total of 58 branches (2 from the Ascending Aorta, 3 from the Aortic Arch, 31 from the Thoracic Aorta, and 22 from the Abdominal Aorta), before reaching the area of the groin, at which point the blood vessel bifurcates into the Right and Left Common Iliac arteries (see Table 1-VI).

The 5 arterial branches of the ascending aorta and aortic arch give rise to 82 arterial tributaries. Those of the abdominal aorta branch into 53 additional tributaries; and those of the thoracic aorta eventually divide themselves into 65 downstream branches, for a total of some 200 *tributaries* that branch from the 58 *arteries* of the main aortic trunk. These tributaries continue to divide further as we move peripherally outward along the vascular tree, so that the total number of major arteries in the body ultimately reaches something on the order of 10^3. These thousand-or-so arteries give rise to 10^8 smaller *arterioles*, which, in turn, distribute themselves via smaller *rami* into a capillary network of some 10^{10} (ten billion) tiny vessels of average diameter 7 to 9 microns and length 1 mm or less. With 10^{10} blood vessels of this size supplying 10^{14} cells of average size 10 microns, it is possible for any given cell to be within a cell size of the nearest capillary.

Capillaries recombine to form 5×10^8 *venules* as blood begins its journey back to the heart. The latter coalesce into some 5×10^3 veins, which eventually join to form two large *vena cava* that empty into the right atrium of the heart. Counting duplicate pairs (e.g., we have two

TABLE 1-VI

MAJOR ARTERIES AND VEINS OF THE SYSTEMIC CIRCULATION

The Ascending Aorta: Branches 1
    and 2 - Coronary Circulation
    (See Table 1-V)
Major Arteries Leading To The
    Head And Neck Region:
The Aortic Arch: Branches:
    3: Brachiocephalic Trunk
        (Innominate Artery)
    4: Left Common Carotid Artery
    5: Left Subclavian Artery
Right Common Carotid Artery
Right Subclavian Artery
Right External Carotid Artery
Left External Carotid Artery
Right Internal Carotid Artery
Left Internal Carotid Artery
Facial Artery
Vertebral Artery
Thyrocervical Trunk
Internal Thoracic Artery: Mammary
Major Arteries Of The Torso:
The Thoracic Aorta: Branches:
    1-3: Bronchial Circulation
            (See Table 1-III)
    4-7: Esophageal Arteries
    8-9: Pericardial Arteries
    10-29:Intercostal Arteries
    30-31:Mediastinal Arteries
Pericardiacophrenic Artery
The Abdominal Aorta: Branches:
    1-2: The Phrenic Arteries
    3 : The Celiac Trunk to the
            Splanchnic Circulation
            (See Table 1-II)
    4 : Superior Mesenteric Art.
    5-6: Suprarenal Arteries
    7-8: Renal Arteries
    9-10: The Testicular, or,
            The Ovarian Arteries
    11 : Inferior Mesenteric Art.
    12-19 : Lumbar Arteries
    20 : Middle Sacral Artery
    21-22: The Iliac Arteries
Colic Arteries (Right, Middle)
Gluteal (Sciatic Artery)
Superior Gluteal Artery
Superior Thoracic Artery
	Major Arteries Of The Upper Limbs
	    Axillary (Armpit) Artery
	    Brachial (Arm) Artery
	    Ulnar (Forearm) Artery
	    Radial (Forearm) Artery
	    Deep Palmar Arch (Hand)
	    Superficial Palmar Arch
	    Common Palmar Digital(Fingers)
	Major Arteries Of The Lower Limbs
	    Femoral Arteries (Thigh)
	    Popliteal Arteries (Knee)
	    Anterior Tibial Artery (Shank)
	    Posterior Tibial Artery
	    Profunda Femoris Artery
	    Dorsalis Pedis Artery (Foot)
	Major Veins Of The Lower Limbs
	    The Femoral Veins
	    The Dorsal Venous Arch
	    Anterior Tibial Vein
	    Small Saphenous Vein
	    Popliteal Vein
	    Great Saphenous Vein
	Major Veins Of The Upper Limbs
	    Superficial Palmar Network
	    The Basilic Vein
	    The Median Cubital Vein
	    The Brachial Vein
	    The Ulnar Vein
	    The Radial Vein
	    The Axillary Vein
	Major Veins Of The Torso
	    Testicular (or) Ovarian Veins
	    The Renal Vein
	    The Portal and Hepatic Veins
	    The Iliac Veins
	    Pelvic and Perineal Veins
	    Glomerular and Tubular Renal V.
	    The Superior Vena Cava
	    The Inferior Vena Cava
	Major Veins Of The Head And Neck
	    Brachiocephalic (Right & Left)
	    Jugular Veins (Four)
	    Right and Left Subclavein Veins
	    The Emissary Veins
	    Left and Right Vertebral Veins
	    Diploic Veins (Five)
	    Inferior Thyroid Vein

arms, two legs, two eyes, two ears, and so on), smaller "rami", and a rather significant network of "potential spaces", i.e., *anastomoses* that provide by-pass connections between any two vascular spaces -- various estimates suggest that if all of the blood vessels of this system were laid end-to-end next to one another, they would stretch around the world some two-and-one-half times, totalling some 60,000 to 100,000 miles of vascular pathways -- the study of which is called *Angiology*.

THE CARDIOVASCULAR SYSTEM IN THERMOREGULATION

The average student of physiology is generally well-versed and trained to understand the *metabolic* and *control* functions of the cardiovascular system. Blood is, indeed, the river of life, bringing nutrients, oxygen and other biochemical supplies to all the cells of the body, while continuously removing carbon dioxide and other toxic waste products. By comparison, the *thermoregulatory* function of this system receives much less emphasis, and the student comes away from a typical physiology course with little appreciation for how important this role really is in terms of the sustenance of life.

And yet, when one envisions the anatomical configuration of the vascular network of pipes and channels through which the blood flows as it courses through the body, and compares this anatomy with, for example, the radiator of an automobile, the potential for thermoregulation afforded by the cardiovascular system becomes immediately obvious. Like engine coolant coursing through the coils of a radiator, blood in the physiologic system can absorb the heat generated by the complex chemical reactions of life; and, like this same engine coolant, blood can carry (convect) this heat to the 3000-square-inch surface area (skin) of the organism or to the 300-million alveoli of the lungs to be expelled to the environment. Indeed, as we shall see when we discuss metabolism and thermoregulation in Chapter 4, various estimates suggest that as much as 95% of the heat generated by the body at any given time *flows* through vascular convection to the cutaneous circulation supplying the outer layers of the skin, and to the pulmonary circulation coursing through the lungs (see Table 1-III), where 60% of it is lost to the environment by radiation, 25% by evaporation (through sweating and latent heat loss), 12% by convection (and sensible heat loss), and 3% by conduction.

Conversely, this thermoregulatory role can be *reversed* when heat is to be conserved by the body. Again, acting like the thermostat in the radiator of an automobile (which keeps the engine coolant *away* from the heat-dissipating surface area when the engine is cold), vascular smooth muscle tissue acting at the arteriolar level (the *control points* of the cardiovascular system) can completely shut off (thus totally by-passing), or significantly reduce flow to the periphery when this becomes necessary to maintain core body temperature at 37°C.

THE LYMPHATIC AND RETICULOENDOTHELIAL SYSTEMS

In general, one may think of the Lymphatic System as the "Sewage System" of the engine. It carries a transparent, slightly yellow, somewhat opalescent, viscous, mucous-like material called lymph, which is to the Lymphatic System what blood is to the Cardiovascular System. The Lymphatic System is responsible for removing waste and toxic materials (by phagocytosis and the action of lymphocytes) from extracellular tissue spaces in the body, which is where the open-ended, almost funnel-like access channels of this system originate. It also allows for re-entry into the vascular system of fluid and materials (mostly plasma proteins) that leak out from the capillaries but, for whatever reason(s), cannot return by the same route. This is due to the fact that the various lymphatic channels ultimately coalesce into lymphatic ducts that eventually drain via the Right Lymphatic and/or Thoracic Duct into the venous system returning blood to the right side of the heart. And finally, intestinal lymphatics are the pathways for the absorption of digested fats from the alimentary canal.

Table 1-VII lists the major anatomical features of the Lymphatic System and one that is closely related to it, the Reticuloendothelial Sys-

TABLE 1-VII

MAJOR ANATOMICAL FEATURES OF THE LYMPHATIC

AND RETICULOENDOTHELIAL SYSTEMS

Lymphatic Vessels And Nodes Of The Head, Face, and Neck Regions:	Parietal Lymphatic Vessels And Nodes Of The Abdomen, Pelvis, And Pelvic Viscera:
Preauricular Nodes above ears	Cisterna Chyli (Stomach region)
Postauricular Nodes behind ears	Common Iliac Vessel
Occipital Nodes (back of head)	External Iliac Vessel and Nodes
Parotid Nodes (cheeks)	Superficial Inguinal Nodes
Submental Nodes (chin)	Inguinal Nodes of the Groin
Submandibular Nodes (jaw)	Preaortic Nodes
Deep Cervical Nodes (neck)	Celiac Nodes (abdominal)
Jugular Lymph Trunk	Lumbar Nodes (lower back)
Parietal Lymphatic Vessels And Nodes Of The Thorax and The Thoracic Viscera:	Other Lymphatic Vessels and Nodes
Right Subclavian Trunk	Apical (shoulder)
Left Subclavian Trunk	Supratrochlear
Right Lymphatic Duct	Palatine Tonsils (pharynx)
Axillary Nodes (armpits)	Nasopharyngeal Tonsils, also known as the Adenoids
Infraclavicular Nodes (collar)	Lingual Tonsils
Lateral Nodes (sides)	The Thymus Gland (near heart)
Pectoral Nodes (chest)	The Spleen (near the stomach)
Scapular Nodes (towards shoulder)	Lymphatic Capillaries
Tracheobronchial Nodes	Macrophages of Loose Connective Tissue (histiocytes & others)
Intercostal Nodes (ribs)	Kupffer Cells Of The Liver
Posterior Mediastinal Nodes	Microglia Of Central Nervous Syst.
The Thoracic Duct	

tem. That latter refers to all physiologic cells that have the ability to ingest (phagocytose, see Chapter 2) and ultimately digest or otherwise dispose of particulate matter, such as bacteria and colloidal particles. Included are macrophages of loose connective tissue (histiocytes, clasmatocytes, resting "wandering" cells); reticular cells of lymphatic organs (such as the Spleen, Liver and Thymus Gland) and myeloid tissue (such as bone marrow); Kupffer cells of the Liver; cells lining the blood sinuses of the spleen, bone marrow, adrenal cortex, and hypophysis; the microglia of the central nervous system; the adventitial cells surrounding blood vessels; and the so-called "dust cells" of the lungs.

INTRACELLULAR TRANSPORTATION

As opposed to the *Intercellular* transportation pathways outlined above, by which mass, energy and momentum are carried to and from the cells of the body, there is also an elaborate *Intracellular* network through which materials get from one organelle ("little organ") within the cell to any other organelle or region within the cell. This connecting network of microcanals (canaliculi) or tubules that course through both the nucleus (control center) and cytoplasm (cellular protoplasm) of the cell is called the *Endoplasmic Reticulum*. In some cases, the endoplasmic reticulum is named according to the particular cell through which it courses, such as the *Sarcoplasmic Reticulum* in the case of skeletal muscles. In any event, the endoplasmic reticulum provides a transportation network among the cellular organelles, as discussed further in Chapter 5.

The Renal-Urological System

While the Alimentary System exhausts fuel that the engine either does not want or cannot use, the Renal-Urological System exhausts the waste products of metabolism as brought to it by the cardiovascular system. This consists of the products of exergonic-endergonic coupled biochemical reactions (except carbon dioxide which is exhausted by the lungs), waste, and other toxic materials that are picked up by capillaries and lymphatic vessels. The Renal (kidneys) System also is involved in fluid and electrolyte balance, and in the overall balance and maintenance of total body fluid volume, since it has the ability to control the exhausting of these materials into the environment outside the physiologic system. Listed in Table 1-VIII are the major anatomical features of this exhaust system.

The Excretory System

Composed of four basic subsystems -- three of which have already been described, and the fourth mentioned briefly -- the Excretory System includes collectively all of those organs that are responsible for eliminating from the body anything it does not want, or anything that should not be there. This includes: (i) indigestible residue, liquid water, dead red blood cells, bacteria, and other waste products that are eliminated through peristaltic *defecation* from the large intestines of the *Alimentary System* (Table 1-II); (ii) carbon dioxide, gaseous water vapor, other gases, and heat that are eliminated through *expiration* from the lungs of the *Respiratory System* (Table 1-III); (iii) liquid water, nitrogenous substances (urea, uric acid, creatine, creatinine), and mineral salts that are eliminated through *urination* (micturition) from the bladder of the *Renal-Urological System* (Table 1-VIII); and (iv) last, but certainly not least, water vapor, certain salts, much more heat, and a small amount of urea that are eliminated through *perspiration* from the previously-mentioned skin surface. The skin and its appendages, including hair and nails constitutes the *Integumentary System*, which is the fourth major subsystem of the Excretory System of organs.

THE INTEGUMENTARY SYSTEM

Being the largest single organ of the body, skin can account for up to 12 to 18% of total body weight in human adults. It envelopes the body in

TABLE 1-VIII

MAJOR ANATOMICAL FEATURES OF THE RENAL-UROLOGICAL SYSTEM

The Two Kidneys (Renal System):	The Renal Circulation:
Nephrons (10^6 per kidney)	Renal Arteries (Branches 7 & 8
Renal Corpuscle (Malpighian	Off Abdominal Aorta)
Body)	Inferior Suprarenal Artery
Glomerular (Bowman's) Capsule	The Lobar Arteries
Proximal Convoluted Tubule	The Interlobar Arteries
(Loop Of Henle)	The Renal Arcuate Arteries
Distal Convoluted Tubule	The Interlobular Arteries
Collecting Duct	Afferent Arterioles
Juxtaglomerular Apparatus	Glomerular Capillary Network Of
The Hilum Opening	The Renal Corpuscles
The Inner Medulla	The Efferent Arterioles
The Renal Pyramids	Tubular Capillary Network Of
The Papillae	The Renal Cortex and Medulla
The Minor Calyx	The Interlobular Veins
The Major Calyx	The Medullary Veins
The Renal Pelvis	The Renal Arcuate Veins
The Outer Cortex	The Interlobar Veins
The Renal Columns	The Lobar Veins
The Urinary System:	The Suprarenal Veins (R. & L.)
Two Ureters	Two Renal Veins empty into the
The Bladder (Urine Storage)	Inferior Vena Cava
The Urethra (Male and Female)	Middle Suprarenal Artery
The Penis (Male)	Capsular Surroundings: Capsule
Micturition	Proper; Perirenal Fat; Fascia

the form of a sheet varying in thickness from 0.2 mm in the eyelids to 6.0 mm in the soles of the feet; and it can stretch out to a surface area from as little as 0.7 m² to as much as 2.0 m² for an average adult. Skin is comprised of basically two layers: the *Epidermis*, and the *Dermis*, or *Corium*. The epidermis (or *Cutis*) is, itself, composed of four layers -- moving from outside in, the Stratum Corneum (10 to 200 micrometers thick), which forms a *protective covering* to shield the body from mechanical, thermal, chemical, radiative, biological, and other environmental insults; the Stratum Lucidum; the Stratum Granulosum; and the Stratum Germinativum (or Stratum Mucosum, or Stratum Malpighi, or Stratum Spinosum). The last of these contains granular melanocytes (1% of all skin cells), which form a pigment called melanin, that is responsible for determining *skin color*. The bottom three layers of the epidermis are sometimes referred to collectively as the Stratum Basale.

The dermis also consists of two layers: the outer Papillary Layer within which is contained the cutaneous microcirculation that is so important in *thermoregulation*; and the inner Reticular Layer. In this layer are located *sense organs* that perceive touch, pressure, pain, heat, and cold (see later in this chapter); lymphatics; sensory and motor nerves and nerve endings; blood vessels; sebaceous, oil-secreting *Holocrine Glands*; sudoriparous, sweat-secreting *Eccrine Glands* (both Merocrine and Apocrine, see page 13); elastic fibers (elastin); collagen fibers; reticulin (a reticulat tissue protein); and fibrocytes (precursors of fiber-producing cells).

Beneath the dermis is a layer of connective tissue called the *Subcutis*, or Superficial Fascia, which contains subcutaneous adipose or fat tissue that is so important in providing *insulation* for the body. Also important in this respect is *hair* (100,000 to 150,000 strands on the head, alone), which can be made to stand-on-end (to increase body surface thickness) by the action of *Arrector Muscles* in a process called "piloerection". And finally, insulation can be increased by a process called "cutis anserina", which is better known to us as the formation of "goose bumps" when we get cold!

If we add to our growing list of skin functions the manufacture of antirachitic vitamin (Vitamin D_3) from 7-dehydrocholesterol (which exists naturally in this organ, and which is activated by the ultraviolet radiation from the sun), the fact that skin also acts as a reservoir for food and water, and that it plays a vital role in preventing dehydration, we see that the Integumentary System plays no small part in the overall function of our physiologic system.

INTRACELLULAR DISPOSAL UNITS

Lying inside of the cytoplasm (cellular protoplasm) of most cells are tiny organelles, approximately the same size as mitochondria (up to 3 or 4 microns), that may be thought of as little "garbage disposal units". This is because these structures have the ability to secrete hydrolytic digestive enzymes which break down fats, carbohydrates, proteins, nucleic acids, and other large molecules, into smaller ones capable of being metabolized by the enzyme systems of mitochondria. The *lysosomes* of white blood cells, for example, can "digest" bacteria and other unwanted microorganisms after the cell envelopes and engulfs them by phagocytosis (see Chapter 2) into self-contained capsules called vacuoles. Even though their importance in health and disease is certain, all of the precise ways that lysosomes can affect changes are not yet clearly understood.

Feature 2: The Human Machine Has the Ability to Perceive and Interact with Both its Internal and External Environment, and its Output is Mechanical

Locomotion of parts or all of the animal body (accomplished by the action of *Striated Skeletal Muscles*) is crucial for its survival. Not only is such locomotion essential for finding, procuring and ingesting food from the *external* environment, but it also plays a major role in getting the fuel to the various body tissues *internally* -- e.g., muscles of mastication for chewing, the diaphragm and thoracic muscles for inspiration and expiration of air, peristaltic visceral smooth muscles for moving the fuel along in the Alimentary System, cardiac and vascular smooth muscle for propelling the blood through the Cardiovascular System and maintaining proper vascular "tone", visceral smooth muscle for maintaining the internal environment of the body, and so on.

The human engine interacts with its environment, both external and internal, by means of five major physiologic organ systems, one of which (the Striated Skeletal Muscular System) comprises nearly half of the total weight of the body in men, and about a quarter in women. The other four are the Nervous System (sometimes referred to collectively with the muscular system as the Neuromuscular System), the Skeletal System (sometimes referred to collectively with the muscular system as the Musculoskeletal System, or, even including the nervous System, as the Neuromusculoskeletal System), the Articular System of Joints, and the Sensory System (containing several subsystems as described further below).

The Nervous System

This may be thought of as one of the electrical systems of our engine, responsible for information transport in electrochemical form (called *Action Potentials*) that activates all moving parts and stimulates glandular function. This is the so-called "motor function" of the nervous system, which is coded into a unique physiologic syntax that can be externally monitored, recorded, and deciphered (see subsequent chapters, especially 6). At the root of this ability to understand and reproduce physiologic signals is our desire to build and install synthetic, man-made, prosthetic organs that can replace the real thing, yet operate the same way when innervated by the remaining tissue that has not been excised (ibid.). The Nervous System includes Cerebrospinal (Table 1-IX), Peripheral, and Autonomic components as subsystems. The latter branches from the former through a series of interconnections that involve intermediate nerves called, collectively, communicating rami or *Ramus Communicans*.

TABLE 1-IX

MAJOR ANATOMICAL FEATURES OF BOTH THE

CEREBROSPINAL AND PERIPHERAL NERVOUS SYSTEMS

Twelve Pairs Of Cranial Nerves:	31 Pairs of Spinal Nerves:
1: Olfactory (Sense of Smell)	8 Pairs of Cervical Branches
2: Optic (Sense of Sight)	12 Pairs of Thoracic Branches
3: Oculomotor (Eye Musculature)	5 Pairs of Lumbar Branches
4: Trochlear (Proprioception)	5 Pairs of Sacral Branches
5: Trigeminal (Three Branches):	1 Pair of Coccygeal Branches
Ophthalamic (Cornea)	Cervical Plexus:
Maxillary (Nose, Face, Scalp)	Lesser Occipital Nerve
Mandibular (Mastication,	Great Auricular Nerves
Tongue, Mouth, Teeth, Chin)	Anterior (Transverse)
6: Abducens, to Lateral Rectus	Cutaneous Nerves
Muscle Of The Eye	Supraclavicular Nerves
7: Facial (Gustation, Taste,	Phrenic (Diaphragm) Nerves
Facial Expressions)	Brachial Plexus:
8: Vestibulocochlear, Auditory	Supraclavicular Nerves
Vestibular Branch (Balance	Infraclavicular Nerves
and Equilibrium)	Musculocutaneous Nerve
Cochlear Branch (Acoustic)	Median Nerve (Arm)
9: Glossopharyngeal (6 Branches)	Ulnar Nerve (Forearm)
Carotid; Tympanic; Lingual;	Radial Nerve (Forearm)
Pharyngeal; Tonsillar; Sinus	Suprascapular Nerve
Nerve of Hering	Subscapular Nerve
10: Vagus, or Parasympathetic	Rhomboid Nerve
11: Accessory (Sternomastoid and	Musculospiral Nerves (Radial)
Trapezius Muscles)	Posterior Thoracic Nerve
12: Hypoglossal (Tongue)	Musculothoracic Nerve
Sacral Plexus: (Legs)	Circumflex Nerves
Great and Small Sciatic Nerves	Intercostal Nerves (Thoracic)
Superior and Inferior Gluteal	Lumbar Plexus:
Tibial (Medial Popliteal) Nerve	Iliohypogastric Nerve
Common Peroneal (Lateral	Ilioinguinal Nerve
Popliteal Nerves)	Genitofemoral Nerve
Deep Peroneal Nerves	Lateral Cutaneous Nerve
Pudic and Pudendal Nerves	of the Thigh
Sural and Sacral Nerves	Obturator Nerves
Superficial Peroneal Nerve	Femoral Nerves (Crural)
Posterior Cutaneous Nerve	Lumbar Nerves
of the Thigh	Genitocrural Nerve
Posterior Tibial Nerve	Accessory Nerves
Saphenous Nerves	Anterior Crural Nerves
Coccygeal Plexus:	External Cutaneous Nerve
Anococcygeal Nerves	Superior, Inferior Gluteal
Single Terminal Nerve (Cranial) to	Hypogastric and Iliac Nerves
Nasal Septum	Communicating Rami

THE CEREBROSPINAL AND PERIPHERAL (OR SOMATIC) NERVOUS SYSTEMS

In general, the cerebrospinal subsystem (part of the Central Nervous System which includes the brain, as well) and the somatic subsystem both receive and transmit information, providing inputs to and responding to outputs from the CNS, or CPU of our engine (see later). The inputs generally come via sensory nerves that are responding to stimulation of internal or peripheral receptors (transducer signals, see Sensory System below and Chapter 6). Outputs generally go via motor nerves to various muscles and glands called effector organs (see Chapter 5). If the entire round-trip path, from peripheral receptor via sensory somatic nerves to the cerebrospinal nervous system, to the central nervous system, and then via motor somatic nerves to the respective effector organs, occurs as an automatic (involuntary) response to a stimulus, the event is called a *Reflex Arc*. Otherwise, it represents voluntary responses to various stimuli, wherein the former may, or may not be under conscious control. The Cerebrospinal Nervous System is broken down further into a Cerebral Part (Cranial Nerves) and a Spinal Part (Spinal Nerves and Plexus), the major anatomical features of which are shown in Table 1-IX.

The 12 Cranial Nerves are so-named because they originate in the brain and emanate from there *directly* out to their respective target organs, without first passing down through the spinal cord. In addition to several other functions, nerves 3, 7, 9 and 10 also give rise to the *Cranial Portion* of the *Parasympathetic Nervous System*, which is described further below. The 31 pairs of Spinal Nerves emanate from the intervertebral spaces between the 33 bony segments of the spinal column, giving rise to 6 basic networks of nerves, each of which is called a *Plexus*. The first four pairs of cervical spinal branches give rise to the *Cervical Plexus* (see Table 1-IX); the next four pair spawn the *Brachial Plexus*. Moving down the spine, the 12 pairs of *Thoracic Branches* yield a network of rami that innervate the intercostal ("between-the-ribs") regions of the thorax; while the first four pairs of lumbar spinal branches give rise to the *Lumbar*

Plexus. Emanating out from the last two lumbar branches and the first four sacral branches is the *Sacral Plexus*; and, to complete this so-called *Peripheral* or *Somatic Nervous System* is the *Coccygeal Plexus*, which issues forth from the last two sacral branches and the remaining coccygeal branches. Going one step further, the 12 thoracic branches also give rise to the *Thoracic Portion* of the *Sympathetic Nervous System* (see below); the first three branches in the lumbar region give rise to the *Lumbar Portion* of the *Sympathetic Nervous System*; and the middle three branches of the sacral region spawn the *Sacral Portion* of the *Parasympathetic Nervous System*, to complete the so-called *Autonomic* or *Visceral Nervous System*.

The functional unit of the nervous system is the *neuron*, the anatomy, physiology and details of which are addressed in more depth in Chapter 6 (see Figure 6-3). In all, there are an estimated 20×10^9 nerve cells, or neurons, per person. Each *spinal* nerve (a *collection* of nerve cells) is attached to the spinal cord by two roots: a dorsal (towards the back) or posterior *sensory* root containing *afferent* nerve fibers that convey action potentials *to* the spine; and a ventral (towards the stomach) or anterior *motor* root containing *efferent* nerve fibers that convey action potentials *from* the spine. On passing through the intervertebral foramen (opening), a typical spinal nerve divides into four branches: a *recurrent branch*, a dorsal ramus or *posterior primary division*, a ventral ramus or *anterior primary division*, and two *rami communicates* (one white, one gray) which pass, via preganglionic efferent nerve fibers, to *ganglia* (masses of nerve cell bodies) of the *Sympathetic Trunk* (see next section).

THE AUTONOMIC OR VISCERAL NERVOUS SYSTEM

The overall function of the autonomic nervous system is to maintain homeostasis and the internal environment of the human machine. It accomplishes this mission by sending peripheral nerve fibers to smooth muscle tissue, cardiac muscle tissue, some exocrine and endocrine glands, and some skeletal muscle tissue. This it does through two basic subsys-

tems working in opposition to one another, namely, the previously-
mentioned sympathetic, and the parasympathetic Systems.

Both sympathetic and parasympathetic nervous systems receive information via visceral afferent sensory nerve fibers (*to* the spinal cord),
and transmit information via visceral efferent (or autonomic) motor nerve
fibers (*from* the spinal cord). The fibers innervate primarily portions of
the circulatory system, the digestive system, the excretory system, the
exocrine system, the endocrine system, and smooth muscle tissue. Their
mechanism of action involves the release from the nerve endings of substances called *Neurotransmitters* (Chapter 6): Acetylcholine from the
parasympathetic (*Cholinergic*) postganglionic nerve endings, and
Norepinephrine (noradrenalin) from the sympathetic (*Adrenergic*)
postganglionic neurons. Thus, autonomic nerves (indeed, *all* nerves)
communicate with their target organs (or other nerves) by "squirting" into
the synaptic space between them (nerves do not actually touch one another or their target organs) a chemical neurotransmitter, which is "picked
up" by receptor sites on the receiving nerve fiber or target organ (see
later).

There are two types of adrenergic receptors attached to most
effector organs: α-receptors (*excited* by norepinephrine and epinephrine)
and β-receptors (*inhibited* by adrenalin and noradrenalin). Among the
actions associated with α-receptors are constriction of blood vessels in the
skin and abdominal region, relaxation of the smooth muscle of the
intestine, and contraction of the radial fibers of the iris (dilating the pupil). Actions associated with β-receptors include dilation of blood vessels
in skeletal muscles and the coronary circulation, increase in the rate and
force of contraction of the heart, and relaxation of the smooth muscle of
the bronchi and their branches. We shall return to the concept of
receptor sites as they relate to neural transmission of information both
later in this chapter (see section on Feedback Control and Homeostasis)
and again in Chapter 6.

As mentioned briefly above, the parasympathetic, or Craniosacral
Division of the Autonomic Nervous System arises from the 3rd, 7th, 9th,

and 10th Cranial Nerves (the *Cranial Portion*), and the 2nd, 3rd, and 4th Sacral segments of the spinal cord (the *Sacral Portion*); whereas the sympathetic, or Thoracolumbar Division of the Autonomic Nervous System arises from all of the thoracic (the *Thoracic Portion*), and the first three lumbar (*Abdominal* and *Pelvic Portions*, or *Lumbar Portion*) segments of the spinal cord. The sympathetic division, however, also has an upper portion that runs parallel to the spine above the thoracic region, although the postganglionic sympathetic nerve fibers that emanate from these *Cephalic* and *Cervical* portions of the autonomic nervous system originally emanate from the 1st, 2nd and 3rd thoracic spinal segments, and then propagate upward towards the neck and head. The parasympathetic nerve network distributes as follows:

From the 3rd Cranial Nerve via the *Ciliary Ganglion* to the eye; from the 7th Cranial Nerve via the *Sphenopalatine Ganglion* to the lacrimal (tears) exocrine glands, and via the *Submaxillary Ganglion* to the sublingual and submandibular salivary exocrine glands; from the 9th Cranial Nerve via the *Otic Ganglion* to the parotid salivary gland; from the 10th Cranial Nerve (the Vagus Nerve) *directly* to the lungs and stomach, via the *Cardiac Plexus* to the heart, via the *Superior* and *Inferior Mesenteric Plexuses* to the intestines, and via the *Celiac* (*Solar*) and *Renal Plexuses* to the liver, kidneys, and gonads; and, from the 2nd, 3rd, and 4th sacral segments of the spinal cord (i.e., the *Pelvic Nerves*) via the *Pelvic Plexus* to the urinary bladder, the Adrenal (Suprarenal) endocrine glands, the genital organs, the uterus, and portions of the large intestine. The Sympathetic nerve network distributes as follows:

From the 1st thoracic vertebra via the *Superior Cervical Ganglion* to the eye, the nose, the sublingual, submandibular and parotid salivary glands, via the *Cardiac Plexus* to the heart, and *directly* to the Thyroid Gland; from the 2nd thoracic vertebra *directly* to the head, heart, eyes, and lungs; from the 3rd thoracic vertebra *directly* to the head, heart, eyes, lungs, and upper extremities of the body (including sweat glands and blood vessels of the skin); from the 4th thoracic vertebra *directly* to the heart, lungs and upper extremities; from the 5th thoracic vertebra *directly* to the liver, gall bladder, and upper portions of the gastrointestinal system; from the 6th through 9th thoracic vertebrae via the *Greater Splanchnic Nerve* to the gastrointestinal system; from the 10th and 11th thoracic vertebra via the *Lesser Splanchnic Nerve* to the gastrointestinal system, the kidneys, and the gonads; from the 12th thoracic vertebra via the *Least Splanchnic Nerve* through the *Renal Plexus* to the kidneys, gonads and lower portions of the intestines; and, from the 1st through 3rd lumbar vertebrae via the *Hypogastric* and *Inferior Mesenteric Plexuses* to the adrenal glands, the kidneys, the urinary bladder, the genital organs, and the lower extremities (including sweat glands and blood vessels of the skin). These are but some of the many distributary pathways by which autonomic control of the body's internal environment becomes manifest. Such control is examined again below and in Chapter 6.

The Muscular System

Perhaps the most obvious characteristic of man is that he (or she) is always *doing* something. Doing involves movement; movement involves musculoskeletal mechanics. Musculoskeletal mechanics involves -- depending on how one counts them (some investigators list as separate muscles what others regard as portions of adjacent muscles) -- at least 639 to as many as 656 specialized transducers that are specifically designed to transform electrochemical energy into mechanical work (see Schneck, D. J., 1984b and Chapter 5). These transducers, accounting for up to 40-43% or more of total body weight in men, and 23-25% in women, are called striated skeletal muscles (the single largest *tissue* in the body); and they transmit work, via connecting tendons (or sinews) to a type of engineering structure composed of levers (bones) that convert muscular work into the maintenance of postural balance, or the locomotion of parts or all of the animal body. The latter is, of course, the 206 bones that comprise the human skeletal system -- and that are held together and in alignment by the ligaments that span at least the same number of joints, called the Articular System (see further below).

The Neuro-Musculo-Articular-Skeletal System gives mobility to the engine, allowing it to seek and utilize fuel, to escape from predators, to perform useful tasks, to protect itself by building sheltors, to reproduce, and to otherwise interact with its environment by means that involve motion. These four subsystems (Nervous, Striated Skeletal Muscle, Articular, and Skeletal) illustrate the mechanical output of the engine. They are also important in giving the body mechanical and structural stability, and for protecting delicate organs (like the heart and lungs) and tissues (like bone marrow). The skeletal system further acts as a reservoir for metals and ions, like calcium and phosphorus. And finally, as shall be discussed in more detail in Chapter 4, the muscular system plays a very important role in temperature regulation. That is, because the biochemical reactions involved in producing contraction of striated skeletal muscles are highly exergonic (energy-releasing), muscles serve as a convenient source of heat during periods of cold-stress. Thus, under the control of the hypothalamus, the *shivering reflex* is but one of several mechanisms by which thermoregulation (or homothermosis) is accomplished in physiologic systems. Others are described elsewhere throughout this text, and especially in Chapter 4. Listed in Table 1-X are some of the major muscle groups of the head, neck, vertebral column, trunk, upper, and lower extremities. They, and others are tabulated and described in more detail in references such as Thomas (1981).

The Articular System

This system, the study of which is called *Arthrology*, includes all joints (pivot points around which bones rotate) and ligaments (strong fibrous connective tissue that binds together the articular ends of bones). The articulations that bones make with one another in the skeleton are of three kinds:

1. Immovable Joints (*Synarthroses*): These include bones that are fused together so firmly that *no* relative movements are possible. The lines at which they join one another may still be visible, however, and

TABLE 1-X

MAJOR STRIATED SKELETAL MUSCLES OF THE HUMAN BODY

Head, Neck, and Trunk:
    Orbicularis Oris and Oculi
    Sternocleidomastoid
    Epicranius (Scalp Muscles)
    Genioglossus, Hypoglossus, and
        Styloglossus of the Tongue
    Occipitofrontalis (Scalp)
    Zygomaticus Major and Minor
    Levator (6 sets)
    Masseter (Chewing Muscles)
    Lateral and Medial Pterygoideus
    Buccinator
    Mentalis
    Infrahyoid (4), Geniohyoid, and
        Suprahyoid (4) Musculature
    Temporalis (Closes Jaw)
    Digastric
    Pectoralis Minor and Major
    Serratus, Anterior and Posterior
    Rectus Abdominis
    Levator Scapulae
    Rhomboideus (2)
    Splenius Capitis
    Sphenosalpingostaphylinus
Lower Extremities:
    Long and Short Biceps Femoris
    Peroneus Brevis and Longus
    Anterior and Posterior Tibialis
    Flexor Digitorum Longus
    Extensor Digitorum Longus
    Plantaris and Soleus Muscles
    Gastrocnemius
    Sartorius
    Gracilis
    Popliteus
    Four Sets of Hamstrings
    Semi-Tendinosus
    Semi-Membranosus
    Four-head Quadriceps Femoris
    Rectus Femoris
    Vastus Lateralis, Medialis,
        and Vastus Intermedius
    Pectineus
    Adductor Longus
    Gluteus Maximus, Minimus, Medius
    Iliopsoas
    Tensor Fasciae Femoris (Latae)
	Superficial And Lateral Cervical
	Muscles And Muscles Moving The
	Vertebral Column:
	    Erector Spinae (Sacrospinalis)
	    Three Iliocostalis Muscles:
	        Lumborum
	        Thoracis
	        Cervicis
	    Three Longissimus Muscles:
	        Thoracis
	        Cervicis
	        Capitis
	    Three Spinalis Muscle Sets:
	        Thoracis
	        Cervicis
	        Capitis
	Upper Extremities:
	    Flexor Carpi Radialis
	    Flexor Carpi Ulnaris
	    Teres Minor and Teres Major
	    Supinator and Pronator Teres
	    Coracobrachialis
	    Pectoralis Major
	    Deltoid
	    Supraspinatus
	    Infraspinatus
	    Latissimus Dorsi
	    Subscapularis
	    Brachialis
	    Three-headed Triceps Brachii:
	        Long Head Triceps
	        Lateral Head Triceps
	        Medial Head Triceps
	    Two-Headed Biceps Brachii:
	        Long Head Biceps
	        Short Head Biceps
	    Anconeus
	    Brachioradialis
	    Pronator Quadratus
	    Extensor Carpi Radialis Brevis
	        and Longus
	    Extensor Carpi Ulnaris
	    Palmaris Longus
	    Trapezius
	    Extensor Digitorum Communis
	    Pollicis (Thumb)
	    Lumbricales Manus

look somewhat like the seams made when pieces of cloth are sewed together -- hence their being called *sutures*. Joints of this kind, also known
as *Synostoses* ("Os" = "bone"), are found between the 8 cranial and 14
facial bones of the skull, between the 3 fused bones (the Ilium, the
Ischium, and the Pubis), lying on each side of the pelvic girdle, that make
up the Innominate ("un-named") hip bone of the pelvic girdle, and between the five fused vertebrae that make up the triangular-shaped Sacrum
forming the base of the vertebral column.

2. Partially Movable Joints (*Amphiarthroses*): These are found especially between the vertebrae of the backbone -- involving fibrocartilage
connections (*Symphyses*); at the articulation between the two bones of the
forearm and those of the leg shank (tibio-fibular, or tibiofibular
syndesmosis joints) -- involving ligament connections (*Syndesmoses*); and
at the junction of the chest bone (sternum) with the rib cage (sternocostal
joints, and the heads, tubercles and necks of the ribs) -- involving cartilage
connections (*Synchondroses*). Such joints, also known as *Arthrodial*
Joints, allow one bone to *glide* over the surface of another, but do not
allow the bones to move freely upon one another. The twisting and
bending movements of the trunk, for example -- especially when the trunk
is not held straight -- are made possible by the sliding of the vertebrae
upon one another at arthrodial vertebral arch joints and amphiarthrodial
vertebral body joints. The small bones in the wrist (carpals) and ankle
(tarsals) also articulate in a partially movable fashion through arthrodial
carpometacarpal, intercarpal, intermetacarpal, tarsometatarsal, inter-
tarsal, intermetatarsal, subtalar, calcaneocuboid, and talocalcaneona-
vicular joints, respectively. The collar bone (clavicle) has an arthrodial
connection with the shoulder blade (scapula) at the acromioclavicular
joint, as does the sacrum with the coccyx at the sacrococcygeal joint, and
the innominate bone (hip) with the sacrum near the lower anterior portion of the hip at the pubic joint (or Pubic Symphysis). Other
amphiarthrodial joints include the costovertebral, costochondral, and
sternoclavicular articulations of the rib cage.

3. Freely Movable Joints *(Diarthroses,* or, *Synovial)* are joints that allow bones comparative freedom of movement. They are further classified in three subgroups depending upon whether the degrees-of-freedom are around one, two, or three axes in space.

Joints allowing movement around only one axis of rotation are of two basic types: those where the axis of rotation is parallel (or nearly parallel) to the long axis of the bone that is rotating *(Trochoid,* or, *Pivot Joints);* and those where the axis of rotation is perpendicular (or nearly perpendicular) to the long axis of the bone that is rotating *(Ginglymus,* or, *Hinge Joints).* The joint between the humerus (upper arm bone) and the ulna (forearm bone) at the elbow is of the latter type. It allows for flexion (reducing the angle between the two bones involved) or extension (increasing the angle between the two bones involved) of the forearm in a manner approximating the action of a hinge (like the spoke of a wheel relative to the axis around which the wheel turns). The ankle (Talocrural) joint, the interphalangeal joints between the first and second, and second and third digits of the fingers and toes, the Temporomandibular joint at the jaw, and some of the metacarpophalangeal and metatarsophalangeal joints in the hands and feet, respectively, are other examples of hinged joints.

In a Trochoid-type uniaxial articulation the rotation is analogous to that of a motor shaft, relative to the shaft-centerline-axis around which it turns. Examples include the Central Atlanto-Axial Joint that exists between the atlas (first vertebra of the neck) and axis (second vertebra of the neck) -- which allows an individual to turn the head from left to right (and back) around an axis lying relatively parallel to the centerline of the spine; and the proximal (Radioulnar), Middle, and distal (Radiocarpal) joints between the radius and the ulna (forearm bones) -- which allow supination (palm up) and pronation (palm down) rotations at the wrist around an axis lying relatively parallel to the forearm.

Joints allowing movement around two axes of rotation are *also* of two basic types: those where the surfaces along which rotation takes place are convex in two mutually perpendicular directions, but with different radii of curvature -- like the surface of a bicycle or automobile tire (i.e., a toroid of revolution) -- are called *Condyloid Joints*; and those where the surfaces along which rotation takes place are convex in one direction, but concave in a mutually perpendicular direction -- just like a saddle -- are called, appropriately enough, *Saddle*, or, *Receptive Joints*. A good (and one of the few anatomical) example(s) of the latter is the *Pollex* Joint at the base of the thumb where it articulates with the palm of the hand.

The articulation of the occipital bone of the skull with the atlas (the first vertebra of the neck, on which the skull rests), known as the *Atlanto-Occipital Joint* between the cranium and the spine, is an example of a condyloid joint. This articulation allows the movement of the head chin-down towards the chest and chip-up towards the back (around one axis of rotation between the shoulders), and, right-ear-towards-right-shoulder or left-ear-towards-left-shoulder rotation around a second axis of rotation that is perpendicular to the chest bone (sternum). These movements take place along two convex surfaces where the occipital bone and atlas articulate. Other examples of condyloid joints are most of the metacarpophalangeal and metatarsophalangeal connections between the fingers and the palm of the hand and the toes with the ball of the foot, respectively; the wrist joint, parts of the ankle joint, and the knee joint.

The knee joint is, in fact, a strange joint to classify. Anatomically, it is *built* like a condyloid joint and hence is normally categorized as such. But mechanically, it *behaves* like a combination of a Ginglymus (hinge) and Arthrodial (partially movable) joint, in that it only allows basically flexion and extension of the lower leg around one axis of rotation hinged at the knee. Furthermore, this articulation between the femur and tibia (as is the case with many joints in the body) does not allow *pure* rotation in the strict sense of the word. Rather, when flexion takes place in the knee joint, the condyles (rounded protuberances at the ends of the bone) of the tibia first *roll* backwards (in rocking-chair fashion) over the surfaces of the femur, after which further flexion takes place by pure tibial rotation about an axis located in the femur. The net movement is thus a combination of rotation plus translation, with the instantaneous center of rotation traversing an arc in space. Joints like the knee, which show both ginglymoid (hinge) and arthrodial (gliding) characteristics are sometimes called an *Amphidiarthrodial* articulation.

Joints allowing movement around three axes of rotation are known as *Enarthrosis* Joints. These, akin to a ball-and-socket joint, restrict action least of all, allowing virtually complete freedom of motion in three mutually perpendicular directions. Included in this category are the junction between the scapula and the humerus at the *Shoulder Joint*, and the articulation between the pelvic girdle and the femur at the *Hip* (Coaxal, Sacroiliac, or Pelvic) *Joint*, which allow movement of the arms and legs, respectively, in almost every possible direction.

The ligaments that keep bones in alignment and bind them together are generally named according to *which* bones they connect (e.g., the Patellar or knee ligament, the Inguinal or groin ligament, the Pubo-Femoral ligament, the Femoral-Lumbosacral ligament, the Iliofemoral ligament, the carpal ligaments, and so on); *where* they are located relative to the joints they span (e.g., ventral or towards the stomach, anterior or towards the front, posterior or towards the back, medial or towards the midline of the body, lateral or towards the sides of the body, proximal or nearest the joint, distal or furthest from the joint, and so on); *what* they are made of (e.g., capsular or fibrous material); *who* discovered and first described them anatomically (e.g., Poupart's ligament in the inguinal region or Gimbernat's ligament in the abdominal region); and/or what they *look* like (e.g., annular, broad, cruciform, falciform, fundiform, rhomboid, round, spiral, trapezoidal, triangular, yellow, and so on). Obviously, then, one finds a rather impressive list of hundreds of ligaments that bind together the 206 bones of the body. Before moving on to the latter, however, we remark that all freely movable joints contain a joint cavity filled with a lubricating fluid called *Synovial* ("Syn" = "like" + "Ova" = "eggs", i.e., having the consistency of egg whites) *Fluid*. Thus, these joints are also referred to as synovial joints, as opposed to the fibrous and cartilaginous nonsynovial joints which are either immovable or partially movable.

TABLE 1-XI

MAJOR ANATOMICAL FEATURES OF THE

206 BONES OF THE SKELETAL SYSTEM

29 Skull Bones:	64 Upper Extremity Bones:
8 Cranial Bones:	2 Clavicles (Collar Bones)
1 Occipital Bone	2 Scapulae (Shoulder Blades)
2 Parietal Bones	2 Humerus (Brachium, Arms)
1 Frontal Bone	2 Ulna, and 2 Radius, making
2 Temporal Bones	up the Antebrachium Forearms
1 Sphenoid Bone	16 Wrist Carpals:
1 Ethmoid Bone	2 Navicular or Scaphoid
14 Facial Bones:	2 Lunate; 2 Trapezium
2 Nasal Bones	2 Hamate; 2 Trapezoid
2 Maxillary Bones	2 Triangular or Triquetrum
2 Palatine Bones	2 Pisiform; 2 Capitate
2 Zygomatic (Malar) Bones	38 Bones of the Hands:
1 Mandible	10 Metacarpals in the Palm
2 Lacrimal Bones	28 Finger Bones, Ossa Manus,
1 Vomer Bone	3 Phalanges (Phalanx) to
2 Inferior Nasal Conchae,	Each Finger Digit, and
or, Inferior Turbinates	2 Bones to Each Thumb
6 Auditory Ossicles Of The Ear:	62 Lower Extremity Bones:
2 Malleus Bones (Hammers)	2 Pelvic or Hip Bones
2 Incus Bones (Anvils)	2 Femurs or Thigh Bones
2 Stapes Bones (Stirrups)	2 Patellas or Knee Caps
1 Hyoid Tongue Support Bone in	2 Tibias (Shin Bones) and,
the Throat	2 Fibulas, comprising the Crus
26 Movable Vertebrae: (33 Total)	or Lower Leg
7 Cervical (including the Atlas,	14 Ankle Tarsals:
C-1; and the Axis, C-2)	2 Calcaneus (Heel) Bones
12 Thoracic (T-1 to T-12)	2 Talus or Astragaloids
5 Lumbar (L-1 to L-5)	2 Navicular (Instep) or
1 Sacral (S-1 to S-5, Fused)	Scaphoid Bones
1 Coccygeal (4 Fused Bones)	2 Cuboid Bones
24 (12 Pairs) Thoracic Ribs, Or,	2 Medial (First Cuneiform)
Costae: 7 Pair (14 Bones) are	or Internal Bones
True Ribs (actually attached by	2 Middle (Second Cuneiform)
Costal Cartilage to the Sternum)	or Intermediate Bones
3 Pair (6 Bones) are	2 Later (Third Cuneiform)
False Ribs (attached to True)	or External Bones
2 Pair (4 Bones) are	38 Bones of the Feet:
Floating Ribs (not attached to	10 Metatarsals in the Foot
either the Breast Bone, or to	28 Toe Bones, Ossa Pedis,
another rib)	3 Phalanges (Phalanx) to
1 Breast Bone, or Sternum	Each Toe Digit, and
Osteoblasts (Make Bone Tissue)	2 Bones to Each Hallux,
Osteoclasts (Resorb Bone Tissue)	Or Big Toe
Osteocytes (Bone Cells)	Two Longitudinal, and
Cartilage (see Text)	One Transverse Arch In Each Foot

The Skeletal System

This system is comprised of the bones (made up of cancellous, or spongy tissue, and compact, or hard tissue) that act as levers, pivoted at joints to provide locomotion, and the entire protective skeleton of the human organism. As shown in Table 1-XI, the sternum (1 breast bone), together with the 12 pairs of ribs (24 bones) and 26 movable vertebrae, comprising the 51 bones of the trunk, are combined with the 29 bones of the head to make up the 80-bone structure known as the *Axial Part Of The Skeleton.* The remaining 126 bones, including the Pectoral (shoulder) Girdle and upper extremities, together with the Pelvic (hip) Girdle and the limbs of the lower extremities comprise the structure known as the *Appendicular Part Of The Skeleton.* In all, the bones of the body account for about 16% of total body weight; and, note further that more than half of them are in the hands and feet! It should also be pointed out, as mentioned already, that the hip bone, although *counted* as one bone, is actually a fusion of the Ilium, the Ischium, and the Pubis, to form the Innominate bone. The latter is united by ligaments with the Acetabulum, the Sacrum, and the Coccyx, along a suture-type joint called a *Synosymphysis*, to form the Pelvis.

The bones of the vertebral column encapsulate and protect the spinal cord, by which sensory nerves enter (afferent) the central nervous system and motor nerves leave (efferent). These enter and leave through intervertebral discs, which are sandwiched between the vertebrae. The vertebral column normally has a double-S-shaped curvature as viewed from the side -- being concave in the dorsal (towards the back) direction in the cervical (neck) region; convex in the dorsal direction in the thoracic (chest) region (an exaggeration of which is called *Kyphosis*, or, hunchback); concave *again* in the dorsal direction in the lumbar (lower back) region (an exaggeration of which is called *Lordosis*); and convex *again* in the dorsal direction in the sacral (tail) region. These curvatures in a sagittal plane (i.e., a plane through the midline of the body that divides it into a right and left half) are normal. Lateral curvature (known as *Scoliosis*) of the spine as viewed from the front or back, however, in a frontal plane (i.e., a plane through the midline of the body that divides it into a front, ventral or anterior, and back, dorsal or posterior half) is *not* normal.

Bones are covered at their sites of articulation by a connective tissue substance called *Cartilage*, which is classified as: (i) Costal (of-the-ribs); (ii) Trachial (of-the-windpipe); (iii) Fibrous (collagen-containing); (iv) Hyaline (articular); (v) Nasal (of-the-nose); and/or (vi) Elastic (Elastin-containing).

The Sensory System of Environmental and Somatic Perception

The Sensory System is responsible for maintaining balance and equilibrium of the human machine relative to its environment, and for transmitting information from the external and internal environment to the Central Nervous System for the purpose of maintaining homeostasis. This is the telemetry system of the engine, by which it keeps in constant touch with both its external and internal environments.

BIOLOGICAL SENSORY RECEPTORS

Central to the operation of this system are biological transducers (*ceptors*) that convert the various forms of energy to which they respond (the *adequate stimulus*) into electrochemical "bits" of sensory information (*receptor potentials*). These are transmitted through sensory nerves (in the form of *action potentials*) to the central processing units, or CPUs of the physiologic organism. The CPU (Central Nervous System consisting of the entire brain and all of the spinal cord) receives, analyzes and evaluates this information and, based on the results, calls for a physiologic response through the motor portions of the various nervous systems. These, in turn, control effector organs that carry out the desired response. The various nervous systems thus have both sensory and motor functions as they seek to maintain homeostasis. Biological transducers which monitor sound energy, light, heat, kinetic energy, potential energy, electromagnetic energy, and so on, are classified further as *Exteroceptors* (i.e., responding to stimuli from the external, or *ambient* environment), and *Interoceptors* (i.e., responding to stimuli from the internal, *somatic* and *visceral* environments). Their mechanism of action is discussed in more detail in Chapter 6. Here, our concern is just to itemize them in terms of the stimuli that they allow the organism to perceive.

DIMENSIONS OF AMBIENSOMATIC PERCEPTION

Our universe, as we know it, is basically made up of "stuff" (matter), various forms of energy (much of which constitutes what we call the electromagnetic spectrum), and interactions among these, which is embodied in the concept of momentum. These entities exist in a space-time environment, within which they manifest themselves in various forms. The *forms*, we identify by *Dimensions Of State* (see Chapter 2). Our *awareness* of these forms is embedded in the concept of *Dimensions Of Perception* (see below). And the anatomical organs that provide us with the *ability* to perceive our universe are what make up the *Sensory System*. That is, to each dimension of perception there corresponds an anatomical organ (or sets of organs). These are designed *specifically* to be responsive to the associated dimension of state that characterizes the particular quantity (mass, energy, or momentum) involved. We have, for example, organs that generate receptor potentials stimulated by:

(i) *inertia* -- i.e., organs which "sense" *mechanical* phenomena, such as linear or translational momentum, as embodied in the perceptual dimension of *mass*, M; and angular or rotational momentum as embodied in the perceptual dimension of *mass moment of inertia*, I_{cg};

(ii) *sound, light,* and other such forms of energy -- i.e., organs which "sense" *electromagnetic* phenomena as embodied in the fundamental perceptual dimension of *electric charge*, q;

(iii) *heat, cold,* and other such forms of energy -- i.e., organs which "sense" *thermodynamic* phenomena as embodied in the fundamental perceptual dimension of *temperature,* T;

(iv) the *Circadian* (daily) and/or the *Circannual* (yearly) cycles of our universe, as well as the *Physiological Rhythms* (e.g., heart-rate, respiration-rate, menstrual-cycle, and so on) of our own bodies -- i.e., organs which "sense" *temporal* periodicities in both our ambient and somatic (including visceral) environments, as embodied in the fundamental dimension of *time,* t, and which also "time" physiologic events accordingly;

(v) *physical contact* and/or *distance* -- i.e., organs which "sense" the *physical* three-dimensional *space* within which we live, as embodied in the dimension of *length,* L, and as also manifest in the sensations of touch, pressure, itch, pain, and vision; and,

(vi) *chemical reactions* -- i.e., organs which "sense" the *biochemical* states of matter which are embodied in its corresponding *taste,* or *smell,* or *color,* or *chemical consistency.* This includes, as well, ceptors that initiate chemical reactions in the body in response to the appropriate signal of adequate stimulus intensity.

Keeping in mind, then, the basic nature of our universe, its manifestation in dimensions of state, and the presence of anatomical biological sensory receptors that enable us to be aware of these states through dimensions of perception, let us take a closer look at some of the more specific organs involved in both ambient exteroception and somatic interoception, i.e., "ambiensomatic" perception.

SENSORY PERCEPTION OF THE EXTERNAL ENVIRONMENT: AMBIENT EXTEROCEPTION

Organs involved in this type of perception include basically those that constitute the *Special Senses* (sight, hearing, smell, taste, and touch); the *Vestibular* (or *Cochlear*) *Kinesthetic Senses* (proprioception, linear and angular acceleration, velocity, displacement, vibration, and impact); and the *Circadian Senses* (including those organs responsible for "measuring" time and synchronizing the body's internal processes with the daily events in its environment, see Moore-Ede, et al., 1982). Each of these is examined briefly below.

The sense of *Vision* (Sight) discriminates energy in the wavelength range from 397 to 723 millimicrons (nanometers, or, from around 400 to 750 Terahertz in frequency). This is accomplished through light-sensitive photoreceptor cells -- some 125×10^6 light-sensitive rods and 7×10^6 color-sensitive cones. The rods-to-cones ratio is generally in the neighborhood of 18-to-1 per eye. Each 10 to 100 rods (and the corresponding number of cones) converge onto a network of some 800,000 to 1,000,000 sensory nerve fibers that ultimately become the Optic Nerve (i.e., the 2nd Cranial Nerve). The anatomic organs from which our sense of sight originates are photo-optic transducers called *eyes*, the major features of which include the: Sclera (outer layer), Cornea, Choroid (middle layer), Iris (colored portion), Pupil, Retina (Inner layer), Ciliary Muscles (control lens), Lens, Aqueous Humor, Vitreous Humor, Fovea Centralis, Macula Lutea Retinae, and Optic Disk.

The eyes are also involved in our perception of the dimension of time. That is, our ability to perceive time seems to have evolved at least in part, from visual "cues" stemming directly from the 24-hour (Circadian) diurnal (day)/nocturnal (night), or light/dark cycles associated with the earth's rotation about its own axis, relative to the position of the sun at any given time. And, although the actual "cues" are no longer absolutely necessary to elicit "timed" biological responses -- i.e., these have become conditioned reflexes that represent genetically-coded adaptations to the inherent cycles of our universe (see Chapter 6) -- it has been shown (see Moore-Ede, et al., 1982) that the *absence* of such cues can alter significantly the timed-sequence of important physiologic processes. Some of these involve sleep/wake patterns, female menstrual cycles, and eating habits.

Finally, the eyes are also intimately connected with our perception of three-dimensional space, i.e., the dimension of length. Being able to "see" something gives immediate relevance to the concept of "size". Thus, of the five major dimensions necessary to describe the physical attributes of our universe, i.e., M,L,t,T, and q, the eye is *directly* involved in perceiving at least three of them (electromagnetic light energy, space, and time), and is peripherally or indirectly involved in perceiving the remaining two (inertia and temperature). No wonder, then, that various estimates suggest that as much as 90% or more of the environmental information received daily by the human organism enters the system through the eyes (the sense of vision); and, that indeed, "Seeing is Believing."

The sense of *Audition* (Hearing) discriminates energy in the frequency range from 20 to 20,000 Hertz (cycles per second), or, in the 17.16 to 0.01716 meter wavelength range (in dry air, at sea level, at a temperature of 68°F, where the speed of sound is 343.2 m/sec). This is accomplished through vibration-sensitive hair-cell receptors -- some 12,000 outer hair cells and 3,500 inner hair cells per ear. The receptors are excited by the ear drums, and ultimately converge upon the Cochlear Branch (i.e., the Acoustic or Auditory Nerve) of the 8th Cranial Nerve (the Vestibulocochlear Nerve). The anatomic organs from which our sense of hearing originates are vibro-auditory transducers called *ears*, the major anatomical features of which include the: Ear Lobes (Auricles), External Ear, External Auditory Canal (Acoustic Meatus), Tympanic Membrane (Ear Drum), Middle Ear (Tympanum), Ossicles (see Table 1-XI), Inner Ear (Labyrinth), Oval Window, Cochlea, Organ of Corti, Eustachian Tube, Tensor Tympani Muscle To Malleus Bones, Stapedius Muscle to Stapes Bones, and Cerumen (Ear Wax).

The ears are *also* involved in our perception of the dimension of time, especially as it relates to the inherent rhythm of music. Indeed, among the most primitive of instruments were those in the percussion family -- the drums; the "beat". Roederer (1975, 1982) and others have postulated that pulsed sound stimuli somehow create a resonance with the "natural clocks" of the brain that ultimately control cyclic physiologic function (like the heart beat or respiration rate) and behavior. "These clocks probably work on the basis of neural activity traveling in closed-looped circuits or engrams, or in any other neural wiring schemes that have natural periods of cyclic response" (ibid., 1975, pg. 165). It is quite possible that many of the therapeutic effects of music (music does, indeed, "soothe the savage beast,") may be traced to the effect of rhythm on sub-cortical structures of the brain as they relate to the perception of time, and to the concept of a "pleasing resonance," (Scartelli, J., 1990, Schneck, J. K., and Schneck, D. J., 1989). Moreover, one cannot deny the importance of rhythm as an element in dance, and the importance of dance and rhythmic body movements as a primitive (perhaps one of *the most* primitive) behavioral characteristics of humankind.

Finally, the ears are also intimately connected with our perception of *each other*, in the sense that auditory stimuli such as speech provide an effective means of communication. One may define speech as the process of uttering articulate words or sounds in order to communicate one's thoughts through verbal expression. The anatomical structures available to generate such cognitive means for expression originate on the left side of the brain (see later) and become manifest through the: Larynx, False Vocal Cords (Ventricular Folds of the Larynx), True Vocal Cords, or Vocal Folds, Vocal Lips of the Larynx, Vocal Ligaments, Rima Glottidis, Thyroarytenoid Vocal Muscle, Vocal Process, Arytenoid Cartilage, Epiglottis, and Thyroid and Circoid Cartilage. Whereas *speech* as a means for *communication* can be traced to meeting one's basic needs for survival, *music* as a means of *expression* appears to have no such comparable function -- at least none that we have thus far been able to identify. This makes the rise of music to such prominence in our society a particularly baffling phenomenon.

The sense of *Olfaction* (Smell) involves some 10 to 20 million chemical receptor cells that lie within a mucous membrane lining the external nose and nasal cavity. These discriminate among volatile substances that dissolve in the thin film of liquid which covers the mucous membrane, extending over an area of about 5 square centimeters. The chemical receptor cells send information to the central nervous system via the 1st

Cranial, or Olfactory Nerve. They, too, are important in giving us a sense of each other -- witness the success and importance in our society of industries that specialize in fragrances and pleasantly-smelling cosmetics. Among the anatomical features that relate to this sense of perception are the: Anterior nares of the nose (Nostrils), Vibrissae (Nasal Hairs), Neuroepithelial tissue containing Olfactory (Chemical Sensory) Cells, Nasal Septum (wall between the nasal cavities), Sinuses (Maxillary, Frontal, Ethmoid, Sphenoid), Nasal Cartilage, Posterior Naris (Choanal or funnel-like) opening into the Pharynx, and the Nasopharynx.

The sense of *Gustation* (Taste) involves chemical stimulation of receptor cells that are embedded in some 3000 to 9000 *Taste Buds* located on the tip, border and base of the tongue, on the soft palate of the mouth, and on certain parts of the throat. There are between 5 and 18 Gustatory Chemoreceptor Hair Cells (and between 30 and 80 non-nervous parenchymal cells) per taste bud. These discriminate among soluble substances that dissolve in the saliva and mucous membranes of the mouth, thereby activating the taste receptors. They, in turn, send information to the central nervous system via the 7th Cranial Nerve (the Facial Nerve). Taste buds can distinguish among four basic chemical sensations: sweet, sour, bitter, and salty. The success of the gourmet industry that caters to epicurean delights clearly reflects the social values of our perception of taste. In addition to the anatomical attributes already mentioned, others related to this physiological sense include: the Vallate and Circumvallate Papillae on the back of the tongue, the Fungiform Papillae on the tip and sides of the Tongue, Supporting Cells, and Taste Pores.

The sense of *Touch* is really several senses collectively grouped in one category. These are the so-called *Somatosensory*, or *Cutaneous* modes of perception embedded in the Integumentary System (see earlier discussion of the Excretory System), which allow the organism to discriminate among external: (i) *concentrated steady loading* (touch, in its traditional context, known as the *Tactile Sense*); (ii) *distributed steady loading* (pressure, or sense of *Baresthesia*); (iii) *concentrated* or *distributed time-varying loads* (cutaneous vibration up to some 200 cycles per second) and *impact* (these being generally called a sense of *Kinesthesia*); (iv) *temperature* (heat, cold, *Thermoesthesia*); (v) *pain* (*Algesthesia* or *Algesia*, including throbbing; terebrant -- sharp, piercing; paresthesic -- tingling, stinging; lancinating -- short, cutting; dull -- continuous, diffuse; cramplike; burning; and aching); (vi) *itch* (*Pruritis*); (vii) Skin *shear* and *irritation* (including the six basic modes of *grasping* things: cylindrical, palmar, tip, spherical, hook, and lateral); (viii) *humidity*; (ix) *resistance*; and (x) various types of *deformation* (including "pinching" and bending).

Tactile receptors for discrete touch are especially numerous in the upper dermis of the hands, feet, lips, and nipples, and in the mucous membrane of the tip of the tongue. They are also found in great numbers in the deepest epidermal layer of the fingertips -- where they allow one to perceive the consistency and shape of things via the *Stereognostic Sense* (also involved in grasping, or gripping), and, in the end organs of hair -- where they connect with basketlike arrangements of sensory nerve fibers around hair follicles. Tactile receptors are known as *Meissner's Corpuscles* and *Merkel's Tactile Discs*. They can be stimulated by direct contact of *Dermal Papillae* (small, nipplelike protuberances or elevations of the skin surface), or, by excitation of the hair-root (*Radix Pili*) endings in hair follicles when the hair shaft (*Scapus Pili*) is disturbed. These hair follicle and skin surface sensors also allow us to perceive vibration.

Pressure receptors for diffuse touch are also particularly numerous in the integumentary system, where they are known as *Vater's*, or *Pacinian Corpuscles*. The epidermis contains virtually all of the receptors that give rise to the sensation of *itch*; while the dermis and deep, lower dermis contain most of the receptors that give rise to the sensation of *pain* -- the former being associated with sharp pain, and the latter with more of the "aching" types of pain. Sensations of pain as perceived by these somesthetic receptors excite the endings of free, or naked sensory nerves, from which ends action potentials subsequently transmit the information to the central nervous system. Tactile and pressure senses are further instrumental in allowing us to perceive and interact with one another. Almost from birth, we require touching as a means of showing affection and of communicating emotions. Hugging, kissing, shaking hands, fondling, caressing, and other forms of physical contact thus play a vital role both socially and physiologically -- as well as giving us a perception of spatial dimensions (to complement the sense of sight). That is to say, we can *feel* whatever we can't *see*, to give us a sense of size and geometry.

The sense of *Temperature* (Thermoesthesia) allows us to perceive variations in ambient thermal conditions from very hot to very cold -- primarily for the purpose of ultimately maintaining body temperature within a very narrow range of 37°C (see Chapter 4). *Ruffini Corpuscles* are a type of thermal receptor located in the skin (i.e., peripheral cutaneous temperature sensors) which responds to heat energy in the infrared portion of the electromagnetic spectrum, from a wavelength of 0.022 cm down to 7×10^{-5} cm (corresponding to a frequency range from about 10^{12} Hertz up to 4×10^{14} Hertz). *Krause's End-Bulbs* are cutaneous temperature sensors that respond to cold. These two types of peripheral cutaneous exteroceptors have intimate neural connections with both the anterior and posterior portions of the hypothalamus in the brain (see later). There are also thermal transducers located in the tongue and nose, where mucosal peripheral sensors provide temperature information to the brain via the Trigeminal, or 5th Cranial Nerve.

We pointed out in our discussion of the day/night (Circadian) cycles as they relate to vision, that our perception of time seems to be directly linked through cues received via our other senses to the naturally occurring periodicities in our environment. So it is that our ability to perceive *temperature* also gives us some sense of time as it relates to the seasonal changes associated with the earth's annual trajectory around the sun. That is, we receive through these changes not only light/dark cues (longer days in the summer, shorter days in the winter, and so on) but also humidity/thermal cues that span the range from very hot and dry to very cold and wet, depending on the seasons of the year. And, again, the evolutionary results of this perception of seasonal (Circannual) time represent a genetically coded adaptation to the cycles of our universe.

In addition to the five "special" senses described above, one also has "exteroproprioceptive" senses which give us an awareness of posture, and movement (locomotion) relative to our external environment, and which provide us with the means to maintain balance and equilibrium. These senses are embedded in the Vestibular (Cochlear) System, which responds to linear and angular displacement, velocity and acceleration, and to impact and vibration -- giving us a kinesthetic perception of inertia. Our ability to perceive *angular* motion derives from hair-like sensory receptors called *Crista Acustica* located in three, semicircular, mutually orthogonal, interconnected arches called *Semicircular Canals* (or, the "Bony Labyrinth"). The three canals -- Lateral (horizontal), Anterior Superior, and Posterior Superior -- are filled with a fluid called Endolymph, and include such anatomical features as the: Endolymphatic Duct, Cupula, Round Window, Oval Window, Ampulla, Utricle, and Utricular Macula.

Our ability to perceive *linear* motion derives from corresponding sensory hair-like receptors located in the *Otolith Organs*. These excite both a *Cupular Membrane* and an *Otoconial Membrane*, which eventually transmit (along with the excitation signals received from the semicircular canals) sensory information to the brain, via the Vestibular Branch of the 8th Cranial Nerve (The Vestibulocochlear or Auditory Nerve, see Table 1-IX). The anatomy of the otolith organs includes, in common with the semicircular canals, the Utricle and Utricular Macula, and, in addition, the Saccule and the Saccular Macula, and the Otoconia. These, together with the semicircular canals, themselves, constitute the anatomy of the inner ear, or *Vestibule*.

SENSORY PERCEPTION OF THE INTERNAL ENVIRONMENT: SOMATIC AND VISCERAL INTEROCEPTION

For virtually every sense organ that allows us to perceive our external environment (or ambient conditions) by means of exteroception, there is a corresponding biological transducer that allows us to monitor our internal environment (somatic and visceral conditions) via interoception.

TABLE 1-XII

MAJOR ANATOMICAL FEATURES OF THE

INTEROCEPTIVE SENSORY TRANSDUCERS

Interoproprioception:	Visceroreceptors:
Neuromuscular Spindles	Thermoesthesia
Spindle Equator (Stretch Recep-	Hypothalamic Thermoreceptors
tors Act Like Strain Gauges):	monitor Blood Temperature
Type I-A Primary Annulospiral	Spinal Cord Cold Sensors
Afferent Excitatory	Visceral Thermodetectors in
Sensory Nerves	muscles; gastrointestinal
Type II Secondary Flower-	tract; kidney; ventricular
Spray-Ending, Afferent In-	wall of the heart; urinary
hibitory Sensory Nerves	tract; Visceral Organs
Nuclear Bag Intrafusal Muscle	Baresthesia
Fibers (Rate-Of-Stretch	Auricular Baroreceptors in
Excitation of I-A Nerves)	the walls of the Heart
Nuclear Chain Intrafusal Mus-	Aortic Baroreceptors in the
cle Fibers (Amount-Of-Stretch	Aortic Arch Sinuses
Excitation of I-A & II Nerves)	Carotid Sinus Baroreceptors
Spindle Poles (Establish Equa-	Vena Caval Baroreceptors
torial Stretch Sensitivity):	Chemoreceptors
γ-1-Plate Dynamic (Rate-Of-	Aortic Bodies in Aortic Arch
Stretch) & γ-1-Plate Static	Carotid Bodies in the Neck
(Amount-of-Stretch) Efferent	Sexual Sensations
Motoneurons Excite Spindle	Hypothalamus: Vegetative Functions
Poles of Nuclear Bag and	Pre-Optic Anterior Thermoregu-
Nuclear Chain Muscle Fibers;	latory Center (heat)
γ-2-Trail Static Motoneurons	Posterior Thermoregulatory
Excite Spindle Poles of only	Center (cold)
Nuclear Chain Muscle Fibers	Suprachiasmatic Nuclei (time)
Neurotendinous Spindles (Golgi	Ventromedial Hypothalamic
Tendon Organs)	Nucleus (Satiety Center)
Type I-B, Spraylike, Granular,	Lateral Hypothalamus (Hunger
Myelinated, Clasplike, Afferent	and Feeding Center)
Inhibitory Sensory Nerves	Supraoptic Nuclei (Osmorecep-
Respond To Muscle Forces	tors, Thirst)
Pacinian (Vater-Pacinian) Corpus-	Posterior Hypothalamus (blood
cles have Elliptic Lamellae that	pressure, pupillary dilation,
function as internal direction-	thermal shivering reflex)
sensitive, rate-sensitive pres-	Medial Preoptic Area
sure Transducers that respond to	(Cardioinhibitory Center)
Joint Position and Orientation	Perifornical Nucleus (Rage)
Bone Tissue Acts As A Piezo-	Mammillary Body (Feeding)
Electric Transducer to convert	Dorsomedial Nucleus (Gastro-
Stress into Electromagnetic	Intestinal Stimulation)
Information in the Functional	Paraventricular Nucleus
Adaptation Servomechanism:	(Conservation of Water)
Osteoblasts; Osteocytes	Posterio-Lateral Hypothalamus
Osteoclasts	(Cardioacceleratory Center)

Table 1-XII lists some of the major anatomical features of the interoceptive sensory transducers that allow us to perceive: (i) the position (orientation in space), weight, and resistance to movement of our limbs in relation to the rest of the body (interoproprioception); (ii) the kinematics (muscle spindles) and kinetics (Golgi Tendon Organs) of striated skeletal muscle function; (iii) factors that tend to cause imbalance (static sense) and other mechanical (or equilibrium) disturbances (dynamic sense); (iv) tissue pressure (Baresthesia); (v) body core temperature (Thermoesthesia); (vi) pain, fatigue, and/or discomfort; (vii) the quantity and distribution of total body fluids (tissue stretch receptors, Osmoreceptors); (viii) chemical imbalances (Chemoreceptors are sensory nerve endings that are stimulated by and react to chemical stimuli, such as carbon dioxide, oxygen, lactic acid generated during fatigue, and other chemical by-products of metabolic reactions); (ix) posture; (x) tissue deformation (stress, strain, bending, shear, torsion, cubical dilatation or contraction, and other types of deformation -- which is sensed by means of strain-gage-type tissue stretch receptors and/or piezoelectric transducers such as bone); (xi) thirst and hunger; (xii) sexual drives; (xiii) time; -- and any other parameters that are relevant to the *internal* environment of the organism (and disturbances of same).

All time cues perceived through extero-and-interoceptors seem to converge eventually on the Suprachiasmatic Nuclei of the Hypothalamus (Moore-Ede, et al., 1982), which plays the role of biological clock for most physiologic processes. That is, just as the Sino-Atrial or SA Node acts as a natural *Cardiac Pacemaker* for the heart, the bilateral suprachiasmatic nuclei, or SCN near the preoptic area of the anterior ventral hypothalamus acts as a natural *Cerebral Pacemaker*. Its function is to "measure" time and to synchronize the body's internal processes with the daily (Circadian), monthly (Lunar), and yearly (Circannual) cycles of our environment. Thus, our perception of time is also linked to our awareness of the biorhythms associated with physiologic processes, which processes include, for example, (i) the cyclic beating of the heart (time-scale on the order of 1 to 2 seconds); (ii) the respiration rate (time-scale on the order of 4 to 6 seconds); (iii) the rhythmic 90-minute sleep-stage cycles; (iv) the 28-day (monthly) female menstrual cycles; (v) the 280-day (approximately nine-month) gestation period; (vi) the 4 to 6 hour cravings for food (hunger cycles); (vii) rhythmic gait (walking) patterns, on the order of 80-or-so steps per minute; (viii) oscillating peristaltic movements in the gastrointestinal system; (ix) approximate 24-hour sleep/wake cycles; (x) diurnal variations of about 1.5°F in body temperature (higher during the day, cooler at night); (xi) the timed-release of endocrine hormones; (xii) the timed-release of urine (bladder function) and fecal wastes (rectal function); (xiii) the 365-day hibernation cycles in some species of animals; (xiv) the process of aging (and our awareness of same) -- which currently has a time-scale on the order of 70-or-so years -- and so on.

Control of the internal environment of the body by the hypothalamus is often referred to as its *Vegetative Function* (see Table 1-XII). Also, all proprioceptors and exteroceptors are frequently grouped collectively in the category of *Somatoceptors*; and the visceral senses are often referred to collectively as the *Seventh Sense*. What, then, is the *Sixth Sense* (after the five Special Senses)? Interestingly, although rats, cats, snails, worms, fleas, and certain plants (like Mimosa) respond actively to the presence of certain forms of ionizing radiation (such as X-Rays), so far as we presently know, our physiologic system contains no sense organ(s) that perceive this form of energy, so that is not the answer. However, you *can* pick from one of two possibilities: on the one hand, the Sixth Sense is often defined to be the somewhat abstract "generalized feeling of normal functioning of the body as a whole," -- a condition known as the sense of *Cenesthesia*. On the other hand, the Sixth Sense is also defined to be the equally abstract "perception of truths, facts, things that are to be, and so on, without the benefit of supporting reasoning," -- otherwise known as *Intuition*, or, if you will, the vague sense of "Hunch". Whichever definition you choose, neither has yet been identified with or otherwise associated with a corresponding anatomical organ, so we shall move on to the third major aspect of physiologic function.

Feature 3:   The Human Machine is Fine-Tuned Through a Variety of Complex Feedback Control Mechanisms That Maintain Within Prescribed Limits Variables Critical to Life, Itself, in a Process Called Homeostasis

Four specific organ systems -- The Central Nervous System (Brain plus Spinal Cord), The Autonomic Nervous System (Sympathetic and Parasympathetic Divisions), The Endocrine System (ductless organs of internal hormone secretion), and The Immune System -- together with a virtual myriad of individual servomechanisms, are responsible for continuously monitoring the "vital signs" of the engine, and for seeing to it that appropriate action is taken, if warranted, through the basic mechanisms discussed so far in this chapter, and in the chapters to follow.   Such monitoring is important not only from the point of view of just keeping the engine *functioning*, but also to keep it functioning *effectively* (i.e., making sure the *right things* get done); and *efficiently* (i.e., making sure the right things get *done right*, within the general limits of quality control); and, to keep it protected from environmental attack through immune responses such as are discussed further in Chapter 2. These are all important for the proper maintenance of life.

Recall that homeostasis is the name given to the steady-state internal environment that is required for the maintenance of life. The concept includes as well those processes which give the organism the ability to maintain within specific (and narrow) limits certain variables which insure a relatively stable condition of life. These include, for example: (i) *body temperature* (see earlier in this chapter, and Chapter 4 for a discussion of the Thyroid Calorigenic Effect, the Shivering Reflex in striated skeletal muscle, peripheral vasoconstriction during cold stress, perspiration and sweat gland secretions during heat stress, and other thermoregulatory homeostatic servomechanisms);   (ii) *blood pressure* (ibid., and see Chapters 2 and 3 for a discussion of baroreceptor homeostatic mechanisms as they relate to the concept of vascular peripheral resistance); (iii) *cardiac output* (e.g., the chronotropic effect, inotropic effect, Frank-Starling mechanism, and reactive hyperemia); (iv) *body fluid volume* (ADH or Pitressin control of renal function, involving tissue stretch receptors and osmoreceptors); (v) *muscle tone* (monosynaptic reflex arc, inverse myotatic reflex, nociceptive reflex arc, crossed-extension reflex, Phillipson's reflex, and others -- see Schneck, D. J., 1984b); (vi) *enzyme synthesis and production* (see Chapters 4 and 5); (vii) *balance and equilibrium* (ibid., Table 1-XII, proprioception and kinesthesis, Labyrinthine reflexes, "righting" reflexes, postural reflexes, nystagmus reflex -- see also Schneck, D. J., 1984b); (viii) *biochemical and metabolic reactions* (chemical and electrolyte balance, p_{CO_2}, p_{O_2}, dissolved gases, chemoreceptor servomechanisms, insulin, respiration); (ix) "Flight-or-Fight" response reflexes; and other processes that the body can control by its own regulatory mechanisms. These intimately involve both the senses discussed earlier, as *input* elements to the regulatory feedback control pathways, and smooth muscle tissue and glands, as *output* elements (or affectors) of the homeostatic responses.

The Central Nervous System

The brain is the main control center for this machine, receiving, processing, storing, recalling, manipulating, discharging, and otherwise handling huge quantities of data on a second-to-second basis. This three-pound physiologic tissue version of man-made computer chips makes split-second decisions, both consciously (the so-called "higher" brain) and subconsciously (the so-called "lower" brain). Working through the Cranial Nerves, the Spinal Cord, the Autonomic Nervous System, and the Endocrine System, the brain maintains complete regulation of all internal and external body functions. This, indeed, is *the* Central Processing Unit (or CPU) of our engine, consisting, with the Spinal Cord, of some 14×10^9 nerve cells and on the order of 10^{11} non-nervous, granular-like parenchymal (or accessory) cells, give-or-take a factor of ten. The brain

plus spinal cord constitute the central nervous system of the body, and
both begin to appear in a growing embryo as early as the third week following conception, when the budding human is but an eighth of an inch
long.

THE BRAIN (ENCEPHALON)

Composed of Substantia Grisea (Gray Matter, including the ten billion
cells that form the 1 mm-thin Cerebral Cortex, the Basal Ganglia, and the
various Nuclei of this soft lump of tissue), and Substantia Alba (White
Matter, including all of the intervening white medulated nerve fibers of
this organ), the brain is anatomically divided into a left and right cerebral
hemisphere, three major *areas* and several *subsections,* some of which are
tabulated in Table 1-XIII, which also includes the major cerebral
circulatory pathways.   The three major areas are the *Forebrain*
(Prosencephalon), the *Midbrain (Mesencephalon),* and the *Hindbrain*
(Rhombencephalon), as described further below.

The   Prosencephalon   includes   all   of   the   *Cerebrum*   (the
Telencephalon, Endbrain, or largest portion of the brain), and the
Diencephalic portion of the *Brain Stem* (see Table 1-XIII).   The latter is
a stem-like portion of the brain which connects the two cerebral hemi-
spheres   with   the   spinal   cord.   It   consists   of   the   Midbrain,   or
Mesencephalon, the Medulla Oblongata, the Diencephalon, and the Pons
Varolii. The Diencephalic portion of the Brain Stem, also known as the
Thalamencephalon, lies *between* the Telencephalon (Cerebrum) and the
Mesencephalon   (Midbrain).   It   includes   the   Epithalamus,   the
Metathalamus, the Hypothalamus, and the Thalamus -- hence its obvious
name (see Tables 1-XII and 1-XIII).

The two cerebral hemispheres (right and left), accounting for about
$\frac{7}{8}$'ths of the whole brain, consist of three primary anatomical subdi-
visions: the *Rhinencephalon,* or Olfactory Lobe, the *Corpus Striatum*
(composed of two basal ganglia, i.e., the Caudate and Lentiform Nuclei,
together with the Internal Capsule that separates them), and the

TABLE 1-XIII

SOME ANATOMICAL FEATURES OF THE HUMAN BRAIN

The Cerebral Circulation:	The Brain Stem:
Circle Of Willis:	Myelencephalon (Myelencephalic
2 Internal Carotid Arteries	Brain Stem; most posterior):
2 Posterior Cerebral Arteries	Posterior Corpora Quadrigemina
2 Anterior Cerebral Arteries	Tegmentum
Posterior Communicating Artery	Posterior Crura Cerebri
Anterior Communicating Artery	Superior Medulla Oblongata
Vertebral Artery	Mesencephalon (Mesencephalic
Ophthalmic Artery	Brain Stem; or Midbrain):
Middle Cerebral Artery	Corpora Quadrigemina
Superior Cerebellar Artery	Crura Cerebri, or Cerebral
Anterior Inferior Cerebellar Art.	Peduncles
Posterior Inferior Cerebellar A.	Continuation Of The Aqueduct
Posterior Cerebellar Artery	Of Sylvius
Inferior Cerebellar Artery	Superior & Inferior Colliculus
Transverse Pontine Artery	Thalamencephalon (Diencephalic
Anterior Spinal Artery	Brain Stem, or Diencephalon):
Posterior Spinal Artery	Epithalamus (uppermost portion)
Anterior Choroid Artery	Habenular Commissure
Basilar Artery	Trigonum Habenulae
Internal Auditory (Labyrinthine)	Pineal Body, and Habenula
Artery	Metathalamus (posterior portion)
Ethmoidal and Lacrimal Arteries	Medial Geniculate Body
Supraorbital Artery	Lateral Geniculate Body
Dural Venous Sinuses	Hypothalamus (lowermost portion)
Cavernous and Sigmoid Sinuses	Optic Chiasma; Infundibulum
Straight and Transverse Sinuses	Hypophysis (Pituitary Gland)
Superior Petrosal Sinus	Tuber Cinereum
Inferior Petrosal Sinus	Thalamus (Thalamic Nuclei):
Superior Sagittal Sinus	Anterior; Lateral; Medial;
Sphenoparietal Sinus	and Ventral
Superior Ophthalmic Vein	Thalamic Medial Geniculate Body
Diploic and Emissary Veins	Thalamic Lateral Geniculate Body
Cerebral and Cerebellar Veins	Pulvinar
Great Cerebral Vein	Ventriculus Tertius (Third Ven-
The Limbic System (collectively re-	tricle); lies between the two
gulates emotional experiences	Optic Thalami, and communicates
and behavior):	with the Fourth Ventricle via
Hippocampus (Cortex)	the Cerebral Aqueduct of Sylvius
Parahippocampal Gyrus	Pons Varolii (together with the
Cingulate Gyrus	Cerebellum, comprises the Met-
Parolfactory Area (Smell)	encephalon, or Metencephalic
Septum Pellucidum (wall)	Region of the Brain)
Mammillary Body (Feeding Reflex)	Cerebral Meninges (Membranes)
Amygdala (group of nuclei)	Dura Mater (hard, outer layer)
Fornix (connects Cerebral lobes)	Arachnoid (intermediate layer)
Thalamus (sensory stimuli)	Pia Mater (soft, inner layer)
Hypothalamus (see Table 1-XII)	Cranial Nerve Nuclei (Table 1-IX)

Neopallium, or Cerebral Cortex. The latter, in turn, is further subdivided into five principal lobes: the *Frontal, Parietal, Occipital*, and *Temporal* Lobes, and the *Central Insula*, or Island of Reil, all enveloping two cavities: the First, or Right Lateral Ventricle, and the Second, or Left Lateral Ventricle -- each connected with the front portion of the Third Ventricle via the interventricular foramina (passage). Connecting the right and left cerebral hemispheres -- the "Right" and "Left Brain", respectively, are the *Corpus Callosum* and the anterior and posterior *Hippocampal Commissures*. Deeply embedded within each cerebral hemisphere are groups of nerves (masses of gray matter) called *Basal Ganglia*. These include the previously mentioned Caudate and Lentiform Nuclei, the Amygdaloid Nucleus, and the Claustrum.

The *Rhinencephalon* (Olfactory Cortex, Archipallium, or Paleopallium) receives and integrates olfactory (pertaining to the sense of smell) sensory impulses. Its anatomical features include the Pyriform Lobe (or area), the Hippocampal Formation, the Olfactory Bulb, the Olfactory Tract, the Olfactory Striae, the Intermediate Olfactory Area, the Paraterminal Area, and the Fornix.

The *Frontal Lobe*, so-named because it lies beneath the Frontal bone of the skull, consists of four main convolutions (Gyri) of the Cerebral Cortex, located in front of the central sulcus (Fissure of Rolando) of the Cerebrum. The front portion of this lobe, called the *Cerebral Sensory Cortex*, is the site of sensory experiences, abstract thought, memory, intelligence, alert consciousness, learning, constructive imagination, higher mental faculties, reasoning, judgment, and emotions. The rear portion of this lobe, called the *Cerebral Motor Cortex*, is composed largely of the *Pyramidal Tract*, and is the site where voluntary movements, striated skeletal muscle function, and conscious management of posture and locomotion originate.

The *Parietal Lobe*, so-named because it refers to the division of the brain lying beneath each parietal bone of the skull, contains a variety of sensory centers. The *Occipital Lobe*, so-named because it lies beneath the occipital bone of the skull, is the *Posterior Lobe* of the cerebral cortex, and is shaped like a three-sided pyramid (hence the terminology, "Pyramidal Tract"). It is the site of visual centers. The *Temporal Lobe*, located laterally and below the frontal and occipital lobes of the cerebrum, is so-named because it lies beneath the temporal bone of the skull. It contains hearing centers and is the site of long-term memory capacity. And finally, the *Insula* (Island of Reil) is a triangular region of the cerebral cortex, lying in the floor of the lateral fissure (Sylvian Sulcus) and constituting the *Central Lobe* of each respective cerebral hemisphere. It also contains sensory centers and is the site of short-term memory capacity (especially in the superficial dendritic layers of the cerebral cortex).

The *Mesencephalon* (Midbrain) is involved in coordinating tracking movements through visual reflexes; in certain aspects of hearing; in motor movement and postural reflex patterns; and, lying literally in the middle of the brain, provides the main motor connection between the forebrain and the hindbrain, and between the cerebrum and the cerebellum.

The Rhombencephalon includes all of the *Cerebellum*, the *Pons*, the *Fourth Ventricle*, the 12 *Cranial Nerve Nuclei*, the *Myelencephalic portion* of the brain stem (which is the most posterior portion of the brain stem, giving rise to the medulla oblongata), and the *Medulla*. The latter is the enlarged portion of the spinal cord in the cranium, just after the cord enters the foramen magnum of the occipital bone, and it constitutes the lowest portion of the brain stem. Within it is contained the fourth ventricle, which communicates with the subarachnoid spaces by the two foramina of Luschka and the foramen of Magendie, and with the third ventricle by the Aqueduct of Sylvius. The medulla has numerous regulatory and reflex centers, including those that control the cardiovascular system, the respiratory system, swallowing, vomiting, coughing, and sneezing.

The *Cerebellum* is the largest portion of the hindbrain, lying behind the pons and medulla, but partially overhanging the latter. Its 20×10^6 Purkinje Cells (large neurons) and 10^{10} granular-like parenchymal (Granule) cells are involved in the coordination of muscular movement and the regulation of balance and equilibrium. This portion of the brain receives afferent impulses from Golgi Tendon Organs, joint receptors, muscle spindles, and Pacinian Corpuscles; and it exercises its control function through efferent motoneurons emanating from *Fiber Bundles* called *Peduncles* (of which there are inferior, middle, and superior ones). The cerebellum consists of two lateral cerebellar hemispheres, and a medial portion, the *Vermis*.

And finally, the ring-like border around the top of the brain stem, formed by regions of the brain that collectively regulate emotional experiences and behavior, is called the *Limbic System* (from the Latin word for "border"). Major anatomical elements of this system are summarized in Table 1-XIII. The Limbic System contains heavy concentrations of a type of neurotransmitter called *Enkephalins,* which appear to have opiate characteristics. Enkephalins may regulate mood by counteracting stimulants that may lead to disappointment and depression, thus leaving the organism in a natural "high" state (similar to the euphoria produced by morphine and other opiates) -- but their mechanism of action is as yet not clearly understood.

CEREBRAL INFORMATION PROCESSING

There is accumulating an abundance of physiologic evidence which confirms the premise that the human brain processes information in two distinct ways: (i) the rational, logical, time-and-language-based *symbolic mode* (which may possibly be "left-brain-centered"); and, (ii) the intuitive, imaginative, perceptive, non-temporal, spatially-based *creative mode* (which may possibly be "right-brain-centered", see Springer and Deutsch, 1981, Schneck and Schneck, 1989, and Edwards, B., 1979, among others). In other words, one might define two "levels of consciousness" in information processing, i.e., a *cognitive level* that functions primarily on verbal and other forms of sequential symbolism (hieroglyphic "cues") as a means for communication; and an instinctive, expressive level (one that is not language-oriented) that functions primarily on a perceptual or sensual means (utilizing intuitive "cues") for communication. Table 1-XIV lists

TABLE 1-XIV

ENCEPHALIC MODES OF INFORMATION PROCESSING

TEMPORAL - RATIONAL - COGNITIVE vs.	SPATIAL - CREATIVE
1. Verbal - Language-Based Logic Which is Inherently Slow	Intuitive-Perceptive-Based Logic Which Is Inherently Much Faster;
2. Sequential, Step-by-Step, Time-Bound, Linear, Progressive Information Processing System	Global, Simultaneous, Non-Temporal Whole-istic, Unified, Instantaneous Information Processing System;
3. Convergent, Deductive, Rational Directed Reasoning (Bound) that has an ultimate goal or purpose	Divergent, Impulsive, Metaphoric, Sensuous Existential Reasoning (Free) that is all-encompassing;
4. Intellectual, Objective Analysis	Imaginative, Subjective Perception;
5. Vertical, Discrete, Realistic and Explicit Definition Of Factual Information Or Data	Horizontal, Continuous, Abstract and Implicit Definition of Factual Information Or Data;
6. Differential, Historical Perspective (always has a "reason", usually based on current dogma)	Integral, Non-Rational Perspective (Does not necessarily require "reasons" to explain everything);
7. Language-Symbolic-Based Functions:	Sensual-Intuitive-Based Functions:
a) Words, Speech, Verbal forms of communication (Hearing-based)	Non-Verbal, Perceptual Communication (wordless-hearing-based);
b) Production & Understanding of Language (Speaking, hearing & Translating Words and Voices)	Awareness & Recognition Of Spatial & Environmental Orientation and Presence (Ubiquity);
c) Reading Comprehension	Sensory Consciousness;
d) Writing and Expression	Music, Art, Fine Arts Skills;
e) Mathematics, Arithmetic, Vision Based Manipulation Of Numbers	Visuo-Spatial Abilities, including Form and Distance Perception
f) Analytic Processes & "Higher" Functions Involving Deductive & Inductive Reasoning Ability	Creative Processes and "Innate" Functions Involving Synthetic "Thing" thinking; Subjective;
g) Naming & Categorizing; "Pigeon-Holing"; Differentiating	Illustrating, Imagining, Abstracting, Coalescing, & Integrating;
h) Intellect (Reasoning Sense)	Intuition; ("Hunch" Sense);
i) Digital, uses Numbers, Symbols, Words, Emblems, and Syntax	Analog, uses Metaphors, Space Figures, Things, Whole Entities;
j) Purposeful (the ability to perform on command movements that may be out of context) and Objective	Introspective (nothing needs to be in any defined context, but a representation of one's own perceptions and emotions)
8. Tends to be specific, problem-oriented and deliberate	Tends to be general, but attentive and reactive;
9. Memory for Facts, Figures and Trivia	Memory for Depth and Spatial Relationships
10. Cannot Accept Anything Out Of The Cause-Effect Mode Of Information-Processing And Always Seeks Justification	Accepts things simply by virtue of their existence, is timeless and needs no explanation to justify an event or occurrence

the essential attributes associated with each mode of information processing. These may be illustrated by a simple example wherein a new-born infant learns to recognize its own mother.

If the child were to identify its parent via the rational, symbolic mode of information processing, it would have to go through the sequential process of observing the individual's height, weight, over-all build, color of eyes, color of hair, complexion, distinguishing facial features, mode of dress, and so on -- building up an arsenal of bits and pieces of factual information that would then be assembled in some logical sequence, to be compared with a corresponding set of information contained in memory stores, in order to arrive at the ultimate conclusion that this individual is, indeed, "mamma".

We know, of course, that this means for identifying an individual is not likely to be the course taken by a newborn infant (indeed, by *anyone*, at *any* age). Rather, it is more probable that the child will take just one glance at its parent, and, based on an integrated, spontaneous assessment of the image created in the visual cortex of its brain, instantaneously identify the individual as "mamma". This is precisely what the perceptual, spatial information processing mode is all about. Note, in particular, how much *faster* this mode of information processing is, how much more *intuitive* it is, how much *earlier* it becomes manifest in the chronological development of human behavioral and learning patterns (implying that it is a more "natural" means for physiologic data-processing), and how much more of a *global* appraisal it is able to make of a given set of facts and details.

The spatial-creative mode of information processing does not necessarily have to put things in any order, or in any time-sequence to have them make sense. It accepts things simply by virtue of their existence; it is timeless; it needs no explanation to justify an event or occurrence; it allows abstract reasoning and it tolerates intuition and "hunches". Therein lies its power to be creative and to transgress the boundaries of current dogma. Creativity knows no boundaries. Boundaries or "Frontiers" are artificial barriers that constrain abstract thought. They are arbitrarily raised, often inaccurately maintained (as in the Galenic tradition in physiology, which persisted for some 1500 years, even though it was based on totally erroneous interpretations of the observations in hand), and intimidating to the creative "free-thinker" or philosopher.

The cognitive-rational mode of information processing, on the other hand, cannot accept anything out of the cause-effect arena. It tends to be specific, problem-oriented and deliberate; it always seeks explicit definition; it insists on everything always making "sense"; it is bound to time, to history, to the scientific method, to rigor, and to symbolism. Therein lies its weakness to stray very far from the course established by precedent. Indeed, it often attempts to *build* on that precedent, rather than to deviate from it, which is not necessarily all that good or desirable -- as illustrated again by the Galenic tradition (see, for example, Schneck, D. J., 1984d, Leake, C. D., 1962, and others). Moreover, the cognitive-rational mode of data-handling is slow. In the traditional "picture-is-worth-one-thousand-words" sense, it takes longer to describe something by attempting to use the tedious, step-by-step methods of verbal or written communicative skills, than it does using the much faster perceptual skills which are so much more innate to the organism.

Because we have become a language-based society that depends so much on symbolism as a means for communication -- which is to say, symbolism (words, numbers, and sequencing of words and numbers) is such an important part of our culture, civilization, and heritage -- information tends to be processed through the language (symbolic) mode, and on functions based on *this* mode (see Table 1-XIV), rather than through the perceptual mode, together with *its* associated functions (ibid.). This, too, may not necessarily be all that desirable for many reasons, some of which are discussed in Schneck, J. K., and Schneck, D. J. (1989).

THE SPINAL CORD AND AUTONOMIC NERVOUS SYSTEM

With the brain representing the *Cranial Portion* of the Central Nervous System, the Spinal Cord -- already addressed earlier in this chapter -- represents the *Cervical-Thoraco-Lumbar Portion* of the CNS. It is an ovoid column of nervous tissue, about 44 centimeters ($17\frac{1}{2}$ inches) long, extending from its origin in the medulla of the brain to the second lumbar vertebra in the spinal canal. In cross-section, it does not fill the entire vertebral cavity. The cord is surrounded by the pia mater, the cerebrospinal fluid, the arachnoid, and the dura mater, which fuses with the periosteum of the inner surfaces of the vertebrae. In all, there are an estimated 12 to 14 billion neurons in the brain and spinal cord that comprise the Central Nervous System. Each neuron is connected by means of synapses with an average 10,000 other neurons, generating a huge communications network. The CNS also spawns the ANS, or Autonomic Nervous System -- which, too, has already been described earlier in this chapter.

The Autonomic Nervous System is concerned with control of involuntary (not under conscious control) bodily functions. Impulses of the ANS which originate in the CNS travel through the spinal cord and terminate in various peripheral "effector organs and tissues" located throughout the body. The parasympathetic division of the autonomic nervous system conserves and restores energy and maintains routine daily functions of the body, such as salivation and digestion. The sympathetic division prepares the body for stressful situations or emergencies, such as fear or anger. Both divisions consist of numerous ganglia and pre-and-post ganglionic nerve fibers or neurons. The electrical activity (action potential) of the nerve fiber is transmitted across the junction (synapse) between the pre-and-post-ganglionic neurons and their respective ganglia, and between the post-ganglionic nerve ending and its respective effector cell, in the form of a chemical agent, which is called a neurohormone, or neurotransmitter (see earlier in this chapter, and Chapter 6).

The neurohormone *Acetylcholine* is released from the ends of the pre-ganglionic neurons (i.e., those directing the action potential *from* the central nervous system *to* the autonomic nervous system) in *both* the sympathetic and parasympathetic divisions of the ANS. This neurotransmitter also carries the nerve impulse across the synapse *from* the respective ganglion *to* the corresponding post-ganglionic neuron (which carries the action potential *from* the autonomic nervous system *out to* the peripheral target organs); and, is the chemical substance released at the ends of post-ganglionic neurons of the parasympathetic nervous system where they synapse with their respective effector organ receptor sites. This is commonly referred to as *Cholinergic Stimulation of Autonomic Receptors*. The Catecholamine substance *Norepinephrine* is the major neurohormone released from the endings of the post-ganglionic neurons in the sympathetic division of the ANS. This is commonly referred to as *Adrenergic Stimulation*.

When a specific nerve is stimulated, the respective neurotransmitter is released from storage granules or synaptic vesicles located in the nerve endings (see Figure 6-3). These "squirt" their chemical contents into the gap, or *Synapse*, between the nerve ending and the effector cell. The neurotransmitter, driven by a combination of the chemical concentration gradient across the synaptic region (c.f., Chapter 2), by the "squirting" action of the nerve endings, and by the "sucking" action of (i.e., its affinity for) the cells of the target organ, crosses the junction, and combines with a chemical receptor site on the surface of the effector cell. It is the union of the neurotransmitter and the receptor which produces the response by the effector cell, as described more fully in Chapter 6. The effector cells are not part of the nervous system, but they do have special receptor properties that enable them to receive and respond to nervous stimulation.

As also mentioned earlier, there are two types of adrenergic receptor sites: α (eliciting an excitatory response), and β (eliciting an inhibitory response). We should, however, point out further that there is one notable exception to this pattern. Stimulation of the β-receptors in the heart muscle causes an *increased* heart rate and *strengthened* force of contraction. The response of the heart is therefore *excitatory* to β-stimulation, and is the single conspicuous deviation to the response one normally expects from adrenergic stimulation of β-receptors. As a result, a theory has been developed that distinguishes between receptors in the heart, and those in sites such as the bronchioles and intestines.

Beta-receptors in the heart are designated β-1, and Beta-receptors elsewhere in the body, such as in the bronchioles and intestines, are designated as β-2 receptors. Beta-1 stimulation elicits an excitatory response, just as does α-receptor-stimulation. Beta-2 receptors elicit an inhibitory response, as is normally expected from β-receptor-stimulation. Cardiac cells in the muscles and nodes of the heart, and the smooth muscle cells found in the bronchioles contain mostly beta-adrenergic-receptors. The intestines and peripheral blood vessels are examples of organ systems that contain *both* α-and-β-receptor sites. For example, stimulation by the sympathetic nervous system of α-receptors in blood vessels evokes excitatory responses, which causes constriction of the vessels. This is an important aspect of the physiologic thermoregulatory response to cold stress (c.f., Chapter 4). The stimulation by the sympathetic nervous system of β-receptors in blood vessels and in the bronchioles evokes an inhibitory response, which causes relaxation and dilation -- an important aspect of the physiologic thermoregulatory response to heat stress (ibid.).

The Endocrine System of Ductless Glands

This is a diverse group of glands that produce and release (as a result of autonomic stimulation) *directly* (as opposed to via connecting ducts) into the blood stream hormones that stay *inside* (i.e., "endo") the body to perform a variety of regulatory functions. This is as opposed to glands that secrete (mostly through connecting ducts) to a free (internal or external) surface hormones that may ultimately *leave* the body -- causing them to be called *Exocrine* Glands. Some of the latter and most of the Endocrine Organs are listed in Table 1-XV. These are discussed briefly below.

Endocrine hormone secretions serve an integrative function, permitting different tissue groups to act as a whole in response to internal or external stimuli. They also serve a homeostatic function that is at once both more *long-term* acting (i.e., effective for a longer period of time), and more *diffuse* (i.e., anatomically more widespread), than are the immediate (more instantaneous), and concentrated (more "focussed") influences of nervous regulation. And finally, hormones serve a growth function that controls the rate and type of growth of the machine. Depending on the specific hormone involved, its respective function may be any, or a combination of these.

The *Hypothalamus*, as has become obvious by now, is an important center for regulating and integrating many vital body functions, among them: (i) sugar and fat metabolism; (ii) body temperature (thermoregulation, including initiation of the shivering reflex); (iii) body fluid volume (osmoregulation and water balance); (iv) the nervous activity of both the parasympathetic and sympathetic nervous systems (postganglionic fibers of which control the secretion of epinephrine and norepinephrine from the Adrenal Medulla); (v) food intake (satiety); (vi) cardiovascular activity (baro-or-blood-pressure regulation, and control of heart rate); (vii) gastric acid secretions (digestion); (viii) timing and synchronization of internal physiologic processes with the natural cycles of our environment; (ix) rage, "flight-or-fight" mechanisms, and other aspects of emotion and behavior; and, (x) the activity of the anterior and posterior Pituitary Gland. The latter is accomplished via the *Hypothalamo-Hypophyseal-Portal System*, an extensive vascular network that the Hypothalamus and Hypophysis share in common. Transported along this network are endocrine secretions of the Hypothalamus, manufactured in the regions listed in Table 1-XV, and including:

a) Thyrotrophin-Releasing Hormone (TRH, from the Paraventricular and Arcuate Nuclei, Preoptic and Anterior Nuclei);

b) Corticotrophin-Releasing Factor (CRF, from the Median Eminence of the Infundibulum, Posterior Nuclei);

c) Growth-Hormone-Releasing Factor (GHRF);

d) Growth-Hormone-Release-Inhibiting Hormone (GHRIH, Somatostatin)

TABLE 1-XV

ANATOMICAL EXOCRINE AND ENDOCRINE GLANDS

ENDOCRINE GLANDS	EXOCRINE GLANDS

Hypothalamus: Endocrine Functions
    Paraventricular Nucleus
    Arcuate Nucleus
    Median Eminence
    Tuber Cinereum
    Ventromedial Nucleus
    Preoptic and Anterior Nuclei
    Posterior Hypothalamus
    Infundibulum
    Anterior Hypothalamus
    Supraoptic Nucleus
    Hypothalamo-Hypophyseal-Tract
Pituitary: Hypophysis
    Adenohypophysis (Anterior)
    Neurohypophysis (Posterior)
    Pars Intermedia
Thyroid Gland (Two Lobes)
Adrenal (Suprarenal) Glands:
    Adrenal Cortex (Glandular)
    Zona Glomerulosa
    Zona Fasciculata
    Zona Reticularis
    Adrenal Medulla (Neural)
Pancreatic Islets of Langerhans
Four Parathyroid Glands
The Gonads: Genital Glands
    Two Ovaries or Two Testes
The Placenta (Female)
The Pineal Gland in the Brain
The Paraganglia:
    Pyloric Gland of Stomach
    Brunner's and Lieberkuhn's
        Glands in Intestines
    Luschka's Gland (Glomus
        Coccygeum) at the Coccyx
    Zuckerkandl Organs (Aortic
        Bodies) in Lumbar Region
	Salivary Glands (Table 1-II)
	Digestive Glands:
	    Cardiac Glands (Stomach)
	    Pancreas & Fundic Glands
	    Crypts of Lieberkuhn
	    Liver (Bile Duct)
	    Gastric and Intestinal Mucous
	        Membranes
	Sudoriferous (Sweat) Glands
	Sebaceous (Oil-Secreting) Glands
	Lacrimal Glands (Tear Ducts)
	Female:
	    Bartholin's (secrete a lubri-
	        cant to ease copulation)
	    Nabothian (Uterine Cervix)
	    Skene's (Para-Urethral)
	    Uterine Glands
	    Mammary Glands (milk)
	    Montgomery's (Areolar, milk)
	    Glans Clitoridis
	    Vaginal (Vaginal Mucosa)
	Male:
	    Cowper's (Bulbo-Urethral, se-
	        crete semen into Urethra)
	    Seminal Glands
	    Prostate Gland (Semen)
	    Tyson's (Sebaceous) Glands
	Ceruminous Glands in the External
	    Auditory Canal (Ear Wax)
	Bowman's Glands (Olfactory Mucosa
	    of Nasal Cavity)
	Mouth:
	    Blandin's (near tip of under-
	        surface of the Tongue)
	    Buccal Glands (in Cheeks)
	    Ebner's (Tongue, Serous)
	    Frankel's (Vocal Cords)

e) Antidiuretic Hormone (Vasopressin, ADH, from the Supraoptic Nuclei, stored in the Pituitary Gland, Paraventricular Nuclei);

f) Oxytocin (Pitocin, from the Paraventricular Nuclei, stored in Pituitary);

g) Gonadotrophin-Releasing Hormone (GnRH, from the Median Eminence for FSH and from the Anterior Nuclei for Luteinizing Hormone, see below);

h) Prolactin-Releasing-Factor (PRF or PRH, probably from the Anterior Nuclei);

i) Prolactin-Release-Inhibiting-Factor (PIH, may be Dopamine);

j) Melanocyte-Stimulating-Hormone-Releasing-Factor (MRF); and,

k) Melanocyte-Stimulating-Hormone-Release-Inhibiting-Factor (MIF).

Lying at the very center of the limbic system, this small (about $\frac{1}{300}$'th of the total brain), multi-functional hypothalamic region of the CPU of the body -- located just below and behind the frontal lobe of the cerebrum -- is linked by many complex nerve pathways to almost all of the other regions of the brain, and it seems to have responsibility for a disproportionate share of the duties of this organ. Indeed, it is a "control center" in the true sense of the word.

Second only to the Hypothalamus in its importance is the *Pituitary Gland*, or, *Hypophysis*, which controls to more or less a complete degree the functional integrity of all the other major endocrine glands, as well as their morphologic development. That is why it is often referred to as the "Master" Gland. Also quite small (having a diameter of about 1.00 cm, and weighing about $\frac{1}{2}$ to 1 gram) in proportion to its physiologic importance, this gland has two identifiable anatomic divisions: An anterior lobe, or glandular portion called the *Adenohypophysis*, and a posterior lobe, or neural portion called the *Neurohypophysis*. Secretions from the Adenohypophysis include:

a) Thyroid Stimulating Hormone (Thyrotrophin, TSH or TTH), secreted in response to TRH from the Hypothalamus;

b) Adrenocorticotrophic Hormone (Adrenotrophin, ACTH), secreted in response to CRF from the Hypothalamus;

c) Growth Hormone (Somatotrophin, GH or STH), secreted in response to GHRF from the Hypothalamus;

d) Gonadotrophin-1, or Follicle-Stimulating-Hormone (FSH), secreted in response to GnRH from the Hypothalamus;

e) Gonadotrophin-2-Type-I, or Luteinizing Hormone (Luteotrophic, or Lactogenic Hormone, LH or LTH), secreted in response to GnRH from the Hypothalamus;

f) Gonadotrophin-2-Type-II, or Interstitial Cell Stimulating Hormone (ICSH), secreted in response to GnRH from the Hypothalamus;

g) Another lactogenic hormone, Prolactin, secreted in response to PRF from the Hypothalamus; and,

h) Melanocyte-Stimulating Hormone (MSH), secreted in response to MRF from the Hypothalamus.

The Neurohypophysis secretes only two hormones: Vasopressin (ADH, or Pitressin), and Oxytocin. Both are synthesized in the Hypothalamus, stored in the Pituitary, and released in response to nervous stimulation from the former.

The two lobes of the *Thyroid Gland* weigh 20 to 30 grams, and contain many 150 to 300-micron follicles that secrete three major hormones: *mostly* (around 90%) *Thyroxin* (T-4) -- which is secreted in response to TSH received from the Pituitary Gland; with lesser amounts of *Thyrocalcitonin* (or Calcitonin, secreted in response to circulating levels of Calcium in the blood stream), and *3,5,3'-Tri-iodothyronine* (T-3). The latter is a hormone also manufactured in the liver and in striated skeletal muscles by de-iodinating Thyroxin from T-4 to T-3. Released from this endocrine organ that lies immediately below the larynx, its two lobes straddling the front part of the trachea, Thyroxin plays a major role in thermoregulation. Its most striking effect in intact homeothermic animals is the stimulation of oxygen consumption and the production of heat for the maintenance of body temperature. This so-called "Calorigenic Effect" is discussed further in Chapter 4.

Like the Pituitary Gland, the two Adrenal Glands also have two identifiable anatomic divisions: the *Adrenal Cortex*, or glandular portion of the *Suprarenal* (above the kidneys) glands; and the *Adrenal Medulla*, or neural portion of the Suprarenal glands. These are triangular-shaped organs, 38 mm in length by 13 mm in diameter (i.e., having an aspect ratio of about three-to-one), and weighing anywhere from 4 to 14 grams -- the mean for adults lying near the lower end of this range. In response to receiving Adrenocorticotrophic Hormone, or ACTH from the Adenohypophysis, or Anterior Pituitary Gland, the Adrenal Cortex manufactures and secretes at least eleven Adrenocortical hormones (or groups of hormones), which are *steroids* made from cholesterol precursors. These hormones are involved in electrolyte metabolism, in the proper metabolism of carbohydrates and proteins, and in the maintenance of the functional integrity of the membranes of many types of cells.

All types of stress -- trauma of any kind, such as cold, heat, pain, fright, infections, or inflammation -- induce prompt activation of the adrenal cortex; as do hypoglycemia (low blood sugar), certain drugs, morphine, ether, nicotine, histamine, and moderate exercise. Thus, responses of the pituitary-adrenal system play a large part in the regulatory systems which help to protect the animal organism from environmental hazards. Secretions of the adrenal cortex include: four Glucocorticoids, or 11-Oxycorticoids (Cortisone, Cortisol, Corticosterone-A, and Corticosterone-B); two Deoxy-corticoids (Deoxycortisol and Deoxycorticosterone); two Mineralocorticoids (Aldosterone and Dehydroepiandrosterone); and three hormones that are important in the physiology of reproduction (Androgens, or 17-ketosteroids, Estrogens, or Estradiol, and Progestins, or Progesterone).

In response to receiving impulses from the nerve fibers of the sympathetic division of the autonomic nervous system, the Adrenal Medulla manufactures and secretes three major adrenomedullary hormones: Epinephrine (Adrenalin, or Adrenine); Norepinephrine (Noradrenalin, or Arterenol); and Dopamine. These are *catecholamines* made from amino acid precursors (principally, Tyrosine), and their major physiologic influence is manifest in cardiovascular and metabolic regulation. That is, the adrenal medulla is intimately related to cardiovascular and metabolic adjustments that take place in the body in response to emotional and anticipatory states; and to the physiological changes that occur in the "flight-or-fight" response to emergency situations. The hormones secreted tend to elicit their response by activating enzyme systems that catalyze biochemical reactions whose products are either excitatory or inhibitory to specific physiologic processes (see earlier in this chapter).

The human adrenal medulla usually secretes four times as much epinephrine as norepinephrine, making the former the principal adrenal medullary secretion. Moreover, there appears to be a synergistic relationship between the thyroid hormones and the adrenal catecholamines, each augmenting many of the responses of the organism to the secretions of the other.

In response to the concentration of Glucose in the blood stream, the *Pancreatic Islets Of Langerhans* (scattered throughout the Pancreas and accounting for 2%, or so, of the whole organ) secrete two hormones: Insulin and Glucagon. Insulin is the major hormone responsible for controlling carbohydrate metabolism and the storage of body fuels. If Insulin is absent (as in *Diabetes Mellitus*), blood sugar levels can rise uncontrolled to above 150-200 mg/100 mℓ blood, or more. Glucagon (also known as the Pancreatic Glycogenolytic Factor) stimulates glycogenolysis (breakdown of glycogen into glucose) and gluconeogenesis (manufacture of glucose de novo) in the liver, but not in muscle tissue. Together with epinephrine, glucagon causes the liver to start releasing glucose into the blood stream. Thus, Glucagon and Insulin, together, appear to constitute a reciprocal system of hormones secreted by the Pancreas for blood glucose regulation.

In response to a drop in serum calcium concentration of as little as 3% from "normal" -- normal being 4.6 to 5.5 milli-equivalents of Ca per liter of serum -- the four 6-mm long by 3 to 4 mm wide *Parathyroid Glands* secrete the hormone *Parathormone*, which has a three-fold effect in increasing calcium concentration: first, acting reciprocally with

Calcitonin (from the Thyroid Gland), Parathormone *stimulates* bone resorption and *inhibits* the deposition of Calcium in new bone tissue; second, it promotes Calcium retention by the kidney by increasing Calcium reabsorption in the kidney tubules; and third, acting synergistically with Vitamin D, it increases the intestinal uptake of Calcium. These activities, and the fact that there are these four endocrine glands located on the posterior aspect of the two lobes of the thyroid dedicated *specifically* to regulating the metabolism of Calcium, reflect the importance of this ion in physiologic processes. Indeed, Calcium may very well be *the* most important electrolyte in the human organism, playing a vital role in at least: (i) neuromuscular excitability; (ii) membrane permeability; (iii) tissue secretory processes; (iv) the activity of enzyme systems; (v) the action of hormones: (vi) synaptic transmission; (vii) blood coagulation; (viii) bone and teeth mechanics, growth and metabolism; (ix) acid-base balance; and (x) photoreception.

The *Pineal Gland* (or Body), shaped like a pine-cone with a 1-cm diameter base and weighing 0.1 to 0.2 grams, is located in a pocket near the splenium of the corpus callosum of the brain. It secretes the hormone *Melatonin*, which is believed to exert an inhibitory effect on the gonads (discussed in a later section of this chapter).

The *Paraganglia* (see Table 1-XV) are groups of chromaffin cells, similar in staining reaction to cells of the adrenal medulla, and associated anatomically and embryologically with the sympathetic system. They are located in various organs and parts of the body, such as: a) near the pylorus opening into the duodenum, where secretions from the *Pyloric Gland* in the lower part of the stomach control the activity of the first part of the small intestine (the Duodenum); b) in the first portion of the small intestine, where alkaline mucinous secretions from the *Duodenal Glands* (Brunner's and Lieberkuhn's Glands) located in the Duodenum and upper Jejunum control the activity of the pancreas, gall bladder, and stomach; c) at the level of the coccyx, where secretions from the *Coccygeal Gland* (Glomus Coccygeum Body, or Luschka's Gland) -- a small arteriovenous anastomosis associated with the median sacral artery and lying ventral to (on the stomach side) or immediately distal to (behind) the tip of the coccyx -- have an as-yet undetermined function; d) in the region of the inferior mesenteric artery, where *Aortic Bodies* (Lumbar Paraganglia, Corpora Para-Aortica Glands, or Organs of Zuckerkandl) lying one on either side of the aorta secrete hormones that appear to have some function early in life, but are virtually non-existent and serve little purpose after puberty; and, e) in ganglionic regions of the sympathetic trunk, where additional little glands seem to be associated with the Celiac, Renal, Supra-Renal, Aortic, Hypogastric, and Abdominal Ganglia of this division of the Autonomic Nervous System.

The Immune System

The engine has developed sophisticated mechanisms to distinguish its own constituents from foreign, potentially destructive or damaging substances. Specialized white blood cells, called Lymphocytes, are "trained" to recognize the unique chemical and geometric identity of cellular polypeptide chains that distinguish the genetic structure of one machine from that of another, or of a different species. These unique chains are called Human Leukocyte Antigens, HLAs, histocompatibility, or, transplantation antigens. The mechanism of identification is actually quite simple, and is discussed towards the end of Chapter 2, in the section on carrier-facilitated transport. Basically, it relies on the "jig-saw-puzzle-like" congruence among the "pieces" that make up our physiologic architecture. If a "piece" (e.g., a foreign protein) "fits" the matrix, it is presumed to belong there, and becomes "biocompatible". If the piece *does not* fit, it is "tagged" for disposal by the body's immune response mechanisms.

    The immune response includes as well participation by Activated T-Cells ("Killer" Cells), Activated Macrophages ("Angry" Cells), Plasma Cells, Memory Cells, Lymphokines, Histamine, and several of the elements listed in Tables 1-IV and 1-VII -- and, is a major aspect of the concept of blood grouping and typing. The role of antibodies in the physiologic immune response is also addressed in the section entitled, *The Concept Of A Codon*, around the middle of Chapter 5.

Feature 4: The Human Machine Has the Remarkable Ability to Make Other Engines Just Like It -- That Is, to Reproduce Itself, or, to Procreate

This function is handled by one, relatively inconspicuous (in terms of weight, size, and so on) organ system of the engine. Inconspicuous, yes, but, oh, so domineering and in control of life! It is said that after the instinct for survival (including reflexes related to hunger, thirst, and "flight-or-fight"), sex is the second most intense drive of humankind. So important is it, that mother nature has created an almost insatiable drive to procreate by giving the experience of conception one of the most enjoyable physical sensations imaginable. Unfortunately, this physiological euphoria has created vice and other problems (such as the "world's oldest profession") associated with situations in which the *joy* of sex has been taken out of context with respect to its intended physiologic purpose -- but, again, this is not of primary concern in this text.

Following the Bible's advice to "be fruitful and multiply", mother nature has provided the human machine with *Gonads*, which are the organs that provide us with the means to reproduce. These are frequently lumped together with the Renal, or Urological System, in what is called the *Urogenital System*. *Gonads*, in fact, refers more specifically to the actual *sex organs*, themselves, which are the two *Testes* (and their respective coverings) in the male, and the two *Ovaries* in the female. The term, *Genitalia*, then, includes the gonads together with the associated architecture required for procreation which, in the male, consists of: two Cowper's (Bulbourethral) Glands, two ejaculatory ducts, the penis with urethra, two seminal ducts (ductus or vasa deferentes), two seminal vesicles, two spermatic cords, the Prostate Gland, the Scrotum, the Epididymis, the Glans, and the Prepuce (foreskin); and, in the female, includes the: Vulva (pudendum, or external genitalia consisting of the mons veneris, labia majora and minora, clitoris, fourchet, fossa navicularis, vestibule, vestibular bulb, Skene's Glands, Bartholin's Glands, hymen, vaginal introitus, perineum, portio vaginalis, and cervix); and the internal genitalia (consisting of the two Fallopian tubes, uterine tubes, or oviducts, the uterus, or womb, the vagina, the endometrium, and, during pregnancy, the placenta and mammary glands).

The two male sex gonads (testes) manufacture sperm cells with an estimated count of over 20 million per mℓ of ejaculate, and an average of 3 to 5 mℓ of ejaculate (60 to 100 million total sperm cells) is required for normal fertilization of an egg (ovum). The testes also have an endocrine function, secreting the hormones Testosterone and Androsterone from interstitial cells. The two female sex gonads (ovaries) manufacture a mature egg cell (ovum) about once every 28 days, from the 400,000 or so primary follicles that are present in the ovaries when the reproductive period of the female begins (i.e., at puberty). Throughout her fertile years (about 40), the adult female will thus produce about 522 ova, give or take a few depending on intervening pregnancies, missed periods, multiple births, and so on. We are definitely not in the cockroach class when it comes to reproduction! The ovaries also have an endocrine function, secreting from their vesicular follicles the hormone Estrogen, and from the corpus luteum the hormone Progesterone. Similarly, during pregnancy, the Placenta has an endocrine function, secreting Human Chorionic Gonadotrophin, Human Placental Lactogen, Progesterone, and Estrogen.

Procreation is not just the simple matter of anatomical propagation, or "cloning" of the species. It is also a matter of passing on from one generation to the next the entire historical physiologic experience of life. In other words, it involves *breeding* (i.e., *improving* on the basic design), as well as producing offspring based on a prescribed pattern. Thus, human reproduction is *sexual*, allowing for improvement, diversification and variety, as opposed to *asexual*, which maintains the status quo (a rose is a rose is a rose ...).

Sperm cells from the male unite with egg cells from the female, each carrying the instructions of generations of history coded into their respective set of 23 double-stranded *chromosomes*, housed in the cell nucleus. The 46 nuclear chromosomes are simply large molecules of the nucleic acid, DNA, arranged in specific sets called *genes*, which are the basic heredity unit. The entire human genetic code is programmed into an estimated 3 billion DNA base pairs which are grouped into some 400,000 genes. That is to say, there are, on the average, 7500 base pairs, plus or minus 2500, per gene, and around 8700 genes per chromosome.

Genes reside in the cell nucleus and never leave. They contain the "recipes", if you will, for all of the mental (psycho-) and physical (somato-) traits that will eventually characterize the individual they are coded to produce. The recipes are executed by Ribonucleic Acid (RNA), which receives its instructions from DNA in the cell nucleus, and then leaves the nucleus to carry out its mission. Part of its mission involves protein synthesis (see Chapter 5), for which it utilizes as "ingredients" some 20 amino acids located in the cytoplasm of the cell. The 20 amino acids are: Alanine, Valine, Leucine, Isoleucine, Serine, Threonine, Glycine, Phenylalanine, Tyrosine, Tryptophan, Cysteine, Methionine, Proline, Histidine, Arginine, Lysine, Aspartic Acid, Asparagine, Glutamic Acid, and Gluthamine.

Protein synthesis takes place in the *Ribosomes* of the cell, which may appear as a single unit, or in clusters called polyribosomes, or *Polysomes*. Hereditary traits are controlled by pairs of genes (called *alleles* -- both dominant, both recessive, or one of each). Cellular proliferation takes place via progressive cell divisions, called *Mitosis*, which involves chromosomal replication at structures called the *Spindle Apparatus*. Forming the poles of the spindle apparatus are the *Centrioles*, or *Centrosomes* of the cell (see Figure 5-2). Mitosis proceeds through four basic stages: the Prophase, Metaphase, Anaphase, and Telophase, these constituting the so-called *cell cycle*. The basic structure and organization of the human cell, as well as cellular processes (including energy conversion, Krebs Cycle, oxidative phosphorylation, aerobic metabolism, anaerobic metabolism, protein synthesis, anabolic reactions, catabolic reactions, and reproduction) are discussed further in Chapter 5 of this text.

Concluding Remarks

Well, there you have it -- a rather brief introduction to the physiologic system as viewed through the eyes of an engineer. This individual sees the conglomeration of systems that comprise the human body as a complex engine designed and built for "who knows what," but, nevertheless, amenable to the techniques of engineering analysis -- using a systems approach and adhering to fundamental laws of physics (see Figure 1-1). As stated at the outset, and summarized in Figure 1-2, the ultimate objective of such an engineering analysis is two-fold: first, to understand and be able to manage effectively engine malfunctions that result from disease (pathology), wear and tear (fatigue), injury (trauma), and/or aging (gerontology, geriatrics); and second, to identify, understand, and be able to expand the envelope of human performance capabilities when the engine is subjected to unusual and/or adverse environments.

The first is a *medical* objective that has as an ultimate goal the development of a health-care technology which addresses both diagnostic procedures and therapeutic (care and treatment) methodologies. Diagnostic procedures address activities such as: (i) physiological monitoring, (ii) non-invasive imaging (Computerized Tomographic Scanning, Magnetic Resonance Imaging, Ultrasonic Imaging, Radioisotope Scanning, Positron Emission Tomography, Thermography, Single Photon Emission Computerized Tomography, Fluoroscopy, and so on), (iii) life-support (left-ventricular assist devices, respirators, cardio-pulmonary resuscitation, renal dialysis, cardiac pacing, defibrillation, artificial hearts, heart valves, blood vessels, and other prosthetic internal organs), (iv) basic physiologic research, (v) high-speed data acquisition and processing, (vi) the quantification of medical information, and (vii) the automation of medical equipment and instrumentation. In addition to many of these same activities, therapeutic methodologies further address such activities

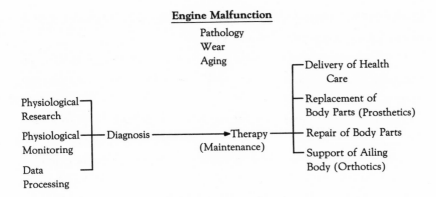

Figure 1-2 The ultimate objective of the engineering analysis of physiologic systems is two-fold: First, to understand and be able to manage effectively engine malfunctions; and second, to identify and be able to expand upon the envelope of human performance capabilities.

as repair of body parts, orthotic support of ailing tissue, and/or prosthetic replacement of body parts and organs.

The second objective is a purely *physiologic* one that includes basic research, design, and development work intended to protect the organism from all (and more) of the hazards listed in Figure 1-2. Such protection may even include *removing* the human body, itself, from exposure to the hazardous environment, and replacing it by robotic telepresence. Thus, a complete understanding of neuromusculoskeletal function and control, as it relates to the kinematics and kinetics of posture and locomotion, becomes a significant aspect of this objective, which is briefly addressed in Chapter 5.

In the material that follows, then, the engineering approach to studying and understanding physiologic function concentrates on three basic aspects of such function: *Transport* of mass, energy, momentum, and information; *Utilization* of mass, energy, momentum, and information (including bioenergetics and metabolism, thermoregulation, optimization, and procreation); and *Control* of the transport and utilization of mass, energy, momentum, and information (including feedback and/or feedforward control and the concept of homeostasis). Transport is addressed in Chapters 2 and 3; utilization in Chapters 4 and 5; and control in Chapter 6. The general philosophy underlying this approach is that, despite its seeming complexity, the physiologic system may be viewed as being basically an isobaric, isothermal, electrochemical engine. It functions in space and time in accordance with fundamental principles of mass, energy, momentum and information transport and utilization. Its output is mechanical; and it may be studied effectively by employing powerful techniques of dimensional analysis and control theory. Toward this end, examples are given in the text that illustrate order-of-magnitude (nondimensional), closed-form analytic (deductive), and empirical (inductive) methods which have been found to be useful in the quantitative investigation of physiologic function. Let us begin, then, by examining some basic principles that govern the transport of mass, energy, and momentum in physiologic systems.

Chapter 2

Basic Principles of Mass, Energy, and Momentum Transport in Physiologic Systems

Introduction

Within the framework of Darwinian theory, it has been suggested (Deutsch, 1981) that evolutionary processes may have been intimately coupled with transport processes. That is to say, phylogenetic advances more than likely favored those species who were successful in developing semipermeable membranes that could:

(a) concentrate *from* the environment desired chemical nutrients such as oxygen, carbohydrates, glucose, proteins, amino acids, lipids, free fatty acids, water, and other raw materials of metabolism; and,

(b) excrete *to* the environment the undesired end products of metabolism, such as carbon dioxide, nitrogenous wastes (urea), non-nitrogenous wastes, water, heat, and other forms of sewage.

It has been further proposed (Schneck, 1981, 1987) that those species who could accomplish their metabolic activities in such a way as to economize on the associated energy expenditure involved, would also have distinct survival advantages. We shall get back to *this* point in a subsequent chapter.

On the basis of these and many other observations, it is clear that among the most important of the processes responsible for the maintenance of life are those that govern the transport of mass, energy, and momentum in physiologic systems. Such processes regulate not only what gets into and out of the system, but also the flow of material, energy, and information (i.e., action potentials, c.f., later chapters) *within* the system. And in this respect, the extent to which air/fuel (oxygen/nutrient) mixtures can be brought to and properly handled at their respective sites of utilization is perhaps less important than is the extent to which the waste products of metabolism can be properly excreted (exhausted).

Indeed, the consequences of an accumulation of waste products are potentially much more damaging than are the effects of a reduced supply or a depletion of fuel and nutrients. This is because the living system can adapt and compensate for insufficient air and fuel supplies, whereas it shuts down completely in the presence of an excessive quantity of waste products. In the former case, compensation is accomplished by operating at a reduced energy level, or by shifting to an alternate set of functional priorities, or even by going to reserve sources of energy which allow the organism to survive for some time in an emergency situation. On the other hand, excessive amounts of waste products, if allowed to accumulate, actually block or significantly interfere with chemical reactions that are vital for the sustenance of life itself.

The same can be said of an accumulation of heat energy that results from the products of metabolism, since we are destined to function as an Isothermal Engine and so must effectively dissipate such heat.

Thus, regardless of whether it be a simple, single-celled organism that relies for its existence on the most rudimentary principles of diffusion, or a large, multicellular, complex organism such as man, in whom survival depends on much more sophisticated, complicated, and diverse transport processes (LaBarbera and Vogel, 1982), the bottom line involves the ability of the organism to move (or to *keep* from moving) some quantity of mass, energy, or momentum from one point to another. It is therefore important for the engineer studying physiologic function to have as well a fundamental understanding of associated transport considerations, and that is the purpose of the material which follows. A broad overview of the means by which mass, energy and momentum transport can be accomplished is first presented below, and this is followed in this and subsequent chapters by the details of some of the more specific aspects of each process.

General Considerations

Because of the immense complexity of the mechanisms that are involved in the plethora of processes associated with physiologic transport, it is extraordinarily difficult (if not impossible) to describe them in a totally comprehensive manner, with complete mathematical rigor and generality. Moreover, because there are so many of them, the approach to studying physiologic transport processes has tended to be inductive, rather than deductive. Thus, the theories that have been formulated are just about as numerous and as diverse as are the very mechanisms that they propose to describe.

Nevertheless, from many of the studies completed to date, there have emerged certain common features that suggest a more general, fundamental principle that governs *all* problems involving transport. Indeed, this principle may be cast in the form of a "Generic Transport Equation." What follows, then, is a discussion of these common features as they are embedded in this generic transport equation, and how this equation may be applied to the exchange of material (or biomass, or energy) among the various "compartments" of the physiologic system (including the environment).

The analysis begins by formulating the Generic Transport Equation in terms of a very fundamental set of physical principles. These are cast within the framework of the following features that are common to all transport:

(a) Some quantity, S, is to be moved from its current location in space to another site. For the purposes of this analysis, S shall represent mass, M, energy, E, or linear momentum, G.

(b) To affect the movement, there exists some transport driving force, F. In physiologic systems, the mechanisms associated with transport driving forces may be collectively grouped into two basic categories: *Passive* (or spontaneous) -- i.e., not normally requiring an expenditure of energy -- and *Active* (or external-energy driven), which are invariably accompanied by an expenditure of energy. Moreover, within each category, one may further distinguish between processes that are driven by *Body* Forces, and those that are accomplished by *Surface* Forces (see discussion later).

(c) Transport must occur along some well-defined path, which may take S across a bounding surface (such as a cell membrane) separating two contiguous regions in space, or which may simply move S intramurally through a suspending medium or solution. Where a boundary is involved, one also distinguishes further between (i) transport through a void, hole, pore, or other discontinuous portion of the boundary (transmembrane *Transudation*), which is a special case of intramural transport within a confined path region, and which usually occurs passively; (ii) transport through the intact, boundary material itself (intramembrane *Permeation*), which may occur by passive or active mechanisms; and (iii) transport by species encapsulation -- resulting from an invagination (involution) and closure of the boundary material around S -- followed by a "ferrying" or "revolving door" permeation mechanism called *Pinocytosis* or *Phagocytosis*, which is always active.

Regardless of which of the above transport pathways applies to a given situation, the *resistance* of the path to movement becomes an important feature of the overall transport process.

(d) And finally, as a corollary to the transport pathway resistance, one must consider as well the *Mobility* of S -- i.e., the ability of the affected quantity to *respond* to the transport driving forces. That is to say, you can put a car on the highway (transport pathway) and give it a good engine with lots of gas (active driving force), but if the brakes on the car are locked, or if it is otherwise constrained in its ability to move (i.e., if it is immobilized), then nothing will happen.

Having derived a generic transport equation from the above set of physical considerations, we shall then address the solution of this equation for some specific physiologic transport mechanisms. Within this context, our attention shall focus on three approaches to engineering analysis of physiologic systems: (i) the purely analytic approach involving the formulation and solution of equations that are derived to describe a specific transport process. This shall be addressed in Chapter 3, using as an example the transport of mass; (ii) the powerful technique of nondimensional analysis, from which one can virtually extract all of the essential physics inherent in *any* transport process by simply scaling the process properly -- and thus derive extremely useful and meaningful information *without* the need to formulate and solve complicated sets of differential equations. This shall be addressed in Chapter 4, using as an example the transport of heat energy; and, (iii) the experimental-empirical approach wherein the laboratory provides solutions to otherwise intractable problems. This shall be addressed in Chapter 5, using as an example the transport of momentum in problems involving human body dynamics.

Whereas purely analytic and nondimensional techniques may be loosely described as being *deductive,* in the sense of going from the general to the specific, experimental and empirical techniques may be thought of as being *inductive,* in the sense that they go from the specific to the general. That is to say, deductive analysis *identifies* specific cases based on a general theory valid for *all* cases, whereas inductive analysis *builds,* or *infers* a general theory *from* the known specific cases, themselves. However, let the reader be advised that the classification of engineering approaches to the solution of biomedical problems as being "inductive", or "deductive", or "analytic", or "numerical", or "nondimensional", or "empirical", or "experimental", and so on, is more as a matter of convenience for purposes of discussion, than as a matter of absolute definition. More often than not, the solution process mobilizes *all* of these techniques in an integrated approach to problem-solving, rather than any *one* of them exclusively. Thus, the ultimate approach may more accurately be termed "hybrid", and it is with this underlying understanding that we proceed.

Derivation of a Generic Transport Equation

Consider the transport of mass, M, energy, E, or linear momentum G. Then, with M already representing a fundamental dimensional measure of inertia, one can derive the dimensions of E and G in terms of M and two other fundamental dimensions that are associated with the physical (space-time) aspects of sensual perception. These are the dimensions of length, L, for space, and time, t, such that $E = \dfrac{ML^2}{t^2}$, and $G = \dfrac{ML}{t}$, as shown in Table 2-I.

Note that (L/t) defines some characteristic transport speed, v, which, since $G = Mv$, can be seen to be a measure of linear momentum (or some corresponding impulse) per unit mass. Similarly, the dimensions (L/t^2) define some characteristic transport acceleration, a, which, since $F = Ma$ (from Newton's Second Law of Motion), can be seen to be a measure of the force, F, per unit mass driving the transport process. And, since energy, E, has the dimensions of a force, F, multiplied by a length, L, i.e., $E = FL = MaL = \dfrac{ML^2}{t^2}$, we see that the dimensions $(L/t)^2 = v^2$ are those of some characteristic energy per unit mass, $e = \dfrac{E}{M} = v^2$. The latter is a measure of an "intensive" property of the system, i.e., one that is either independent of the mass of the system, or is expressed per unit mass. Henceforth, the terminology "is a measure of", or "is measured by" shall be designated mathematically by the symbol $\overset{m}{=}$, as suggested by Becker (1976).

The intensive properties of a physical system (collection of "$e's$", if you will), are those attributes that define the "State" or "Configuration" of the system at any given time. Becker (1976) has coined the word "charges" to designate these attributes, which include such quantities as enthalpy, internal energy, Gibbs Free Energy, Helmholtz Free Energy, electrochemical potential, and so on. The set of all numerical values that can be "legally" assigned to such intensive properties uniquely determines the domain containing all of the possible physical manifestations of the system. The word "legally", included in quotes relates to values of the intensive properties that are capable of being *physically* realized, as op-

TABLE 2-I

FUNDAMENTAL QUANTITIES AND DIMENSIONS DEFINING

THE PHYSICAL CONFIGURATION OF ANY SYSTEM

QUANTITY, S	DIMENSIONS*	CONCENTRATION $= \dfrac{S}{L^3}$	FLUX $= \dfrac{S}{L^2 t}$
MASS, M	M	$\dfrac{M}{L^3} = \rho$	$\dfrac{M}{L^2 t}$
ENERGY, E	$\dfrac{ML^2}{t^2}$	$\dfrac{M}{Lt^2}$	$\dfrac{M}{t^3}$
MOMENTUM, G	$\dfrac{ML}{t}$	$\dfrac{M}{L^2 t}$	$\dfrac{M}{Lt^2}$

* L = Dimension of Length

t = Dimension of Time

q = Dimension of Electric Charge

T = Dimension of Temperature

M = Dimension of Mass

v = Transport Speed $= \dfrac{L}{t}$ = Impulse Per Unit Mass

a = Transport Acceleration $= \dfrac{L}{t^2}$ = Force Per Unit Mass

e = "Charge" $= \dfrac{E}{M} = \dfrac{L^2}{t^2}$ = Energy Per Unit Mass (Intensive Property)

α = Kinematic Transport Coefficient $= \dfrac{L^2}{t}$ = et = vL = aLt

posed to *mathematically* assigned, so that the state of the system predicted
by the numerical values of the various e 's can *actually* be realized (not
just hypothetically conceptualized). For example, it may be possible to
assign numerical values to a matrix of e's that, on paper, allows the system to violate the Second Law of Thermodynamics; but in real life the
physical state predicted by such e-values could never be actually realized
(or, the system could never manifest itself in that predicted state).

When the spatial and temporal distribution of charge in a system is
uniform and homogeneous throughout, the system is in a state of *Equilibrium.* In this state, no net transport of S occurs, and the system is not
in the process of transitioning from one state or configuration into another. Transport, then, results from mechanisms that disturb the state
of equilibrium of a system, generating *gradients* (designated mathematically by the Greek symbol, ∇) in its charge distribution. Indeed, such
gradients manifest themselves as the very forces, F, that actually *drive* the
transport process, so that $a = F/M$ is, in some sense, related to ∇e, and
there results from such a gradient a *Flux* in S as discussed more fully below. When this flux occurs "down" the gradient in e, i.e., driven by and
in the direction of the force generated by ∇e, the transport is said to be
Passive, in the sense that it tends to occur "naturally" as a consequence
of the gradient in e, and thus does not normally require an external source
of energy to drive it.

Active transport, then, is said to occur when the flux is in a direction
opposite to that associated with the existing gradient in e; indeed, almost
in spite of it and certainly in opposition to the natural tendencies of the
system. Such transport generally will not occur spontaneously, and thus
is driven or accompanied by an expenditure of energy. Examples include
carrier-molecule driven mass transport mechanisms, white blood cell
phagocytosis, mechanical pumping such as is accomplished by peristalsis
in the gastro-intestinal system or by contractions of the heart in the
cardiovascular system, and certain enzyme-driven chemical reactions.

There is an in-between transport category wherein the flux *does* occur down a naturally occurring gradient in charge, but it is enhanced or

accelerated by active mechanisms. Thus, it is aptly categorized as *Facilitated* transport.

Next, define the "Concentration" of S, designated by placing brackets around the quantity, $[S]$, as the amount per unit volume of mass, energy or momentum that is available to participate in the transport process. Since volume has dimensions L^3, we conclude from the dimensions of M, E, and G given in Table 2-I that concentration has the corresponding dimensions given in the second column of the same Table, i.e, (M/L^3), (M/Lt^2), and (M/L^2t), respectively. Let mass concentration, $\left[\dfrac{M}{L^3}\right]$ be designated by the Greek letter, ρ, which shall stand for mass density. Then, writing $(M/Lt^2) = \left[\dfrac{M}{L^3}\right] \times \left(\dfrac{L^2}{t^2}\right)$, and $(M/L^2t) = \left[\dfrac{M}{L^3}\right] \times \left(\dfrac{L}{t}\right)$, we see that concentration may be written, in general, as: $[S] \equiv \left(\dfrac{S}{L^3}\right) \stackrel{m}{=} \rho v^j$, where $j = 0$ for mass, $j = 1$ for momentum, and $j = 2$ for energy.

Finally, define more formally the "Flux" of S as the quantity of mass, energy or momentum transported per unit time (t), per unit area (L^2), as a result of the driving force generated by ∇e (or by the active transport mechanisms that oppose ∇e). Letting the flux be designated by the Greek letter, Φ, we observe that it has the corresponding dimensions given in column three of Table 2-I, i.e., $\left(\dfrac{M}{L^2t}\right)$ for mass, $\left(\dfrac{M}{t^3}\right)$ for energy, and $\left(\dfrac{M}{Lt^2}\right)$ for momentum. Again, writing $\left(\dfrac{M}{L^2t}\right) = \left[\dfrac{M}{L^3}\right] \times \left(\dfrac{L}{t}\right)$, $\left(\dfrac{M}{t^3}\right) = \left[\dfrac{M}{L^3}\right] \times \left(\dfrac{L^3}{t^3}\right)$, and $\left(\dfrac{M}{Lt^2}\right) = \left[\dfrac{M}{L^3}\right] \times \left(\dfrac{L^2}{t^2}\right)$, we see further that flux may be written, in general, as:

$$\Phi \equiv \left(\frac{S}{L^2t}\right) \stackrel{m}{=} \rho v^{j+1} = (\rho v^j) \times v \stackrel{m}{=} [S] \times v, \qquad [2\text{-}1]$$

where: $j = 0$ for mass,

                               $j = 1$ for momentum,

and $j = 2$ for energy;

or, in words: Flux $\stackrel{m}{=}$ [Concentration] × (Transport Velocity). Note that *momentum flux* has the same dimensions as *energy density* (which is defined to be energy per unit volume), and that *energy flux* has the same dimensions as *intensity* (which is defined to be power per unit area).

To convert the $\stackrel{m}{=}$ sign in equation [2-1] into an exact equality, one may introduce an appropriate dimensionless scaling coefficient, A_j. That is to say, the generalized functional form of equation [2-1] is as far as dimensional considerations can take you, and while the equation does inherently account for all of the physics associated with transport, it does so to within an undetermined constant of proportionality that must be determined by other means. In some cases (one of which is discussed in Chapter 3), by a judicious choice of units for M, L, and t, A_j may be set equal to unity without a loss in generality. In other cases, A_j can frequently be estimated by a more complete mathematical analysis of the system or by experimental techniques, as discussed in Chapter 5.

Before proceeding, let us perform an index transformation that lets $j = 2 - i$, so that $i = 2$ for $j = 0$ (mass), $i = 1$ for $j = 1$ (momentum), and $i = 0$ for $j = 2$ (energy).   Then, we may write a Generic Transport Equation based on [2-1] that says:

$$\Phi = A_i \times [\rho v^2] \times \frac{1}{v^i} \times v \qquad\qquad \text{[2-2]}$$

The convenience of this form of the generic transport equation shall become clearer later on. Again, recognizing that dimensions are a qualitative concept by which we perceive of an entity as having a particular size, and that dimensions are made quantitative by introducing arbitrary units of measurement relative to carefully defined reference values, it is often possible to choose such units in a way that will make $A_i = 1$; but we leave it in equation [2-2] for the sake of generality.

The three key components defining the generalized concept of a physical configuration, or realization of state, are thus seen to be charge (e), geometry (L), and flux (Φ) -- these being of the nature of basic "dimensions of state", and, as a set, being equivalent to mass (M), length (L), and time (t), which represent "dimensions of sensual perception" (Becker, 1976). Note further that length and momentum are conjugate pairs, as are time and energy, in the sense that the ratio of energy to momentum yields the dimensions of a velocity, which is the same as the ratio of length to time. Or, viewed another way, the product of energy and time is dimensionally the same as the product of momentum and length, both yielding the result $\frac{ML^2}{t}$. The latter product of mass and a kinematic transport coefficient is called Action. Thus, as a set, mass, length, and time on one scale are equivalent to mass, momentum, and energy, respectively, on another scale. Furthermore, from our point of view here, the set $e, L,$ and Φ are considered to define uniquely *any* transport process, provided the variables involved can be somehow measured or calculated. This is addressed in general in the sections which follow, and more specifically in subsequent chapters.

Solution of the Generalized Transport Equation

Concentration is simply the amount (per unit volume) of S available to participate in the transport process, and, as such, it is usually not very difficult to measure or calculate. The transport velocity, however, is a somewhat abstract kinematic variable that, except in specific convective processes, is not always easy to determine either from direct measurements or analytically. This makes it difficult to use the flux equation in the specific form [2-1] or [2-2] derived above. What is more convenient in practice is to evaluate charge, e, and charge distribution in a system, so we proceed to express v as a function of e using the following reasoning -- which, incidentally, also puts the equation in a more meaningful physical context.

By definition, acceleration, a, is equal to the time rate of change of velocity, v, i.e., $a = \frac{dv}{dt}$, so that $v = \int a\, dt$, where a, in general, is some function of time, i.e., $a = f(t)$. If $f(t)$ is continuous when t spans some interval τ, then the mean-value theorem for integrals may be applied to get:

$\int_o^\tau f(t)dt = \tau f(t^*)$, where $f(t^*) = a^*$ is some characteristic transport acceleration (having dimensions $\frac{L}{t^2}$) calculated at some specific time t^* in the elapsed interval, $0 \le t \le \tau$. Then v may be written as the product of a^* and some characteristic transport time scale, or time constant, τ, (having dimension t), i.e., let $v = a^*\tau$. Then the flux equation [2-1] takes the form: $\Phi \overset{m}{=}$ Concentration $(\rho v^j) \times$ some characteristic Driving Force per Unit Mass $(a^*) \times$ some characteristic "Mobility" Time Constant (τ), which corresponds to a similar expression first proposed by Teorell (1953) for the free diffusion of mass. Recall that a^*, being a measure of the force per unit mass, or "thrust" that propels mass, energy, or momentum from one point to another, is in some sense related to ∇e (c.f., previous section). Let this relationship be designated by some function f_1, so that $a^* = f_1(\nabla e)$.

The time constant, τ, is a measure of the "mobility" or "ease" of transport of the quantity S, and it may be expressed dimensionally as the product $\left(\frac{t^2}{L^2}\right) \times \left(\frac{L^2}{t}\right)$. Once again, $\left(\frac{L^2}{t^2}\right)$ are the dimensions of energy per unit mass, or charge, e, so τ, too, is in some sense a function of the reciprocal of e. Let this function be designated f_2, so that $\tau = f_2\left(\frac{1}{e}\right) \times \left(\frac{L^2}{t}\right)$. Going one step further, the dimensions $\left(\frac{L^2}{t}\right)$ are recognized to be those of a kinematic transport coefficient, α -- such as, for example, the kinematic viscosity of a fluid, the mass diffusion coefficient, the thermal diffusivity of a substance, and so on -- which can usually be calculated for a given transport process based on the physical properties of the system involved (c.f., Chapters 3, 4, and 5). The product of mass, m, and α, again is known as *action,* and has the same dimensions as angular momentum (or "moment-of-momentum"). Thus, summarizing all of the reasoning developed to this point in the analysis, one can convert the general flux equation written in terms of transport velocity, into a generic mathematical form written as a function of charge:

$$\Phi = A_j \times \rho \times \left\{ \alpha \times f_1(\nabla e) \times f_2\left(\frac{1}{e}\right) \right\}^{j+1} \qquad [2\text{-}3]$$

The generic flux equation [2-3] is a generalized transport equation that contains within it all of the essential physics that govern *any* transport process, i.e., the *amount* of S available to participate in the process, the *force* driving it, and the *path* and *mobility* of the quantity being transported. A complete description of any transport process thus reduces to defining for it a corresponding transport coefficient, α, the functions $f_1(\nabla e)$ and $f_2\left(\frac{1}{e}\right)$ -- which include, as well, provisions for Passive, Facilitated and/or Active Transport, Intramural Motility, Permeation, Transudation, and Pino-or-Phagocytosis, -- the mass density, ρ, and the appropriate value of j (0 for mass, 1 for momentum, and 2 for energy).

The functions f_1 and f_2 depend on the nature of the force systems involved in driving the transport process. For *conservative* force systems, the driving force, F, per unit mass (i.e., some characteristic a^*) can often be written exactly as the negative gradient of an associated *scalar* intensive potential energy function per unit mass, which we shall designate as ϕ; so that $f_1 = -\nabla\phi$. Moreover, because energy is conserved in such systems, potential energy is converted directly into kinetic energy, and vice versa, so that ϕ, having dimensions $\left(\frac{L^2}{t^2}\right)$ is also in some sense a measure of the square of the characteristic transport velocity, v; i.e., potential energy, ϕ, per unit mass $\overset{m}{=}$ kinetic energy, $\frac{1}{2}v^2$, per unit mass. This allows us to write, $\tau = $ mobility $\overset{m}{=} \left(\frac{t^2}{L^2}\right) \times \left(\frac{L^2}{t}\right) = \frac{\alpha}{v^2} = \frac{\alpha}{2\phi}$, so that $f_2 = \frac{1}{2\phi}$, and $\alpha = \tau v^2 = 2\phi\tau$. Substituting these results into equation [2-2], one can write:

$$\Phi = A_i \times [\rho v^2] \times \frac{1}{v^i} \times v = A_i \times [\rho v^2] \times \frac{1}{v^i} \times (a^*\tau)$$

$$= A_i \times \rho(\tau v^2) \times f_1 \times \frac{1}{v^i} = -\frac{A_i \rho \alpha (\nabla\phi)}{(2\phi)^n}$$

[2-4]

where $a^* = f_1 = -\nabla\phi$ is the force per unit mass driving the transport process, and now, $n = 0$ for energy, $n = \frac{1}{2}$ for momentum, and $n = 1$ for mass. When the amount of some species, S, is expressed as a concentration, $[S]$, in, say, moles per liter, we note that $\rho = [S] \times MW$, where

MW is the molecular weight of S (its formula weight expressed in mass units).

For *dissipative* force systems (non-conservative), the functions f_1 and f_2 can often be written in terms of the negative gradient of an associated *vector,* or *complex* (rather than scalar) potential, $\vec{\Omega}$, where the arrow over a quantity designates that it is a vector, and thus $\nabla\vec{\Omega}$ yields a nine-component, second-order tensor, Ω_{ij} $(i, j = 1, 2, 3,$ respectively). The diagonal terms of this tensor are associated with one component of the vector potential, the energy-conserving, real part which corresponds to the scalar potential, ϕ, discussed earlier. The off-diagonal terms are associated with an orthogonal component of the vector potential, the energy-dissipating, imaginary part, which we shall label ψ. Thus, $\vec{\Omega} = \phi\hat{i} + \psi\hat{j}$, where $\hat{i}$ and $\hat{j}$ represent, respectively, unit vectors designating mutually orthogonal directions in space.

Alternatively, the vector $\vec{\Omega}$ may be written as a *complex* potential function, $\Omega = \phi + i\psi$, where $i = \sqrt{-1}$. This comes from the geometric depiction of the vector in what is called an "Argand" Diagram, or complex plane, as a ray of length: $\mathscr{R} = \sqrt{\phi^2 + \psi^2}$, making an angle: $\theta = $ Arc Tan $\dfrac{\psi}{\phi}$ with the direction in space corresponding to the positive direction of $\hat{i}$. Then, $\mathscr{R}$ is called the *Modulus* of Ω, and θ is called the *Argument* (or phase angle in the language of control theory) of Ω. Using Polar notation, we can thus write:

$$\Omega = \mathscr{R}e^{i\theta} = \mathscr{R}\cos\theta + i\mathscr{R}\sin\theta = \phi + i\psi,$$

and the corresponding derivatives of Ω are in some sense related to f_1 and f_2 (or, vice versa). Examples of the functions ϕ, ψ, and α, together with methods for obtaining them are discussed below, in Chapters 3, 4, and 5 which follows, and in Schneck, D. J. (1989).

Where appropriate, the generic transport equation [2-4] may be further modified by the addition of "Correction" Coefficients, such as "activity" factors (c.f., Chapter 3, for particle-to-particle interference in concentrated solutions); "dielectric" constants (for the lack of a vacuum in electromagnetic transport); "Staverman" Factors (c.f. more, below to account for the semipermeability of a membrane); "Energy-loss" coefficients (at valves, sharp bends or curvatures in tubes, sudden pipe constrictions or dilatations); and so on. With equation [2-4] and the preceding discussion in mind, we proceed now to examine some transport processes involving the movement of mass, energy, and momentum, using as illustrations some physiologically-relevant cases.

Transport Driving Forces and Associated Intensive Potentials

As mentioned earlier, the forces that drive transport processes may be grouped into two basic categories -- *body* forces and *surface* forces -- for which associated potentials or related energy functions may be defined.

Body forces are those long-range (at-a-distance) forces having a magnitude proportional to the mass or volume of the body upon which they act. These forces may exert their influence without coming into actual contact with the body (i.e., they may act "at-a-distance"), but their influence decreases (generally quite dramatically) as the body upon which they act moves further and further away from them. Body forces are classified as being of the *Field, Convective,* or *Inertial* type. For the most part, they tend to be conservative, and so each has associated with it a corresponding intensive scalar energy potential, ϕ. Body forces that may affect physiological transport include:

Field-Type Forces: These can usually be written in terms of the negative gradient of the corresponding field potential as follows:

1) Forces generated by a *gravitational* field are proportional to a gradient in height, z, above the center of the earth ($\phi = g_o z$), or, more generally, to a gradient in gravitational potential as defined from Newton's Law of Gravity. (*Note*: all quantities used herein, together with their corresponding dimensions, are defined in the List of Symbols at the beginning of the text).

2) Forces generated by a *magnetic* field are proportional to a gradient in magnetic field intensity, H, $\left(\phi = \frac{1}{8\pi} H^2 \frac{\mu_e}{\rho}\right)$, which is de-

fined to be the force that acts upon a unit positive pole when the latter is placed at a point in the field.

3) Forces generated by an *electric* field are proportional to a gradient in electric field intensity, E_e, $\left(\phi = \dfrac{1}{8\pi} E_e^2 \dfrac{1}{\rho k}\right)$, which is defined to be the force that acts upon a unit positive charge when the latter is placed at a point in the field. Such forces lead to an important biological transport process called *Electromotive Diffusion*, wherein the lack of a uniform distribution of positive and negative charges on both sides of and across the membrane separating ionized or electrical particles causes them to migrate in an effort to neutralize the charge.

4) Forces generated by both an electric and a magnetic field acting simultaneously, i.e., *electromagnetic* forces are proportional to a field potential whose magnitude is equal to the square-root of the sum of the squares of the component electric and magnetic field potentials: $\phi = \sqrt{\phi_{\text{electric}}^2 + \phi_{\text{magnetic}}^2}$. This is due to the fact that electric field vectors and magnetic field vectors are mutually orthogonal to one another.

5) Forces generated by a *nuclear* field (such as natural radio-activity), are proportional to a gradient in the quantum energy $\left(\phi = 2\left\{\dfrac{q^2\pi k}{h_o i}\right\}^2\right)$, as defined by Planck's Law, that is associated with particular atomic states ($i = 1, 2, 3, \ldots$).

Note that, in addition to M, L, and t, the dimensions of the above-referenced body force field potentials (as well as Table 2-I) also sometimes include the dimensional quantity electric charge, q, to account for electromagnetic perception. In the convective potentials to be considered next, we shall add still a fifth dimension, temperature, T, to account for thermodynamic perception. These five dimensions: M, L, t, q, and T, normally suffice to measure and define all of the quantities of physics, or of physiological processes.

Convective-Type Forces: These can generally be expressed as the negative gradient of thermodynamic potentials as follows:

1) Forces generated by a nonuniform distribution of mass are proportional to a gradient in *concentration* potential energy

$(\phi = \dfrac{RT[S]}{\rho}$, where $[S]$ is molar concentration), and lead to a process called *Ordinary Diffusion*. In ordinary diffusion, the lack of a uniform distribution of *uncharged* chemical constituents, either intramurally or on both sides of and across a membrane causes them to migrate by random Brownian motion until the concentration gradient is dissipated. This shall be addressed at length in Chapter 3.

2) Forces generated by a nonuniform distribution of *charged* particles -- i.e., by the combination of an electric field together with a nonuniform Brownian motion field -- are proportional to a gradient in *electrochemical potential* $\left(\phi = \dfrac{1}{MW}\{\overline{F}ZB + RT\ell n[S]\}\right)$, and lead to a process called *Electrochemical Diffusion*. This, too, is an important physiological phenomenon that is discussed more in Chapter 3.

3) Forces generated by a nonuniform distribution of mass across a membrane barrier that is impermeable to S are proportional to a gradient in *osmotic pressure* $\left(\phi = \dfrac{RT\Sigma_i[S_i]}{\rho}\right)$, where the latter results directly from the concentration potential energies of the various S_i's distributed nonuniformly across the membrane (see Chapter 3). Osmosis refers specifically to the passive movement of water across a membrane. This results both from passive or ordinary diffusion due to a concentration gradient, and from inertial properties exhibited by the bulk mass of fluid (see later).

4) Forces generated by a nonuniform distribution of mass density (which may result from an associated thermal or temperature gradient) are proportional to a gradient in *"Buoyancy" Potential*, $\phi = \left\{\dfrac{\rho V_B g_o}{\mu\ell}\right\}^2$.

5) Forces generated by a nonuniform distribution of temperature -- i.e., those originating in a thermal field, are proportional to gradients in *Thermodynamic Potentials*, such as: Enthalpy $(\phi = c_p T)$, Internal Energy $(\phi = c_v T)$, Entropy $(\psi = sT)$, Gibbs Free Energy $(\phi = c_p T - sT)$, Helmholtz Free Energy $(\phi = c_v T - sT)$, and Thermodynamic Pressure $\left(\phi = \dfrac{p}{\rho}\right)$. Also included in this category are processes involving a phase

change, where $\phi = c_L$ represents either the heat of vaporization or the heat of fusion of the respective quantity involved.

Processes involving phase changes are particularly important in physiologic systems from the point of view of thermoregulation, which shall be addressed in greater detail in Chapter 4. Recall from Chapter 1 that some 25% of the heat generated as a product of metabolism is lost through evaporation from the skin surface. This can be significantly higher during strenuous exercise, when much sweat is produced to absorb heat in the process of transforming from the liquid state to the vapor (gaseous) state.

Inertial-Type Forces: These can generally be written as the negative gradient of kinematic potentials. This is because the forces originate on mass elements when their motion is referred to an *accelerating set of axes,* or when they are travelling at very *high speeds,* or then their motion is otherwise *time-dependent.*

The kinematic variables involved include displacement, velocity, acceleration, "jerk" (the rate-of-change of acceleration), strain, and strain-rate, gradients of which define:

1) *Centrifugal* (or Centripetal) forces ($\phi = \frac{1}{2} k_r^2 \omega^2$), which are caused by the change in direction of the velocity vector, which is tangent to the curvilinear path of a moving particle.

2) *Coriolis* acceleration forces ($\phi = 2v_r \ell \omega$), which are caused by radial particle motion v_r along an arbitrary curvilinear path, referred to a coordinate system rotating with an angular velocity, ω.

3) *Relativistic* forces ($\phi = c^2$), which are caused by a change in mass experienced during motion at very high speeds (approaching the speed of light, c), and governed by the principles of Einstein's Theory of Relativity.

4) *Kinetic Energy Inertial* forces $\left(\phi = \frac{1}{2} v^2\right)$, which are body forces inherent in a particle as a result of the work done to generate movement (at an associated speed, v).

5) Other types of *Unsteady Temporal* (time-dependent) forces (such as vibration, $\phi = \ell^2 f^2$, turbulence, $\psi = \dfrac{\overline{v_i v_j}}{2}$, $i, j = 1, 2, 3$ -- the bar over the quantities indicating a spatial average over all directions i and j in space -- and so on). These arise from the unsteady motion of particles, which results in such phenomena as secondary streaming, turbulent dispersion, and other effects related to unsteady physical processes. They have particular relevance in the cardiovascular system (Schneck, D. J., and Ostrach, S., 1975, Walburn, F. J., and Schneck, D. J., 1976, 1978, 1980, Schneck, D. J., 1977, and Davis, R. B., III, Schneck, D. J., and Gutstein, W. H., 1984), and these effects are actually manifest *both* through inertial responses and surface responses, some of which are discussed below.

As opposed to body forces, surface forces are short-range forces which have a direct molecular origin. They decrease even more rapidly than do body forces when the distance between interacting elements increases, and they are appreciable only when that distance is of the order of the separation of molecules in the medium. Indeed, surface forces are negligible unless there is direct physical contact between the interacting elements, because without that contact none of the molecules of one of the elements would be sufficiently close to a molecule of the other element for their interaction to be manifest (i.e., for a force to be experienced).

Short-range forces result primarily from the transport of momentum by molecules migrating across a common boundary between two interacting species. Therefore, it is convenient to divide surface forces into those that are transmitted *normal* to the common boundary (or surface of contact) and those that are transmitted *tangent* to the common boundary. Examples of surface forces that may affect physiological transport are:

Normal Forces: These are manifestations of hydrostatic pressure and various forms of compressive, tensile, and secondary stresses. Note that pressure is a *body* stress in the thermodynamic sense discussed in the earlier section on convective-type forces, but a *surface* stress in the mechanical sense addressed here. That is to say, as defined by an equation such as the perfect gas law, $pV = NRT$, pressure is in some thermodynamic sense related to Temperature, and so is a measure of translational molecular kinetic energy per unit volume (c.f., Chapter 3 and Schneck, 1985). On the other hand, as defined in continuum mechanics, pressure is the negative mean of the three normal components of the generalized stress tensor, i.e., $p = -\frac{1}{3}(\tau_{ii})$, where the double-subscript notation implies summation on $i = 1, 2, 3$.

It can be shown (see, for example, Jeffreys, 1965) that the two separate definitions for pressure reduce to the same quantity in a material that is in a state of thermodynamic equilibrium, wherein expansions or contractions take place reversibly with no energy loss. That is to say, when thermodynamic equilibrium prevails, the *work done* by normal stresses in expanding or compressing some material is transferred *reversibly* (isentropically) into molecular translational kinetic energy, without generating any heat -- and it acts to alter the *pressure* (i.e., "potential" energy) in the material, rather than the *temperature* (i.e., "entropy") of the material. Thus, rather than being accompanied by the generation of *heat* (i.e., *dissipation* of energy), which raises the temperature, such work is *stored* elastically in the form of pressure energy, which can be retrieved when the process is reversed.

For this special case, normal forces in fluids (i.e., hydrostatic mechanical pressure forces) caused by a nonuniform distribution (or gradient) in the negative mean of the normal stresses acting at a point in a solute-solvent mixture, can be written as the negative gradient of the same associated scalar potential that was defined previously in our discussion of convective-type body forces, i.e., $\phi = \frac{p}{\rho}$. For elastic solids, normal forces can be written as the negative gradient of the potential, $\phi = \frac{E_s \varepsilon^2}{\rho}$, where E_s is the conventional static Young's Modulus of Elasticity and ε is the (nondimensional) normal strain in the material. An analogous potential proportional to E_d/ρ can be written for viscoelastic materials, where, in this case, E_d represents the Dynamic Elastic Modulus.

Apropos to our discussion on page 103, it is of interest to note that in each of the cases cited above, the characteristic speed, v, associated with the propagation of a disturbance through the respective material involved is, indeed, related to the square-root of the corresponding intensive scalar potential, ϕ. That is to say, scaled by the dimensionless specific heat ratio, $\frac{c_p}{c_v}$, the speed of sound in gases is given by $\sqrt{\frac{c_p}{c_v}\frac{p}{\rho}}$. Similarly, the speed of sound in liquids is equal to: $\sqrt{\frac{\Xi}{\rho}}$, where $\Xi = V\frac{dp}{dV}$ is the bulk modulus of the fluid; the speed of sound in solids is equal to $\sqrt{\frac{E_s}{\rho}}$; and, scaled by the dimensionless ratio of tube thickness, ℓ, to inner tube radius, r_i, or, more precisely, diameter $2r_i$, the speed of propagation of a pressure wave through an inviscid fluid contained within a thin-walled, linearly elastic tube of dynamic modulus E_d is approximately given by $\sqrt{\frac{E_d\ell}{2r_i\rho}}$. The latter, of course, is the famous Moens-Korteweg Formula which gives a remarkably reasonable estimate of the pulse-wave velocity in the human cardiovascular system. Various advanced forms of these equations also take into account damping (Womersley, 1957), axial strain (Poisson's Ratio), and wall mass (Lamb, 1898), but, each effect simply adds an appropriate scaling factor to the basic functional form, which is $\sqrt{\phi}$. Thus, in *general,* one can write for the velocity of transport of a disturbance through any given material: $v_{transport} = \sqrt{A^*_j\phi_j}$, where A^*_j is the proper scaling coefficient that depends on the physical properties of the material involved, or the geometric configuration of the corresponding physical system, and ϕ_j is the associated scalar intensive potential for species j.

Normal forces acting on deformable fluids can produce movements of large masses of solvent across a membrane-type barrier separating two contiguous regions in space. Such movement, in turn, may carry with it dissolved particles -- frequently moving them in an active sense against passive forces that might normally drive them the other way, were the solvent flow not present. This type of mass transport is generally termed *solvent drag,* or *flow-induced transport.*

In addition to solvent drag, a further consequence of flow due to normal forces is that movements of large masses of solvent also disturb the concentration of solute particles on both sides of the membrane -- irrespective of whether or not these particles are dragged along with the flow. Such disturbances may thus *generate* ordinary diffusion forces where there might not have been any to begin with. The combination of flow-induced *solvent drag*, together with flow-generated *ordinary diffusion* is called *bulk flow*, and is believed to play a major role in mass transport in and out of the cardiovascular system at the capillary level (c.f., later and Chapter 3).

When normal forces are dissipative, such as is the case, for example, in situations where there arise *secondary* normal stresses that are attributable to the nonhomogeneous, anisotropic, non-Newtonian characteristics of biological solutions -- or, -- where there arise *turbulent* normal stresses (so-called "mixing forces") attributable to flow instabilities and other disturbances in streamline curvature -- then these normal forces usually *cannot* be expressed in terms of the gradient of a scalar potential alone. Instead, they must be written together with a conjugate orthogonal function, ψ, in the form of a vector or complex potential, as discussed previously. More often than not, energy dissipation is associated with tangential forces, however, rather than with normal ones.

Tangential Forces: These are manifestations of surface tension and various forms of shear, viscous, and secondary tangential stresses. Surface tension is one of the *few* tangential forces that have associated with them a conservative potential function, $\phi = \dfrac{\zeta}{\rho \ell}$. The others all have associated with them dissipative functions that can usually be written in terms of kinematic-variable gradients that are occurring within any given material.

For example, *shear* or *viscous forces* are caused by relative motion (flow) taking place tangential to the common region of contact between interacting species, thus causing them to "rub" against one another. In this case, then, the forces result from a nonuniform distribution of particle velocities, and so are proportional to spatial *gradients in velocity* ($\dot{\gamma}$), called *shear rates,* that exist in the flow, i.e., $\psi = \frac{\dot{\gamma}\mu}{\rho}$. In the case of solids, the forces involve gradients of displacement, i.e., $\psi = \frac{\sigma\varepsilon}{\rho}$, while in the case of viscoelastic materials, the forces involve gradients of *both* velocity *and* displacement, i.e., $\psi = \frac{\eta}{\rho}$, where η represents the viscous modulus.

The significance of viscous shear forces in affecting *any* membrane mass transport processes may be addressed from three different aspects. First, there is a characteristic time-scale associated with such transport requiring (among other things) that the quantity involved remain at the designated transport site *long enough* for the transfer to take place. If viscous-induced flow tangent to the boundary causes the material to go by too quickly -- relative to however long it takes for the gradients driving the transport process to accomplish getting the quantity across the membrane -- then there simply may not be enough time for the material to interact with the membrane long enough to pass through from one side to the other. That being the case, transport is impeded.

Second, the speed with which a quantity flows by tangent to a boundary also controls the gradients, themselves, of those potentials that may be responsible for driving materials across the membrane from one side to the other. A faster moving flow, for example, while allowing less *time* for mass transfer to occur, also leads to *larger concentration gradients* across the membrane. This is due to a *wash-out effect* that does not allow mass equilibration to occur across the membrane. That is to say, mass may be continuously removed from one side due to viscous flow, so the concentration on that side never stabilizes. Indeed, wash-out effects may even *generate* transmembrane gradients where none might otherwise have existed. In a balanced situation, it may turn out that wash-out effects in high speed flows lead to increased transport-driving gradients, which, in turn, amplify the transport driving forces, which, consequently *decrease* the time *required* for transport to be affected, which may just balance the decrease in time spent by the quantity at a transport site along the membrane.

Going one step further, and third, a faster moving flow on one side of a membrane *decreases* the hydrostatic pressure on that side -- a

Bernoulli Effect -- relative to a slower moving flow on the other side of the membrane. Thus, a *lateral*, or transverse (normal to the direction of flow) pressure gradient is established that results in bulk flow transport *across* the membrane, although the mechanisms causing it are occurring *parallel* to the membrane. Thus, the *wash-out effect* enhances transmembrane gradients in potential, the *Bernoulli effect* generates an induced transport pressure gradient, and the two effects together generate the same bulk flow phenomenon discussed in the previous section. The difference here is that the bulk flow is being induced by *shear effects* occurring tangent to the membrane surface, as opposed to those in the previous section that were assumed to be occurring *normal* to the membrane surface, i.e., in the direction of the actual transport.

As was the case for normal forces, the nonhomogeneous, anisotropic, non-Newtonian characteristics of biological materials also generate secondary *shear* stress forces that need to be contended with, and, where relevant, *turbulent shear* stresses, as well. The presence of turbulence adds greatly to the energy of a molecule in a solvent and thus has significant effects on mass and other forms of transport. These go under the general term, *turbulent dispersion*, which can be somewhat quantified by a parameter known as the *Eddy Diffusion Coefficient*. Turbulent shear stresses are due to double and triple auto-and-cross-correlations $\left(\psi = \frac{\overline{v_i v_j}}{2} \right)$ that define velocity component interactions leading to additional momentum transport between interacting species. These are *in addition* to shear stresses that exist in "normal" laminar flow, and they have significant effects on transport because of the dispersion phenomenon. They, too, however, as are virtually all tangential forces, are dissipative. That is to say, the kinematic gradients discussed above produce a shearing action that, through friction, *dissipates* (converts to heat) rather than *conserves* (converts to a potential form) energy.

To summarize, then, depending on the particular transport process involved, one can write:

$$\vec{\Omega} = \phi \hat{i} + \psi \hat{j} \qquad\qquad [2\text{-}5]$$

with the realization that, in some sense, the force driving the transport process is related to $\vec{\nabla\Omega}$, and, for the case of conservative force systems, $\psi = 0$. Moreover, the characteristic transport velocity, v, in this case can also be written as $\sqrt{A^*_j | \vec{\Omega}_j |}$, where, $| \vec{\Omega}_j | = \mathcal{R}_j = \sqrt{\phi_j^2 + \psi_j^2}$.

Kinematic Transport Coefficients

Kinematic transport coefficients, α, are in some sense a measure of both the *mobility* (ability to *react* to a driving force) of the transported quantity, S, and the *permeability* (*resistance* to the passage of S) of the transport pathway. They may be grouped according to whether they define the transport of mass, energy, or momentum -- and -- within these groupings, whether they address active, passive, or facilitated processes involving conduction (i.e., diffusion), convection (i.e., flow), or radiation (i.e., emission). The latter three: conduction, convection or radiation have to do with the *scale* of the transport process; that is, whether it is of a molecular, differential, or continuum order-of-magnitude (see later).

Kinematic transport coefficients can often be calculated from the *properties* of a given system, in conjunction with certain fundamental constants or parameters of physics. Properties may be: (i) chemical (including molecular mean free path, λ, molecular weight, MW, and valence, Z); (ii) electrical (including electrical conductivity, σ_e, and electrical resistance, R^*); (iii) magnetic (including magnetic permeability, μ_e, and h_p); (iv) nuclear (including the mass, m_a of an atomic particle, and the various stationary states of an electron); (v) material, or physical (including specific weight, w, static Young's modulus, E_s, dynamic elastic modulus, E_d, bulk modulus, Ξ, viscous modulus, η, Poisson's ratio, dynamic coefficient of viscosity, μ, surface tension, ζ, and mass density, ρ); and (vi) thermodynamic (including thermal conductivity, k_t, temperature, T, convection coefficient, h, specific heat at constant pressure, c_p, specific heat at constant volume, c_v, latent heat of fusion or vaporization, c_L, and radiation emissivity, κ). Fundamental constants or parameters of physics may include the universal gas constant, R, the electric Coulomb constant, k, the Faraday constant, $\overline{F}$, the magnetic Coulomb constant, k_m, Planck's constant, h_o, the gravitational acceleration constant, g_o, Boltzmann's constant, k_b, and the Stefan-Boltzmann constant, k_s. All of these quantities, together with their associated dimensions are defined in the list of symbols at the beginning of the text.

Conduction-Type Transport: Conduction is a molecular concept that attributes the transport of S to events (such as Brownian motion and random thermal vibrations) taking place at the sub-microscopic level. It assumes molecule-to-molecule transmission (or redistribution) of mass, energy, momentum, information, and so on, due to localized gradients in charge that exist in an otherwise quiescent macroscopic environment.

Thus, for example, the molecular diffusion coefficient, $\lambda\sqrt{\dfrac{RT}{MW}}$, to be discussed further in Chapter 3, addresses the transport of *mass* by conduction due to a concentration gradient (Fick's Law); the thermal diffusivity, $\dfrac{k_t}{\rho c_p}$, to be discussed further in Chapter 4, addresses the transport of heat *energy* by conduction due to a temperature gradient (Fourier's Law); and the kinematic viscosity, $\dfrac{\mu}{\rho}$, addresses the transport of *momentum* by conduction due to a velocity gradient (Newton's Law). For this reason, it is often called the "momentum diffusivity."

In physiologic systems, gradients in charge are mainly the result of metabolic processes that consume raw materials and generate waste products and heat, causing gradients in both concentration and temperature. These gradients are generally of sufficient magnitude to drive *short-range* transport processes, on the order of the mean-free-path of molecular motion, but not to affect long-range, bulk transport. To accomplish the latter, the physiologic system depends on convective transport. This is what LaBarbera and Vogel (1982) call the "first design principle" that characterizes the transport systems of organisms, i.e., that:

> "While diffusion is always used for short-distance
> transport, it is augmented by bulk flow ... for any
> long-distance transport."

Convection-Type Transport: Convection is a continuum concept that attributes the transport of S to events (such as "natural" passive or "active" forced flows) taking place at a substantially more macroscopic level. It assumes region-to-region transmission of mass, energy, momentum, information, and so on, in a bulk-conveyance process that results from a disturbed environment.

Thus, for example, the kinetic flow coefficient, $v\ell$, addresses the transport of *mass* by convection due to a pressure gradient; the thermal convection coefficient, $\frac{h\ell}{\rho c_p}$, addresses the transport of heat *energy* by convection due to both density (free convection) and pressure (forced convection) gradients; and the eddy diffusion coefficient, $\overline{v_i \ell_j}$, $i, j = 1, 2, 3$, addresses the transport of *momentum* due to tubulent dispersion. The bar over the quantities indicates a spatial average over all directions i and j of the respective fluctuating variable.

In physiologic systems, "disturbed environments" can result from several mechanisms. One such mechanism is the semipermeable membrane (discussed later) which, by selectively letting certain materials across, but not others, generates convective currents such as are characteristic of osmotic processes (see Chapter 3). These convective currents play an essential role in getting things in and out of the cardiovascular system at the capillary level.

Another convective mechanism is active transport which, by physically creating concentration gradients where none might have otherwise occurred naturally, establishes both osmotic pressure and electrochemical gradients that drive transport across cellular membranes. These are essential for propagating information in the form of action potentials through nerves and muscles.

Then, too, there are pumping devices such as arrays of cilia (finger-like processes that wave back and forth to generate movement of anything in contact with them), muscular chambers, such as the stomach and diaphragm, which churn" or otherwise undulate to generate movement; peristaltic wall movements of, for example, the intestines, which move food and other material along in the Alimentary canal; and, the action of the human heart. By establishing a hydrostatic pressure gradient ($\alpha = \dfrac{\rho \pi r_i^4 f^2}{8\mu}$, where r_i is a blood vessel radius and f is the heart rate), it generates pulsatile flow through the cardiovascular system, which also illustrates LaBarbera and Vogel's Second and Third Design Principles for transport systems:

> "Transport systems use both large and small pipes
> -- small pipes at exchange sites and large pipes for
> moving material from one exchange site to another;"

and,

> "Total cross-sectional area of smaller pipes greatly
> exceeds that of larger pipes, so that velocity in
> smaller pipes is less than that in larger ones."

Cardiovascular flow is crucial for accomplishing the proper transport of mass into, out of, and within the physiologic system fast enough to maintain metabolism at a rate sufficient for survival. It is also vital for accomplishing the thermoregulatory processes that are essential for keeping the internal environment of the system functioning at an approximately isothermal 37°C setting. The latter depends, too, on a significant amount of radiation.

Radiation-Type Transport: Some 60% of the heat generated by physiologic metabolic processes is dissipated by mechanisms associated with radiation (Cooney, 1976). Radiation can be either at the molecular or the continuum level. This concept attributes the transport of S to some central source that is continuously "manufacturing" and "emitting" it to a transporting medium (which may even be a vacuum). Although there may be identifiable mass and momentum sources and sinks, the concept of radiation is most often associated with energy, especially in physiologic systems. Thus, one speaks of "Emissivities" that are associated with thermal $\left(\alpha = \dfrac{k_s \kappa T^3 \ell}{\rho c_p} \right)$, electromagnetic $\left(\alpha = \dfrac{1}{\sigma_e \mu_e} \right)$, and nuclear $\left(\alpha = \dfrac{h_o}{2\pi m_a}\right)$ transport of energy. From a physiologic point of view, the most important of these is the thermal emissivity of the system.

Transport Pathways

As mentioned earlier in the section on *General Considerations*, S may move intramurally *within* a confined region in space, or, it may move extramurally *across* a bounding surface (such as a cell membrane) between two contiguous regions in space.

Intramural Transport includes: (i) getting S from one point to another entirely within the interior of a cell (i.e., ICF transport); (ii) getting S from one location in the organism to another via the "river of life" that courses through thousands of miles of blood vessels (i.e., intravascular ECF transport); (iii) getting S *from* the exterior of a blood vessel *to* the exterior of a cell or vice versa (i.e., extravascular ECF or interstitial fluid transport); (iv) moving S along in the Alimentary System (peristaltic transport); (v) transporting S through the Lymphatic System; (vi) getting S to leave the organism from the bladder to the environment (i.e., urination or micturition); (vii) getting S to enter the organism from the environment to the lung sacs (diaphragmatic inspiration); (viii) getting S to leave the organism from the lung sacs to the environment (diaphragmatic expiration); and (ix) in general, moving S along any path that does *not* require it to pass through a membrane or some other barrier that stands between it and its intended destination. The mechanisms of intramural transport are many and varied, and, for the case of *mass* transport, in particular, are affected by:

(1) The drag characteristics between S and the solvent or suspending medium. This relates to the degree to which dissolved or suspended particles of a given species S interact with the solvent or suspending medium *through which* or *with which* they are travelling. It is a function of the *consistency* or *viscosity* of the fluid (mostly water) or other suspending medium that constitutes the solvent, its degree of *ionization*, its density (or, more specifically, the *difference* between its

density and that of S), its *chemical* composition, and its *thermodynamic* and/or *physical state* (e.g., solid, liquid or gaseous), among other things.

The drag characteristics between S and the solvent may be quantified by defining a frictional coefficient, τ_{sw}, which addresses the mutual drag of solvent and solute on each other (more on this coefficient later); and, by defining a corresponding intramural drag transport coefficient, α. As a simple example, let S be a spherically shaped (of radius r_o) particle settling with velocity v in a medium of viscosity μ_f. Stoke's law then asserts that the total drag force on such a sphere is equal to $6\pi\mu_f v r_o$. Moreover, the net buoyant force per unit volume acting on the sphere is equal to the gravitational constant, g_o times the difference in density, $\rho_S - \rho_f$ between the particle, S, and the medium, f. In the steady-state, one can equate this buoyant force to the total drag force to derive a characteristic transport velocity,

$$v = \frac{2}{9} r_o^2 \frac{g_o}{\mu_f} (\rho_S - \rho_f) \qquad\qquad \text{[2-6]}$$

and kinematic transport coefficient,

$$\alpha = \frac{\mu_f}{\rho_S - \rho_f} \qquad\qquad \text{[2-7]}$$

Equation [2-6] is commonly used to calculate the "Sedimentation Rate" of red blood cells in plasma, a clinical parameter that has some value in characterizing various disease states. Typically, with r_o of order 2.75 - 3.00 microns, μ_f of order 1.8 millipascal-seconds (centipoise) for plasma, $\rho_f = 1.01$ gms/cm^3, and $\rho_S = 1.08$ gms/cm^3 for red blood cells, sedimentation rates calculated from equation [2-6] are of order 2.5 millimeters per hour. This compares with clinically measured values of 1-3 mm/hr for men, and 4-7 mm/hr for women, so the equation gives a relatively good analytic estimate of this variable, despite the simplifying assumptions that have been imposed in its derivation.

(2) The drag characteristics between solvent or suspending medium and the confining boundary (for example, a blood vessel wall). This relates to the degree to which a confining boundary offers resistance to the flow of a contained transport medium. It is a function of many of the same variables as are listed in (1) above, plus, the *isotropic* (or anisotropic) characteristics of both the boundary and the transport medium, their *homogeneity* (or lack thereof), their *Newtonian* (or non-Newtonian) characteristics, their *viscoelastic* (i.e., time-dependent) deformation behavior as it relates to the stress-history of the materials involved, and, the *smoothness* (or roughness) of the bounding surface, among other things.

The drag characteristics between solvent and confining boundary may be quantified by defining a wall frictional coefficient, τ_w, to be addressed further later, and, by defining a corresponding kinematic transport coefficient, α, for this interface phenomenon. The phenomenon affects bulk flow and solvent drag transport processes, and has to do with, or is affected by the size, symmetry, configuration, and other geometric and chemical properties of the molecules that constitute both the boundary and the transport medium materials. As a simple example, consider the steady (time-independent), laminar, fully-developed flow of a homogeneous, isotropic, incompressible, Newtonian fluid (like pure water), having a constant viscosity μ_f and density ρ_f, through a smooth, rigid, circular tube of internal radius r_i and length ℓ. If the flow is being driven by a difference in pressure, Δp, across ℓ, one can derive a radially-dependent characteristic transport velocity,

$$v = \frac{\Delta p}{\ell} \frac{r_i^2}{4\mu_f} \left[1 - \frac{r^2}{r_i^2} \right] \qquad \text{[2-8]}$$

and kinematic transport coefficient,

$$\alpha = \frac{\mu_f}{\rho_f} \qquad \text{[2-9]}$$

Integrating equation [2-8] across the tube cross section, one can solve for the total volumetric flow rate through the tube, which, of course, is the famous Poiseuille Equation. Written in terms of mass flux, the equation takes the form,

$$\Phi = \frac{\Delta p}{\ell} \frac{\rho_f}{\mu_f} \frac{r_i^2}{8}$$ [2-10]

Equation [2-10], converted to volumetric flow rate, i.e., in the form $(V/t) = (\Delta p/\ell)(1/\mu_f)(\pi r_i^4/8)$, is commonly used to calculate the "Peripheral Resistance" of parts or all of the cardiovascular system. This clinically-useful parameter is a ratio of the pressure drop, Δp, necessary to generate a given volumetric flow rate through the particular part of the cardiovascular system being analyzed. Thus, peripheral resistance $= (\Delta p/V)t = (8\ell\mu_f/\pi r_i^4)$ can be seen to vary directly with fluid viscosity and tube length, and inversely with inside tube diameter. For the cardiovascular system taken as a whole, volumetric flow rate is approximately 5 to 6 liters per minute in an average individual. Furthermore, the pressure drop from the output of the left ventricle to the input of the right atrium of the heart -- i.e., through the systemic circulation -- is about 95 - 100 mm of mercury, on the average. Thus, the peripheral resistance of the systemic circulation is on the order of 20 mm Hg per liter per minute of flow generated. This is the pressure that the heart must generate in order to drive that amount of fluid through the entire systemic circulation under normal conditions.

(3) Intramural mass transport is also influenced by the relative magnitude of dissolved particle interactions with the boundary material itself. This relates to the fact that when a moving species S collides with or otherwise comes in contact with the boundaries of a region within which it is moving, one or a combination of several things may occur:

(a) S may stick to the boundary permanently; that is, undergo a specific *adhesive* physico-chemical reaction.

(b) S may adhere to the boundary momentarily, i.e., become *adsorbed*, followed by a subsequent disassociation (called *surface diffusion*) from the boundary.

(c) S may rebound from the boundary according to the simple mechanical laws of impact and reflection -- a process called *specular*, or *mirror-like reflection*.

(d) In the case of significant surface roughness, S may rebound from the boundary by some other more complicated mechanism (including *dispersion*) that is somewhat more random than simple reflection -- for example, *Clausing reflection* following the cosine law of emission. And,

(e) The movement of S may be significantly impeded by *frictional* interactions with the bounding surface. These may be quantified by defining a frictional coefficient, τ_S, describing the shear interactions between solute and boundary material (see later).

The extent to which surface diffusion becomes important in mass transport varies rather widely with the nature of the boundary-solute-solvent system, since it depends upon the existence of an adsorbed layer of material. However, where it is a factor, one may account for it by assuming that surface diffusion flux is proportional to the gradient of the surface concentration, and to a surface diffusion coefficient, i.e., a surface kinematic transport coefficient. The latter has been found to be a decreasing function of the nondimensional parameter, $\dfrac{c_A}{jRT} (MW)$, where c_A is the latent heat of adsorption, $j = 1, 2$, or 3, depending on the nature of the adsorption bond, and R, T, and MW have been previously defined. For temperatures below 100°C, this inverse relationship for α yields useful approximate values for the surface diffusion coefficient, which is small (10^{-2} to 10^{-13} cm^2/sec) (Sherwood, Pigford, and Wilke, 1975).

(4) Last, but certainly not least, intramural mass transport is affected by the relative magnitude of dissolved particle interactions with each other. Most important in this respect are the size and configuration of S, the number of molecules of S in a given quantity of solvent, and, for charged particles, their degree of ionization and hydration.

Obviously, the ultimate path that S takes intramurally in getting from one point to another will depend upon how many times it collides with obstacles in its path, and how serious those collisions are. Collisions with solvent molecules were addressed in (1) above. Collisions with bounding surface molecules were addressed in (3) above. Now, we address collisions with solute molecules, which becomes increasingly important as the concentration of S becomes such that the solution or suspension can no longer be considered to be dilute. That is to say, typically, particle-to-particle, or molecule-to-molecule interference is neglected in mathematical formulations that seek to describe mass transport. Thus, by inference, the equations that result from such formulations are strictly valid only for very dilute solutions, where the probability of one particle of S encountering another particle of S along its path is quite small.

To compensate for such interactions, one may introduce a so-called "activity factor", α^*, which multiplies $[S]$ to account for concentration variations resulting from particle interference. The quantity $\alpha^*[S]$ is called the *activity* of the solution. It may be thought of as the "effective concentration," as opposed to the actual concentration of species S. The activity factor, α^*, depends, itself on concentration and approaches unity for very dilute solutions (defined to be less than about 0.01 molar, i.e., 0.01 moles of S per liter of solution). That is to say, for more concentrated solutions particle interference slows down the process of transport due to intramural mechanisms, making it "appear" that there is less of a driving force (∇e) propelling the process than there really is. The activity factor is an empirically determined coefficient that also depends on S, the solvent in which S finds itself, and an electrical phenomenon known as *molecular association*.

Molecular association has to do with the degree of ionization of S, and, as a result, the extent to which charged particles of S are hydrated. That is to say, the charge distribution on ionized particles has a tendency to attract and bind polarized water molecules to S, thereby increasing its effective molecular dimensions and promoting more intramural transport interference. The attraction of water molecules to S is called hydration, and the resulting transport interference is called molecular association disturbance. It stems from the fact that hydration of ionized particles "encumbers" them with more effective mass, causing them to become less agile and more sluggish in their random Brownian motion. Indeed, molecular association may involve the formation of large aggregates among *any* number of charged, randomly coiled macromolecules -- where the presence of well-defined, discrete concentrations of charge in a complicated molecular configuration (which is, itself, neutral as far as *total* charge is concerned) -- or, where the existence of complementing molecular invaginations in the chemical structure of solute particles -- all cause them to interact in ways that increase their effective size, and so, significantly change their mass transport characteristics.

Molecular association implies less than a permanent chemical bonding between particles. These types of molecular interactions are frequently short-lived, are weak in the sense that the aggregations are easy to break apart, and thus tend to have transient, episodic influences on intramural transport. They are also affected by the pH of the medium through which the transport is taking place.

Extramural Transport includes (i) getting S (oxygen) into the blood stream from the alveolar air sacs to the pulmonary capillaries -- or, in reverse, getting S (carbon dioxide) out of the blood stream from the pulmonary capillaries into the alveolar air sacs; (ii) getting S (nutrients) into the blood stream through the microvilli of the small intestines to the mesenteric capillaries; (iii) getting S (urea and other metabolic wastes) out of the blood stream from the glomerular capillaries into the Bowman's capsule of the kidney nephron -- or, in reverse, getting S (electrolytes and other ions) back into the blood stream from the ascending loop of Henle in the kidney nephron to the tubular capillaries; (iv) getting S into or out of a cell across the cell membrane; (v) getting S into or out of a capillary across the thin endothelial tube wall; (vi) getting S into (wastes) and out of (water, electrolytes) the large intestines through the inferior mesenteric capillaries; (vii) getting S into or out of the lymphatic system across the thin endothelial tube wall (especially involving intestinal lipid absorption); (viii) getting S (hormones) into or out of glands; and (ix) in general, moving S along any path that *does* require it to pass through a membrane or some other barrier that stands between it and its intended destination.

As mentioned earlier, extramural transport may involve transudation through a discontinuous portion of the barrier (such as a pore), permeation through the boundary material, itself, or Pinocytotic or Phagocytotic membrane invagination/encapsulation/evagination of S. In order to address each of these mechanisms, one must first examine some of the structural, anatomical and physiological features of the typical cell membrane.

THE CELL MEMBRANE AND PORE GEOMETRY

On the basis of chemical analyses, electron microscopic observations, x-ray diffraction studies, optical polarization investigations, and other means that have been used to identify and describe a "permeability barrier" between physiologic cells and their environment, the structure of the cell membrane is believed to be a bimolecular lipid layer covered on both sides by a protein coat and, presumably, interrupted periodically by long(ish) narrow pores (see Figure 2-1).

The reason we say "presumably" interrupted by pores is that the cell membrane *behaves, functionally,* as though it also contains pores which have a diameter slightly larger than that of a urea molecule, i.e., some 5 to 10 Angstrom (10^{-10} meters) units. Other than that, the existence of pores is actually based more on circumstantial evidence than on specific visual observations -- but the overwhelming evidence leaves little doubt that certain types of transport across cell membranes could not possibly occur at recorded rates if such transport had to take place through the membrane material itself. It is both plausible, and, indeed, sometimes *necessary* to postulate the existence of pores in order to reconcile measured mass transport characteristics of certain types of species S with the

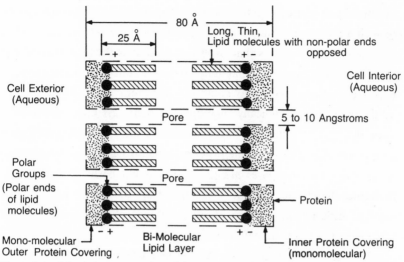

Figure 2-1 Schematic representation of the bi-lipid configuration, periodically interrupted by pores, believed to represent a model of the cell membrane.

presumed anatomical structure of the cell membrane. A schematic model of such a presumed configuration is shown in Figure 2-1.

One theory suggests that Brownian motion of the lipid molecules in the plane of the membrane might cause them to undergo some oscillatory movement. Such oscillations could generate "temporary" pores cyclicly, as the molecules periodically "line-up" just right. At other times in the cycle the pores are just not there because of an overlapping molecular configuration in the membrane material. The further suggestion has been made that when the pores *are* there, they might be lined with protein molecules that extend into the pore lumen from the surface protein coating, but little is really known about what type of "material" (if indeed there is such a thing) actually fills the pore itself.

Be that as it may, the membrane protein coating *is* known to consist generally of a high molecular weight fibrous molecule, outweighing the lipid molecules by nearly two-to-one. The lipids consist mainly of cholesterol and various phospholipids. Although they *weigh* much less than the protein, they *outnumber* them nearly one-hundred-to-one. Various estimates claim that there are enough lipid molecules in the 10^{14} cells that comprise the human organism to cover the surface of each cell two to three times.

EXTRAMURAL TRANSPORT THROUGH PORES

Since most pores in biologic systems are filled with aqueous solutions, and since fats and/or fat-soluble substances are hydrophobic, it is believed that transport through pores in membranes is restricted to materials that are basically water-soluble. In general, the more polar a substance is the more soluble it is in water. The more non-polar a compound is, the more soluble it is in lipids or lipid solvents. Thus, lipid-soluble substances are believed to penetrate a biologic membrane by dissolving into the membrane material itself, rather than by passing through pores, whereas ions and other polarized substances traverse the membrane through water-filled pores. This -- i.e., the solubility of particles in the solvent, is still another in the list of considerations that address solute-solvent inter-actions as determinants of ultimate mass transport characteristics.

Moreover, since pores provide a continuous pathway from ICF to ECF (or vice versa) through an essentially common solvent, extramural transport through pores may actually be thought of as a special case of intramural transport, where the latter takes place through a well-defined and *confined* route. Thus, many of the considerations discussed earlier with respect to transport pathways in intramural transport apply equally well to extramural transport through pores. These include, for example, (i) the interaction (i.e., adhesion, cohesion, adsorption, specular re-flection, dispersion or scattering) and drag (τ_S) characteristics between S and the pore walls (pgs. 122-123); (ii) the resistance to transport (τ_{sw}) re-sulting from interactions between S and the pore material that provides continuity from one side of the membrane to the other (pgs. 119-120); (iii) the resistance to flow (τ_w) resulting from interactions between pore fluid and pore wall (pgs. 120-122); (iv) the degree of ionization of the material which fills the pore (pgs. 124-125); and (v) the physical properties (such as density, viscosity, isotropy, homogeneity, and degree of Newtonian behavior), chemical composition, and thermodynamic state of the mate-rial which fills the pore (ibid.).

In the specific case of transport through a pore, however, several additional features must be considered, to wit:

(1) The *dynamic state* of the material which fills the pore, i.e., whether it is flowing or not, and, if it is, in what direction. The signif-icance of the whole concept of hydrostatic flow as it applies to transport across a membrane is that, when water moves through the membrane, it may also carry with it dissolved particles that are small enough to get through the pores. If the flow is strong enough, these particles may ac-tually be transported *against* electrochemical gradients that might be tending to drive them back the other way. This phenomenon is called *counterflow*, wherein species S being passively driven in one direction gets caught up in flow coming through the pore in the opposite direction.

Counterflow can seriously impede transport (as in the case of male sperm cells fighting the female menstrual flow to get to an egg cell); or, it can act as a means of *active* transport to carry S into a region where it would either not normally be transported passively, or, be transported too slowly by passive mechanisms to do any good (as is the case with the dissipation of heat energy). Because of the method by which such transport takes place, it has come to be known as *solvent drag,* and it may be the very basis for all mass transport that takes place at the capillary level (see Chapter 3).

The influence of solvent drag on particle motion depends on at least τ_S, τ_{sw}, and, τ_w, and it is possible to show (Stein, W. D., 1967) for intramural transport through a pore that these three frictional coefficients are related to the phenomenological kinematic transport coefficients by the following equation:

$$\alpha = \frac{\pi d^2 \left\{ \dfrac{\tau_w}{\overline{V}_w} + \tau_S(1-\beta)\left(\dfrac{[S]_{\text{inside}} + [S]_{\text{outside}}}{2} \right) \right\} \delta^2}{4\rho_f \Psi_w} \qquad [2\text{-}11]$$

where Ψ_w is the volume fraction of water in the membrane (i.e., total volume of water divided by total volume of water plus membrane material), $\overline{V}_w$ is the partial molar volume of the solvent, water (i.e., the volume occupied by one mole of the solvent), the last fraction in the numerator is the mean of the solute concentrations on the two sides of the membrane, d is some mean diameter of an average pore of length δ, and the other quantities are as previously defined. Furthermore,

$$\beta = 1 - \frac{4\rho_f \alpha \Psi_w \overline{V}_S}{\pi \delta^2 d^2 (\tau_{sw} + \tau_S)} - \frac{K\tau_{sw}}{\Psi_w(\tau_{sw} + \tau_S)} \qquad [2\text{-}12]$$

where $\overline{V}_S$ is the partial molar volume (i.e., volume occupied by one mole) of the solute, S, K is the distribution coefficient for the partitioning of solute between aqueous phase and the membrane, and the other quantities are as previously defined. And finally, with the kinematic transport coefficient in equations [2-11] and [2-12] representing the flow of pore fluid, and the corresponding transport of $[S]$ at zero initial concentration difference (i.e., solely due to solvent drag) -- the concentration of $[S]$ being presumed to be uniformly distributed at some mean value, $\frac{[S]_{inside} + [S]_{outside}}{2} = \overline{[S]}$, we may add a third equation,

$$\Psi_w = \frac{\pi d^2}{4} \delta^2 [\tau_{sw} + \tau_S] \overline{[S]} \left\{ \frac{\alpha_S}{RT\delta^2 \Delta[S]} - \frac{\beta^2}{\rho_f \alpha} \right\} \qquad [2\text{-}13]$$

where α_S now represents a kinematic transport coefficient for species S, assumed to be moving through the pore due solely to a driving force generated by a transmembrane concentration difference, $\Delta[S] = |[S]_{inside} - [S]_{outside}|$ -- with no other pressure gradient driving the pore fluid to flow.

Equations [2-11] through [2-13] can be derived analytically from statistical mechanical considerations (Stein, W. D., 1967). Thus, knowing or experimentally determining the phenomenological coefficients, α (for the pore fluid driven by an applied pressure gradient) and α_S (for species S driven by an applied concentration gradient), and further measuring the geometry of the pore (d, δ), $\overline{V}_w$, $[S]_{inside}$, $[S]_{outside}$, ρ_f, Ψ_w, $\overline{V}_S$, T, and K, one can theoretically solve equations [2-11] to [2-13] simultaneously for the frictional coefficients, τ_w, τ_{sw}, and τ_S -- *provided*, we can somehow define and say something further about β.

The quantity β is called the Staverman Factor, or Reflection Coefficient (Stein, W. D., 1967). It is a measure of the *semipermeability* of the membrane or boundary involved in the transport, and it takes on values between zero and unity. If the reflection coefficient is equal to unity, the membrane selects absolutely and allows *only* water (not S) to pass through the pores. If the Staverman Factor is equal to zero, the membrane does not distinguish between solute and solvent, and it is *totally* permeable to both. Equilibration then follows in the steady state and there is no subsequent flux of material in either direction across the membrane.

Going one step further, the quantity $(1 - \beta)$, as given in equation [2-12], is a measure of the relative rates of flow of the volume occupied by solute to the flow of total volume under the influence of a *hydrostatic pressure gradient alone,* whereas the quantity

$$\frac{4\rho_f \alpha \Psi_w \overline{V}_S}{\pi \delta^2 d^2 (\tau_{sw} + \tau_S)} ,$$

where Ψ_w is defined by equation [2-13] is a measure of the relative rates of flow of solute to the flow of total volume under the influence of a *chemical gradient alone.* Thus, if these two quantities are identical, the rate of flow of the solute from one side of the membrane to the other will be the same whether the transport is caused by a chemical concentration gradient *or* by flow due to a hydrostatic pressure gradient. That is, for this case equation [2-12] yields the result that τ_{sw} is essentially negligible, such that the mutual drag characteristics between solute and solvent will not be of great significance in affecting mass transport.

On the other hand, if $(1 - \beta)$ is substantially greater than

$$\frac{4\rho_f \alpha \Psi_w \overline{V}_S}{\pi \delta^2 d^2 (\tau_{sw} + \tau_S)} ,$$

then more solute will be transported when a pressure gradient is imposed (i.e., when a resulting flow of solvent takes place) than under the influence of just a chemical gradient. This would indicate that the bulk flow of water drags along a significant amount of solute. As shall be addressed further in Chapter 3, the latter is probably the case for transport from capillaries to cells, and vice versa, through the interstitial fluid. In this case, equation [2-12] reveals that the solute-solvent interaction frictional coefficient, τ_{sw} is a dominating factor in controlling mass transport.

Whether or not solvent drag plays an important role in mass transport across a membrane is one piece of evidence that has been used to suggest that in some cases the solute and water cross the membrane by different routes -- the solute perhaps directly through the bi-lipid layer of membrane substance and the water through aqueous pores or channels. The dynamic state of the material which fills the pore is a function of the stress-strain-rate constitutive characteristics of the fluid, as well as the hydrostatic pressure gradient across the membrane, and τ_w. The dynamic state of the solute is a function of its mobility, τ_{sw}, τ_S, α^*, the geometry of the pore, certain characteristics of the particles being transported, and the patency of the pore (see below).

(2) The *geometry of the pore* is a key determinant of the success or failure of extramural transport by this route. Obviously, if the mean *diameter, d*, of the pore is considerably smaller than the average size of the species trying to get through, then the membrane is effectively impermeable to that species. This is one explanation for why certain biological membranes are permeable to certain species of particles, but not to others.

Plasma proteins, for example, with molecular weights in excess of 65,000 - 70,000, and mean diameters on the order of 100 angstroms or more, are considerably larger than the 5 to 10 angstrom diameter of a cell-membrane pore, and slightly greater than the 6 to 7 nanometer diameter of capillary wall intercellular slit-pores. They thus generally get trapped inside the walls of capillaries, unable to permeate either the wall material itself (since the proteins are not lipid-soluble) or "voids" (pores) in the wall surface. Their inability to get out of the circulation is an important aspect of the maintenance of osmotic equilibrium in the circulatory system (see Chapter 3). This, in turn, is an important determinant of the proper transport of nutrients, waste products and dissolved gases into or out of the circulation at the capillary level.

On the other hand, chemical species such as water, electrolytes, and smaller nutrients, having mean diameters in the range 1 to 5 angstroms and not being readily lipid-soluble, are believed to traverse pores with little or no difficulty -- this being their major route for transport across cell and other membranes.

In addition to the diameter of the pore, its *cross-sectional contour,* compared with the geometrical configuration of the species trying to get through, may enhance or impede transudation. This connotes the traditional "square peg" trying to fit into the "round hole". Although the cross-sectional area of the latter might be significantly greater than that of the former, the discrepancy in shape could very well preclude any degree of fit between the two, thus impeding transport.

For example, consider a simple circular pore of diameter d. It presents a cross-sectional area $(\pi d^2/4)$ suitable for transport. Now consider a simple species S having a square-shape of side length $\frac{d}{2}\sqrt{2}$. The cross-sectional area of the square is $(d^2/2)$ -- less than two-thirds that of the pore -- and yet, if the side-length of the square is just *slightly* larger than 0.707d, the square will not fit through the hole. Here is a case of a pore having a cross-sectional area one-and-one-half times that of S, and yet not letting S pass through because of a geometrical incompatibility.

And speaking of cross-sectional areas, one must also address the latter in terms of the *number* of particles of species S that are trying to get through the pore. A physiologic example of this point involves blood flow through capillaries. Red blood cells have a mean diameter of 6 to 8 microns, whereas the average diameter of a capillary is on the order of 8 microns or less. Thus, these cells must literally line up and pass through the capillary in single file -- one at a time. Indeed, many blood cells actually have to deform in order to "squeeze through" the capillary lumen. This obviously restricts not only the number of cells that can get through, but also the transit time through the capillary from one end to the other. Flow through capillaries is quite similar to transport through pores -- many particles trying to get through a small pore will find it much more difficult that if the pore had a larger total cross-sectional area. We will emphasize this point again in section (3) below.

Still considering the pore cross section, one needs also to address its *symmetry* characteristics. This is not unrelated to the effects of d and pore cross-sectional contour on transudation, but it highlights a different aspect of the same problem. That is, considerations of cross-sectional contour actually include symmetry characteristics as well, but the latter are singled out here because so many physiological molecular species are typically asymmetric. Thus, *one part* of the molecule might fit through one part of a corresponding portion of the pore contour, but, if the latter is symmetric and the molecule is not, the other part of the molecule might not fit through -- or, vice versa, and so transport could be impeded. This jig-saw-puzzle-like dilemma of fit can be further complicated if the size and geometry of the *opening* to the pore differs from the size and geometry of the *interior* of the pore.

Consider that a relatively large table, for example, may be made to fit through a relatively small doorway into a good-size room if it is turned on end and the legs are "fed through" first. The table can then be "rotated" through the doorway so that its smallest contour is gradually manipulated into place in the narrowed opening. The key here, as it is in the corresponding situation involving mass transport through the pore, is that if one follows exactly the sequence of steps described, the table will, indeed, fit through the doorway. Otherwise, it will not, even though the space on either side of the doorway can accommodate with ease the geometrical configuration of the table.

Turning now to the *profile* of the pore as a factor in transudation, consider its *length*, which is equal to the thickness, δ, of the membrane if the pore walls are comparatively smooth and straight and its centerline is perpendicular to the walls of the membrane. This parameter becomes important when viewed in terms of the likelihood that frictional interactions between the species S being transported and the interior wall of the pore will significantly impede or otherwise affect transport from one side of the membrane to the other. That is, viewed in terms of the frictional coefficients, τ_w, and τ_S, defined earlier, the length of pore that a transported species needs to negotiate could significantly affect net mass transport.

Since work can be loosely thought of as the product of transport driving force and transported distance, the work done against frictional resistance will be directly proportional to δ. Thus, longer pores require more work in order to get S through from one side to the other; and more work means the need for larger gradients in the driving forces already described. Note that this consideration has to do with the *distance* part of force times distance (work). The *smoothness* of the walls of the pore addresses the *force* part of force times distance.

Smoother walls mean less friction, i.e., reduced wall roughness decreases frictional effects that act to impede the movement of S through the pore. Less friction means less work required to transport S by transudation from one side of a membrane to the other. Less work required means greater transport may be accomplished for the same driving force, because less of the force is dissipated in overcoming friction, leaving more of it available for the transport itself.

Of course, the exact opposite is the case for walls which are rough. Not only does that increase friction, but, getting back to the asymmetry of many physiological molecules, their "jagged edges" and convoluted surface shapes may actually cause them to "catch" or "snag" on the rough surface of the pore wall, thereby trapping them in voids and cavities that may keep them from getting through the membrane. Indeed, such "trappings" of asymmetric molecules may also lead gradually to the proliferation and thickening of the wall of the pore, and so, its ultimate *patency* may be further affected.

The patency of a pore has to do with whether or not there is anything located within the pore cavity, or at either end, that acts to impede the movement of material through it. The obstruction could be due to: (i) an accumulation of particles on the wall surface of the pore as described in the previous paragraph; (ii) the formation of an adsorbed layer of S as discussed in the earlier section (3) under the heading of intramural transport; (iii) the existence of microscopic "one-way" pivoted doors (see later), like any door that opens one way into a room, or revolving-type doors, or "gates"; or, (iv) the existence of a variety of molecular species that effectively act to block or otherwise occlude the lumen of the pore. A category of pharmaceuticals that accomplish just that, i.e., that act to obstruct the patency of pores or other transport passageways, is termed, generically, *Channel Blockers.*

In physiologic systems, it is generally acceptable to assume that an adsorbed layer of transported species is so thin that the cross section of the pore available for mass transport is not greatly reduced. However, if a material is strongly attracted to the surface of a pore -- as measured by the *Surface Diffusion Coefficient* (a kinematic transport coefficient for adsorption) between S and the pore wall material -- the adsorbed layer may actually be thick enough to reduce mass transport significantly, perhaps even to fill the pore entirely.

The existence of "gates" in pores is another explanation frequently given for the semipermeable characteristics of certain biological membranes. Consider the schematic model shown in Figure 2-2, for example, where the *patency* of the pore is obstructed by "one-way hinged doors", all of which open in the same direction, like the one-way leaflet-type valves located in the veins and heart of the cardiovascular system. Suppose, further, that these "microdoors" swing open and closed either by their own thermal movement, or by the force exerted on them as a result of bulk flow or the random thermal motion of particles of species S. We have, then, a situation wherein the probability of a molecule of species S passing from left to right through the barrier is much greater than that for its passage in the reverse direction, all other conditions being the same.

Even if the effect were very small -- say only 0.1% -- then placing 1000 of these barriers in series from one end of the pore to the other would produce something in excess of a two-fold osmotic pressure difference across the whole system of 1000 units (Deutsch, D. H., 1981). Thus, this "swinging-door" model for the semipermeable characteristics of a cell or biologic membrane illustrates how the geometry of the pores themselves can influence both the forces that normally dictate mass transport, and the concentrations of S on either side of the membrane barrier. The one-way "valves" illustrated in Figure 2-2 may take the form of long-chain molecules that are attached to the wall at one end and swing freely into

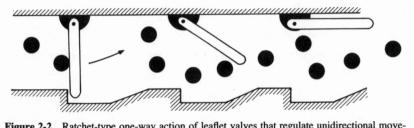

Figure 2-2   Ratchet-type one-way action of leaflet valves that regulate unidirectional movement through a pore.

the pore cavity at their other end, being capable of interacting with the opposite wall of the pore.

In a more general sense, of course, pore patency also involves whether or not there is anything physically located within the pore cavity to impede the movement of material through it. Thus, pharmaceutical channel blockers are administered specifically to alter membrane permeability in a wide variety of clinical situations.

There are at least two other aspects related to the geometry of a pore that determine its semipermeable characteristics. These are the *orientation* of the pore, and the *configuration* or *shape* of the walls of the pore.

By the configuration or shape of the walls of the pore, we mean specifically whether they are straight and parallel to each other, or whether they are straight but taper to a smaller diameter on the inside of the membrane compared to the outside, or vice versa, or whether they have some tortuous complicated shape like winding S-curves or worse, or whether they otherwise facilitate or impede the movement of particles across the membrane -- as in a maze.

Consider, for example, the situation depicted in Figure 2-3. In accordance with principles of Brownian motion, particles of average size $2r_o$ will tend to drift randomly from one side of the membrane to the other and, at equilibrium, there should be an equal concentration of S on both sides of the barrier. However, with d_2 significantly larger than d_1, and, depending on the relative sizes of d_1 and d_2 compared with $2r_o$, it is clear that the "funnel-type" geometry of the pore makes it easier for particles to move from right to left, than from left to right.

In fact, the probability is that particles moving from right to left will most likely get trapped on the left side of the membrane, and this probability is quite close to 100% for d_1 exactly equal to $2r_o$. In the latter case, only those particles attempting to travel from left to right who happen to hit the opening of diameter d_1 *squarely on center*, and with a velocity vector lying virtually along the centerline of the pore would stand a chance of getting through from left to right -- and the probability of this happening in a purely random motion is quite low. Thus, membranes

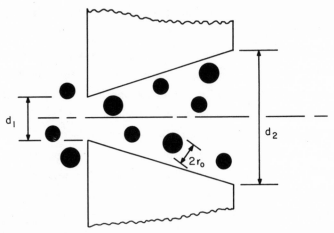

Figure 2-3   Funnel-type configuration of pore that causes species S to become more concentrated on the left side 1 of the membrane, than on the right side 2.

endowed with such funnel-type pore configurations have the ability to be unidirectionally semipermeable, i.e., to allow the passage of a species S from inside the cell out, but not from outside in, or, vice versa.

Going one step further, suppose the concentration of species S on both sides of the membrane started out to be the same. That is, suppose $[S]_1 = [S]_2$ to begin with, such that there was originally no concentration gradient causing a net movement of mass across the membrane. Since random thermal activity would still cause particles to move *across* the membrane (normally, in an equilibrium situation as many particles would move from left to right as from right to left), and since the membrane configuration shown in Figure 2-3 would tend to *cause* particles to accumulate on the left side 1, one concludes that the *geometrical* configuration of a semipermeable membrane can, per se, *produce* a gradient in chemical potential by selectively allowing particles that *started out* in equilibrium across the membrane to pass through in only one direction.

Associated with this gradual build-up in chemical potential, there would, of course, be an associated gradual build-up in the osmotic pressure gradient tending to drive water, too, the same way. Thus, a semipermeable membrane can cause the net flow of fluid from one side of the membrane to the other by mechanisms that initially appear to defy the second law of thermodynamics at the micro-continuum level (see, for example, Deutsch, D. H., 1981).

Closely related to pore wall configuration (i.e., how the pore walls are oriented with respect to *each other*), is the concept of pore wall orientation, by which we mean more specifically the relation of its centerline to the inside and outside surfaces of the *membrane*. This has to do with whether or not the pore is oriented at right angles to the membrane, and whether or not there is some other orientation that might enhance or retard transport, depending on other variables such as those already addressed.

If the pore centerline is exactly at right angles to the membrane surfaces, if its walls are smooth, straight, and parallel, and if it is basically quite patent, then particles of species S have the least obstructed, shortest-distance path from one side to the other. Any deviation from this configuration slows down transport and may otherwise significantly impede it.

For example, if a particle enters a pore that has its centerline inclined at an angle (i.e., "slanted") relative to the membrane surface -- even though these walls may themselves be parallel -- the particle will most likely suffer collisions with the slanted walls (c.f., earlier discussion on pgs. 122-123 and 128-132). Such collisions will cause the particle to lose momentum and kinetic energy, and thus it will find it more and more difficult to negotiate the path from outside the membrane in, or vice versa. Indeed, the more tortuous the path and the greater the distance that has to be travelled, the less likely it is that successful mass transport will transpire through any given pore. For the case of ordinary diffusion through "tortuous pores", Sherwood, Pigford and Wilke (1975) define an *effective diffusion coefficient,* or *effective kinematic transport coefficient,*

$$\alpha_e = \frac{\alpha_S A_p}{\xi} \qquad \qquad \text{[2-14]}$$

where A_p is the fractional free-cross-section of the membrane (i.e., the total cross-sectional area of all pores in the membrane divided by the total surface area of the membrane), α_S is the "usual" ordinary diffusion coefficient for species S (defined earlier), and the dimensionless factor, ξ, called the *tortuosity factor,* is introduced to account for the fact that the diffusion path is greater in tortuous pores than it is when the distance travelled through the pore is simply normal to the surface. Where applicable, the tortuosity factor also corrects for the varying cross sections of the pores, which may not be straight round tubes.

In addition to correcting for the greater path length, ξ also accounts for the possible existence of tiny orifice restrictions in membranes of complex structure. This factor is usually obtained, as are most of these coefficients, experimentally for a given membrane. Indeed, many of the considerations related to the geometry of pores involved in mass transport are handled empirically by all of the various coefficients, correction factors, and rate constants introduced into the relevant equations governing these processes. This is a quantitative way of saying that we are not entirely sure of what, exactly, is going on, but we can account for it!

The last (though certainly not the least) consideration to be addressed in this section is almost obvious. The more pores one has per unit area of membrane surface, the greater the mass transport one can expect to transpire across the membrane. Thus, the *number* of pores affects extramural transudation in basically two ways: first, a particle must *find* a pore in order to get through it. The more pores, then, the greater the probability that a particle will find one to enter and pass through. Second, the more pores, the greater the *number of particles* per unit time that will find and pass through the membrane per unit surface area. The net result is that *many pores per unit area enhance transport.*

The geometric characteristics, patency, and number of pores are frequently lumped together in some sense to define an abstract quantity called the transudative permeability of a membrane to species S. This permeability generally applies to a basically water-soluble S (i.e., a polar molecular species), and it assumes that S passes from one side of the membrane to the other via distinct passages or tunnels. The geometric considerations associated with the transudative permeability of a membrane also provide a convenient means for explaining its semipermeable or "selectively" permeable characteristics to certain particular species, as explained above. In this sense, however, one must further realize that transport through a discontinuous portion of membrane material depends also on certain characteristics of the particles being transported, the most important of which are addressed in the next section.

(3) The ease with which particles can penetrate a membrane barrier certainly depends on their size, relative to the size of the pore opening (i.e., will the particles *fit* through the pore?). Referring back to Figure 2-3, consider a spherical particle of radius r_o attempting to clear the edge and enter a pore of radius $(d_1/2)$. If the particle is to get through, its center must fall within the circle defined by the radius $\left(\dfrac{d_1}{2} - r_o\right)$ about

the axis of the pore. The "effective area", then, through which mass transport can occur via the pore route is $\pi\left(\dfrac{d_1}{2} - r_o\right)^2$. When $(d_1/2) >> r_o$ (pore size large relative to particle size), the effective area for mass transport is essentially the entire cross-sectional area of the pore. This is generally the case when the diameter of the pore is greater than 100 times the diameter of the particle (Deutsch, 1981). Otherwise, the effective area for mass transport is considerably less than the cross-sectional area of the pore -- a phenomenon which has come to be known as the *Edge Effect*.

When $(d_1/2) = r_o$, the effective area has been reduced essentially to zero because a particle must hit the pore squarely and dead on center in order to get through -- an event which is statistically highly unlikely. Indeed, when $(d_1/2)$ falls *below* r_o, the particle can no longer fit through the pore at all, and the effective area for transport is precisely equal to zero. The membrane is now totally impermeable to S, which, as discussed previously, will collide with it and adhere, become adsorbed, reflect in a specular fashion, reflect in Clausing fashion, or be otherwise retarded by friction. Regardless of which combination of the above events occur, the particle does *not* get through the membrane via the pore route.

In this respect, it is of interest to observe that physiological pore sizes have been estimated to be on the order of 5 to 10 angstroms. This is of the same order of magnitude as the average size of a Sodium (2-5 angstroms) or Potassium (2.5-4 angstroms) ion, but considerably larger than water molecules, which average 1-2 angstroms or less in diameter. Of equal significance is the fact that the pore size is considerably *smaller* than plasma protein molecules, which, while being water-soluble, have diameters on the order of 100 angstroms or more. Thus, water, urea, oxygen, carbon dioxide and various ions can move freely across biological membranes, but large protein molecules cannot.

The size and charge of dissociated molecules also affects their degree of hydration. Thus, depending upon the extent to which a solute is ionized, the charge of the species involved, and the consequential number of H_2O molecules it attracts, its "effective" r_o may be such that the Edge

Effect might become a relevant factor in the transport of S via pores. Small, univalent ions, such as Sodium, for example, will bind *more* water than will larger univalent ions, as a general rule.

Taking the above reasoning still one step further, when the diameter of a pore is large in comparison with the size and mean free path of the molecules undergoing transport, then the processes involved are essentially independent of the presence of the pore. That is, the transport proceeds according to ordinary molecular diffusion principles and others already discussed, as if the events taking place were entirely intramural. In this case, appropriate intramural diffusion coefficients and rate equations apply, as shall be described more fully in Chapter 3.

On the other hand, if the mean free path, λ, or "size", r_o, of solute particles is large compared with the pore diameter, d, then the molecules will collide much more frequently with the *pore walls* than with each other. In this case, the resistance to transport will be due more to collisions of S with the wall, than collisions of S with each other, which is the case for ordinary diffusion. Since the criterion for determining which of the two scenarios above actually governs the transport is the *Knudsen Number*, $\frac{\lambda}{d}$, this effect has come to be known as the *Knudsen Effect* (Sherwood, et al., 1975).

The theory of Knudsen Diffusion through straight, nonadsorbing round pores can be developed according to kinetic theory, from which one obtains the following expression for transport with diffuse wall reflection in a pore of radius $\left(\frac{d}{2}\right)$ centimeters:

$$\Phi = -\frac{d}{3}\sqrt{\frac{8RT}{\pi MW}}\,\frac{d[S]}{dr} \qquad [2\text{-}15]$$

where Φ is (molar) mass flux based on the pore cross section, R is the universal gas constant (82.06 cm^3 $-$atm/mole $-$ °K, or, 8.314 Joules/°K $-$ moles), T is absolute temperature in degrees Kelvin, MW is the molecular weight (in grams per mole) of the diffusing species S (which is being transported in the r-direction), and $[S]$ is its concentration in moles per cubic centimeter. Equation [2-15] states that the flux is proportional to the concentration gradient, to the free molecular velocity (which depends on temperature), and to the ratio of pore volume to pore surface. The volume-to-surface ratio of a straight round pore is $\frac{d}{4}$.

Knudsen Diffusion is a factor to be contended with in very dilute solutions and, even more so, in rarified gas diffusion -- where the probability of a particle of S encountering another particle of *any* kind is much lower than is the probability of that particle colliding with the walls of the pore once it enters.

Note that the Knudsen Diffusion coefficient, $(d/3)\sqrt{8RT/\pi MW}$ in equation [2-15] is a kinematic transport coefficient having the dimensions of a length-squared divided by time. Note further the dependence of this coefficient on temperature, T. The latter appears over and over again as one of the variables in kinematic transport coefficients -- and for good reason! The higher the temperature, the greater the average kinetic energy of any molecular species (see Chapter 3); and, the greater the average kinetic energy, the more transport processes are enhanced. Temperature can also have a first-order effect on the geometrical configuration of pores, thereby affecting mass transport by thermally warping, expanding, contracting or otherwise altering the transport channels. Moreover, Temperature affects such physical variables as dynamic viscosity, mass, density, kinematic viscosity, and viscous modulus, which are, themselves, determinants of kinematic transport coefficients (see earlier section on this topic). And the list goes on. It should therefore come as no surprise that T is of prime importance when considering transport of *any* kind.

As was the case when we considered the geometry of *pores* earlier, it should also come as no surprise that the geometry of S is equally important as a factor in transport through pores. To put it simply, the two geometries must *complement* each other if the transport is to be a success. Thus, if one views a particle of species S as a piece of a jig-saw puzzle, and the biological membrane as the rest of the puzzle, then clearly the void in the puzzle (i.e., the pore in the membrane) must have a shape and configuration that is congruent with the puzzle piece (i.e., the particle of species S) if the two are to fit together (i.e., mass transport is to take place through the pore). This is especially true when the pore size is small compared with the particle dimensions, so that "fit" becomes a rather critical factor.

Going one step further in terms of the *symmetry* of the particles involved, consider the situation depicted in Figure 2-4. Suppose that, instead of spheres, the particles of species S are in the shape of long rods, where the length to cross-sectional diameter ratio of the rods is rather large (say, at least five-to-one). Entering the pore from the right side, through the larger diameter, the rod-shaped particles will collide with the walls of the tapered holes and they will thus be manipulated so that there will be a tendency for the major axis of these particles to become aligned with the axis of the tapered hole. This phenomenon, loosely termed the *Beveling Effect*, works to the advantage of mass transport from right to left, since, as the rod-shaped particles approach the small hole from the right, they are guided to become aligned with the pore opening such that they can pass through relatively easily.

By contrast, when the rod-shaped particles approach the small hole from the left side of the membrane the pore will not act to assist them in passing through the small opening. Assuming the most general case of purely random orientation of these particles in space, then, the probability of their becoming properly aligned to pass through the pore opening is extremely low. Moreover, even if, by pure chance, the particles *should* approach the hole with their centers of mass moving *directly* toward the center of the small opening, the probability is still very low that these rod-shaped particles actually would enter and pass through the small end of the tapered opening. Thus, the rod-shaped molecules, after passing through the asymmetric barrier from right to left, will, as a result of their first collisions on the left side of the membrane, become randomly oriented and thereby preferentially trapped on the left side of the semipermeable membrane.

Not only the geometry of the *pore*, then (i.e., Figures 2-2 and 2-3), but also the geometry and symmetry of the *species*, itself (Figure 2-4) may contribute to the "auto-induced" generation of chemical concentration

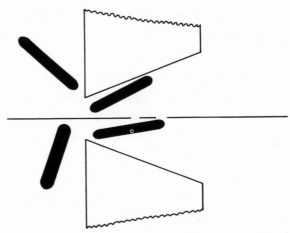

Figure 2-4   Large length-to-length-to-cross-sectional-diameter ratio, rod-shaped particles of species S cause it to become more concentrated on the left side of the membrane, than on the right side.

gradients where none existed otherwise. This discussion also points to the importance of the *orientation* of the particle when it hits the pore opening. Depending on its symmetry, or lack thereof, the particle might fit through the pore if it hits the opening in one particular spatial orientation, but not otherwise. Of course, this assumes that the particle has a fixed geometry and does not deform to "squeeze" through the opening.

Here, again, we point out what happens in the case of blood transport through the microcirculation. Where red blood cells are smaller than the blood vessels through which they have to pass, the transport is no problem. But in very many instances, capillary diameters (d) are smaller than mean red cell sizes (r_o). In this case, the cells can *deform* and alter their shape to squeeze through the blood vessel in single file -- and, indeed they do! Similarly, the deformable characteristics of particles of species S may allow them to penetrate membrane pores that might otherwise have presented an impermeable barrier. Globular species, for example, may be able to elongate in such a way that they can be *extruded* through tiny pores -- much like the method by which certain white blood cells (most notably, Neutrophils and Monocytes) enter and leave capillaries as a result of their "Ameboid Motion" through endothelial intercellular clefts. The ability to thus deform implies that there is more than one equilibrium configuration for a species of particle, such that it can transform from one to the other with relatively little energy expenditure.

In other cases, highly polarized molecular configurations can form secondary internal structures due to the formation of weak, electrostatic bonds. This allows the molecule to "coil-up" where intra-molecular uneven charge distributions cause parts of the long-chain structure to attract one another. This occurs when the *distribution* of charge is not uniform, even though the *total* net charge on the molecule is zero. In its "coiled-up" state, S thus appears to be denser, and, due to its secondary structure, to offer a greater resistance to deformation. The latter, however, is subject to change in situations that cause the molecules to open up. This partially explains the time-dependent deformability characteristics of certain materials that exhibit *Thixotropic* behavior.

Thixotropic and other physical properties of particles (i.e., density, specific gravity, degree of ionization, molecular configuration, and so on) also affect the drag characteristics between solute and solvent (τ_{sw}), sedimentation rates (c.f., equation [2-6]), inertial responses, solvent drag, bulk flow, surface tension effects, and other processes that are intimately associated with mass transport, not the least of which relates to the *solubility* of S in the pore solvent. And, just as the number of pores per unit area of membrane surface *enhances* transudation, the number of particles trying to get through the pore at any one time *impedes* transudation.

In this sense, one can envision a situation analogous to hundreds of people trying to get out of a burning building through the same door. Particles may "pile up" at the entrance to the opening in such great numbers that the crowding itself will cause interference that will preclude *any* particle from getting through. This might be termed *Stereometric Interference* -- wherein particle-to-particle interference in highly concentrated solutions significantly impedes mass transport through pores because there are simply too many particles trying to get through too few pores at the same time. The "bottle-neck" is exactly similar to highway traffic jams where many lanes of traffic (particles of S) converge into a one-lane bridge (the pore), or to any similar situation where there are not enough avenues to handle the flow.

EXTRAMURAL TRANSPORT THROUGH MEMBRANES

Regardless of whether or not pores do exist, another way in which substances can cross a bi-lipid membrane is by dissolving into the lipid film on one side, diffusing through it, and dissolving out of it on the other side. Thus, the abstract concept of "membrane permeability" includes as well transport through the intact membrane material *itself*. In this sense, S is assumed to be primarily a non-polar, lipid-soluble material. If it is not, then membrane permeation is generally not passive, since energy is required both for the solute to leave the water or aqueous phase that is definitely present on either side of the membrane, in order to *enter* the lipid phase, and, energy is further required to *actively* transport S across the membrane lipid layer from one side to the other (see later section on active transport). In general, somewhere in the neighborhood of 5000 thermodynamic calories per mole are required to break hydrogen bonds between polar compounds and water, in order to get them to leave the aqueous phase and enter the lipid phase.

Many of the factors that control transport through a pore apply equally well to transport through a bi-lipid layer. For example, the characteristics of the transport pathway, the transported material, and the membrane material all enhance or impede transport by many of the same mechanisms that have already been discussed. In examining transport through the intact membrane material itself, however, certain additional factors must be considered, such as the degree of oil solubility of S.

The oil solubility of a molecule can be expressed as its oil-to-water partition coefficient, or dimensionless distribution coefficient, K. The latter is defined to be the ratio of the solubility of S in oil, i.e., $[S]_{oil}$ for a *saturated* oil solution of S, to its solubility in water, i.e., $[S]_{water}$ for a saturated aqueous solution of S:

$$K = \frac{[S]_{oil}}{[S]_{water}} \qquad \text{[2-16]}$$

In general, the more non-polar a compound is, the greater is $[S]_{oil}$ and the less is $[S]_{water}$, so that its partition coefficient K is high. Studies have shown that there is a definite and well-defined *increase* in membrane permeability as the partition coefficient increases, suggesting rather strongly that medium and large size lipid particles penetrate biological membranes by actually dissolving through the lipid film. Moreover, species with the smallest molecular dimensions (especially volume) seem to penetrate the membrane much more rapidly than do larger species, when both have the same degree of oil solubility (i.e., the same partition coefficient). As compounds become more and more polar in nature, K approaches zero. Thus, in the generalized transport equations [2-3] or [2-4], as applied to extramural transport through intact membranes, the coefficients A_j or A_i may be considered to be functions of K. Alternatively, the kinematic transport coefficient, α, for intra-membrane permeation may be scaled by the partition coefficient.

Like the concepts of adsorption and surface diffusion addressed earlier, a related aspect of transport through the membrane material has to do with the degree to which there are sites along the membrane surface that obstruct or otherwise immobilize solute particles -- preventing them from *penetrating* the membrane to begin with; or, having been successful in the latter respect, preventing them from *leaving* the membrane at the other end. Furthermore, since S has to *get* from one side of the membrane to the other once it *does* penetrate, one has to consider here, as well, all of the electrochemical and other gradients that are available *within* the membrane to drive lipid-soluble solutes, as well as membrane resistance to transport, solute-solvent interactions, solute-solute interactions, and so on. And, finally, one must examine the degree to which the journey of the solute particles through the membrane can be *facilitated* by special "carrier molecules" (facilitated transport), active transport mechanisms, pinocytosis, and/or phagocytosis. This last point leads us to the subject of the general *mobility* of S, which concludes the list of *General Considerations* presented at the beginning of this chapter for the discussion of basic principles of transport in physiologic systems.

Species Mobility in Physiologic Transport

The abstract concept of "mobility" has to do with how well S *responds* to transport driving forces. It encompasses many considerations, several of which have already been addressed, to wit (i) the *kinetic energy* of S, as it relates to both its temperature (T) and the level of turbulence intensity $(\overline{v_t \ell_j})$; (ii) the degree to which S interacts with the transport medium, as it relates to inter-molecular association (hydration), solvent drag (τ_{sw}), counterflow, bulk flow, degree of ionization, solubility ($[S]_{saturated}$), and secondary bond formation; (iii) the degree to which S interacts with transport boundaries, as it relates to adhesion, adsorption, surface diffusion ($c_A MW/jRT$), specular reflection, dispersive reflection, wall shear (τ_S), transudation and permeation; (iv) the degree to which S interacts with S, as it relates to concentration ("activity" factor, α^*), stereometric interference, inter-molecular association, Knudsen diffusion, and secondary bonding; and, (v) the geometry of S, as it relates to size, symmetry, configuration, cross section, profile, complexity, tortuosity, intramolecular association ("coiling up"), secondary internal structure, and "jaggedness."

Thus, to a great extent, one cannot divorce particle mobility from the discussions that have preceded this point. Indeed, all of the physical principles addressed coalesce to encourage or discourage S from moving from one point to another, and many of them can be classified as factors that immobilize or enhance the mobility of S. What we are concerned with in this section are several *additional* considerations that have *not* yet been addressed, and that exacerbate or impede transport *directly* through their influence on mobility. These include: (i) Gibbs-Donnan effects; (ii) carrier-facilitated transport (passive, active, metabolic substrate-control, chelation); (iii) transport by directional sites; (iv) Pinocytosis; and, (v) Phagocytosis.

Gibbs-Donnan Effects

In addition to hydration and molecular association, charged particles in solution pose a special set of mobility problems when they encounter an impermeable barrier. That is to say, consider a biological membrane which has semipermeable characteristics such that one of the nonpenetrating particles (say, a high-molecular-weight protein) happens to be an electrically charged ion or highly polarized long-chain polymer.

The condition of electroneutrality across the membrane requires that for every charged macromolecule that cannot penetrate the membrane, an equivalent oppositely charged particle must stay behind with it. This essentially immobilizes otherwise agile charged particles which, themselves, might normally have no trouble at all getting across the barrier. Thus, the resulting concentration of such agile, oppositely charged parti- cles on the side of the membrane where the impermeable macromolecules are "trapped" is higher than it would be normally, and the equilibrium concentrations that are eventually approached differ from those that would be predicted from purely passive electrochemical considerations. In this case, particle mobility is impaired by *electrostatic binding,* which constrains its ability to be transported. This is as opposed to *adhesive binding, adsorptive binding,* or other types or particle-to-wall interactions that constrain mobility.

The phenomena involving fixed ions were investigated exper- imentally by F. G. Donnan and analytically by W. J. Gibbs, hence, they have come to be known as *Gibbs-Donnan Equilibrium Effects.* The partic- ular case involving Sodium ions, Chlorine ions (both of which pass easily across cell membranes) and *large* plasma protein ions -- which carry a predominantly negative charge and which, because of their large size, normally do not penetrate the walls of mammalian capillaries -- is of im- mense physiologic importance. For this reason, we shall examine the Gibbs-Donnan phenomenon in much more detail in the next chapter, and so we postpone any further discussion here.

Carrier-Facilitated Transport

Included under this topic are a group of transport processes that share a common feature: they all involve *carrier molecules.* Call them "taxi- cabs", call them "facilitators", call them "chelating" agents (a pharmaceu- tical term) ... call them whatever suits your purposes ... the bottom line is that these carrier molecules actively participate to "hand-carry" ("Fed- eral Express", if you will) S to its intended destination, or, to otherwise increase the probability and speed with which a given transport process is intended to occur. The mechanism of action of carrier molecules is to secure S to a "binding site" on the molecule long enough for the desired effects to take place, and then to release it at an appropriate time and destination. The binding usually involves a certain degree of geometrical compatibility or *affinity* between the carrier molecule and *S*, such that the former must in some sense be *activated* before it can do its thing. The "thing" is then accomplished through molecular association by many of the same mechanisms already described.

A physiologic example of such molecular association, to be discussed further in Chapter 4, is the electrostatic or stereo-specific binding of enzymes with substrates, inhibitors, coenzymes, and activating ions. These act to catalyze biochemical reactions by bringing raw materials together faster and more efficiently than would be the case if they were allowed to simply roam about free, and it was left to the laws of chance and probability to determine whether or not the materials would ever meet and successfully combine chemically. Thus, physiologic enzymes act as "match-makers" to speed up chemical reactions that are vital to the well-being of the organism. They bring interacting species into proximity with one another, and then let nature take its course. But they do so by being in the right place at the right time, and they do so primarily by affecting particle mobility.

Another type of geometric carrier-molecule-*S* interaction involves a "C-clamp-type grip" by one particle on the other, and has come to be known as *chelation*. This word comes from the Greek, "chele", which means "claw", and from the derivative, "chela", which describes the nipperlike organs terminating certain limbs of crustaceans (e.g., lobsters and crabs). Thus, chelation implies that one particle is trapped, or held in the claw-like grip of another through geometric interaction, enhanced by electrostatic attraction. That is to say, the polarized molecular configuration produces an electrostatic attraction that causes these particles to interact faster than they would in the case of a mere chance encounter. This interaction then forms more permanent aggregates by chelation or molecular "entanglement." Many pharmaceutical agents, such as some blood anti-coagulants, work through chelation. By immobilizing certain blood-clotting factors through a grip-like hold, these chelating agents prevent those factors from contributing to the formation of a clot.

And yet another example of carrier-mediated molecular association involves the body's immune system response to foreign invasion, discussed briefly in Chapter 1. Each cell in the organism contains long, polypeptide chains, called *human leucocyte antigens* (HLAs) embedded in its outer membrane. One part of the antigen molecule extends into the interior of the cell and another part extends out from the cell surface, forming a one-of-a-kind pattern. Due to the astronomical variability of genetic possibilities, the geometric configuration and structural characteristics of these HLAs can be made so complicated that the tissues of any given individual may be (and are) endowed with a unique physical and chemical identity. Now, specialized white blood cells, called Lymphocytes, are the "quality control engineers" of the physiologic system. They contain the geometric "template" that is *exactly* compatible (congruent) with the individual's unique HLA code, and they circulate freely through the body, "checking" everything with which they come into contact for a proper "fit". If the lymphocyte contacts a foreign antigen -- which, by definition, is anything containing an HLA that does not mesh properly with the coded template -- then it (the lymphocyte) becomes activated and triggers an immune response.

The latter brings into play a category of globular proteins called *antibodies*. These contain binding sites for antigens. Each binding site is a pocket or cleft in the globular molecular configuration that can form bonds with two antigen molecules, thereby holding the antigens together in large clumps through a process called *agglutination*. The antigens thus agglutinated are subsequently engulfed by large white blood cells called phagocytic macrophages, or Polymorphonuclear Granulocytes (PMNs). The PMNs exemplify the transport process called Phagocytosis (see later). The reader is referred to the literature (e.g., Cunningham, 1977, Guyton, 1981, and Roesel, 1978) for a more detailed description of immune system processes. Our intent here is simply to show that it further illustrates the concept of carrier-mediated transport.

And the list goes on, and on. One can cite example after example of physiologic events or transport processes wherein "mediators" or "facilitators" are involved. The essential mechanisms by which me-diation is accomplished are addressed further below and in subsequent chapters of this text. Furthermore, we distinguish between those mechanisms that are external-energy-driven (active) and those that require lit-tle or no outside energy to accomplish the transport (passive). In general, additional energy will be required anytime S is being moved from one point to another under circumstances wherein such transport would oth-erwise not occur, such as *against* a concentration gradient. If the medi-ation acts to *facilitate* transport that has a tendency to take place anyway (but perhaps too slowly for metabolic purposes), then such mediation tends to be passive, or, at worst, require much less energy than active processes.

INTRAMURAL FACILITATED TRANSPORT

Oxygen does not readily dissolve in water (hence, in blood). Its solubility in aqueous fluid, as measured by its "Henry's Law Coefficient," k_h is only 0.024 at 37°C. Henry's Law is given by:

$$[O_2] = k_h P \qquad \text{[2-17]}$$

where $[O_2]$ is the number of milliliters of gas (indeed, *any* S) dissolved in one milliliter of blood (or *any* solvent) at a partial pressure of P atmos-pheres. Now, moist alveolar air contains 13.6% oxygen, 5.3% carbon dioxide, 74.9% nitrogen, and 6.2% water vapor; and it is at a pressure of 760 mm Hg (one atmosphere) at sea level. Blood travelling through the lungs thus tends to equilibrate its gaseous partial pressures at: $p_{O_2} = 0.136(760) = 103$ mm Hg, and $p_{CO_2} = 0.053(760) = 40$ mm Hg. Thus, for $P = \frac{103}{760} = 0.136$ atmospheres and $k_h = 0.024$, equation [2-17] reveals that only about 0.003 milliliters of dissolved oxygen can be carried intramurally per milliliter of whole blood. This is usually expressed as 0.30 mℓ O_2/100 mℓ of arterial blood (i.e., the unit "volume percent"), and is nowhere near the quantity required to meet the metabolic needs of the organism under even the simplest of circumstances.

To alleviate this dilemma, mother nature has invented the red blood cell, whose single primary purpose is to transport oxygen to living tissues at a rate sufficient for survival. This it does by binding the gas to a non-protein, iron-containing group called, "heme", which is folded into and carried by a larger protein called, "globin", to form the oxyhemoglobin complex in the intact interior of the erythrocyte. Hemoglobin has a molecular weight of 68,000 a viscosity of 6.00 milli-Pascal-seconds (centipoise) -- about twice that of whole blood -- and occupies about one-third of the red blood cell. The remaining two-thirds consists mainly of water (65%), with some proteins, lipids, gluthathione, Adenosine Diphosphate (ADP), Adenosine Triphosphate (ATP), and various ions making up the rest.

One molecule of hemoglobin can combine with four molecules of oxygen, and each red blood cell is estimated to contain some 280 million hemoglobin molecules. Therefore, every erythrocyte can carry (or "ferry") about 1.12 billion (10^9) oxygen molecules if all of the hemoglobin binding sites are occupied (i.e., if the oxyhemoglobin complex is "saturated" with O_2). If one further assumes an average of five million red blood cells per cubic millimeter of whole blood, then the latter can carry 5.6×10^{15} oxygen molecules at full saturation. A cubic centimeter (1000 cubic millimeters), or one milliliter (cc) of whole blood can then carry 5.6×10^{18} oxygen molecules, and 100 milliliters of whole blood (i.e., one deci-liter, dℓ) can carry 5.6×10^{20}, or, 0.0056×10^{23} molecules of oxygen at full saturation. Since the Avogadro Number of molecules, 6.0225×10^{23}, represents one mole of a substance (in the case of oxygen, 32 grams), 0.0056×10^{23} molecules of O_2 represents 0.00093 moles, or about 0.03 grams.

Now, the mass density of oxygen varies with temperature and pressure, but is in the range 1.43 - 1.48 grams/liter for most physiologic conditions. Thus, using an average, 1.455 grams/liter, 0.03 grams of oxygen converts to just over 0.020 liters, or, around 20 milliliters. Actual measured values quote approximately 20.1 mℓ of oxygen per dℓ of blood, commonly expressed as 20.1 volumes percent at full saturation. Compared with 0.3 mℓ O_2/dℓ blood calculated from Henry's Law, we see, then, that the hemoglobin-carrier-mediated mechanism described above increases the oxygen-carrying capacity of blood by a factor of 67! Going one step further, since one mole (68,000 grams) of hemoglobin can combine with 4 moles (128 grams = (128/1.455) = 88 liters) of oxygen, one gram of hemoglobin can combine with approximately $\frac{88,000}{68,000}$ = 1.29 cubic centimeters of oxygen to form the oxyhemoglobin complex. Thus, the usual concentration of hemoglobin is approximately $\frac{20.00}{1.29}$ = 15.5 grams per 100 milliliters of whole blood, which compares favorably with measured values of 14-18 grams/100 mℓ for men and 12-16 gms/dℓ for women.

Hemoglobin saturation is very sensitive to a wide range of factors, the discussion of which is not immediately relevant to the intent of this chapter (see, for example, Cooney, 1976 or Guyton, 1981). Suffice it to say that the partial pressure of oxygen in blood leaving the lungs is sufficient to almost saturate hemoglobin completely. That is, hemoglobin leaving the lungs is about 97.5% saturated, such that arterial blood leaving the lungs contains approximately $0.975 \times 20.1 = 19.6$ milliliters of oxygen bound to each 100 cc of blood (19.6 volumes percent), at normal body temperature and pH. At the level of the tissues, at rest, for an average tissue partial pressure of 40 mm Hg, hemoglobin releases to the cells about 5 milliliters of oxygen per 100 cc of blood. These values can change significantly with exercise and altitude, beyond around 10,000 feet above sea level.

EXTRAMURAL FACILITATED TRANSPORT

Some sugars (most notably, Glucose) and amino acids (the end result of the digestion of proteins) gain entry to the cell across its membrane by carrier-mediated mechanisms, not all of which are yet as clearly understood as is the oxyhemoglobin connection. As pointed out by Guyton (1981), a protein having a molecular weight of approximately 45,000, and being capable of combining specifically and reversibly with the same types of monosaccharides that are typically transported, has been discovered in the cell membrane, and is a likely candidate to be the carrier substance for glucose transport. Furthermore, in the presence of the Pancreatic hormone, Insulin, this carrier-mediated transport process can increase by a factor anywhere from 7 to 10 the rate of penetration of glucose and amino acids into the cell by permeation through the membrane material. This is one of many similar processes that are hormone linked. It is believed that the hormones may, in some sense, "activate" the carrier molecules, as suggested earlier.

EXTRAMURAL ACTIVE TRANSPORT

The key difference between extramural *active* transport and extramural *facilitated* transport is that in the latter case mobility is enhanced by special carrier molecules *in the same direction* as that of an already existing gradient in charge which is driving the transport process *anyway,* in a passive sense. The word, "active" is reserved, therefore, to describe the transport of S when it cannot be accomplished passively by any of the mechanisms discussed thus far, or, when S is to be delivered to a location that involves moving it *against* a gradient in charge that would normally tend to drive it back the other way. In these types of situations, the transported quantity can sometimes be physically *carried* by active processes that involve significant expenditures of energy.

One such process is the *Sodium-Potassium electrolyte pump,* which operates in a manner analogous to facilitated diffusion, but which moves sodium *against* an electrochemical gradient that attempts to drive it in exactly the reverse direction. This is one of the reasons that much more energy must be expended to accomplish this type of transport. In effect, S is being carried "up-hill". Active transport may also be used to *generate* gradients in charge in an otherwise equilibrated situation.

A mechanism that has been proposed to describe how active transport is accomplished across a physiologic membrane is discussed at some length in Chapter 3, and so we shall not dwell on it further here. Nor shall we say anything further at this point regarding *Intramural Active Transport*, except to point out previous examples of this which have already been cited, i.e., enzyme-substrate interactions (see Chapter 4), molecular chelation, and antibody-antigen immune responses. Before continuing, however, let us address an active transport process that proceeds through pores, but does *not* involve carrier molecules. This is so-called *Active Transport by Directional Sites.*

Active transport by directional sites postulates the existence of binding sites fixed in the membrane structure, itself (rather than on mobile carrier molecules). These fixed binding sites "guide" particles of S in a preferred direction from one side of the membrane to the other. In a manner somewhat analogous to surface diffusion, certain particles entering the pore in a membrane are assumed to undergo a reversible chemical reaction with specific "binding sites" along the walls of the pore. The rate constant for the *formation* of species-binding-site bonds is much higher than that for the *dissociation* of such bonds, so the reaction tends to go essentially in the direction of association.

Having thus been "trapped" at a binding site momentarily, S is subsequently "released" in such a way that it is propelled along the direction of desired transport. The propelling force is actually presumed to be the force exerted on S by *another* binding site located adjacent to the one which caught S to begin with. In this manner, S is "handed along" from binding site to binding site, each attracting and catching the species as it is released in an energy-consuming (active) process, until it gets from one side of the membrane to the other.

Active transport is harder to accomplish the longer it progresses. This is due to the fact that as more and more particles get delivered to the other side of the membrane, there develops a progressively increasing passive *concentration gradient* that tends to drive them back the other way. As this gradient increases, so does the force that resists active transport, so the transport mechanism finds itself having to work harder and harder with each subsequent particle that it tries to carry across. If the membrane happens to be permeable to the species being transported (as is the case for sodium and potassium ions), then an asymptotic limit defining an equilibrium condition may ultimately be reached. At this limit, the concentration gradient is such that for every particle of species S that is transported actively across the membrane, another one drifts back passively the other way, so the whole process becomes self-limiting. This, in part, is how a resting membrane potential is generated in excitable tissue, as we shall see in Chapters 3 and 6.

Also limiting the effectiveness of active transport is the availability of the energy source that drives the whole process. This is referred to as *metabolic substrate control* and includes, as well, the source of enzymes or hormones that may be required to catalyze the various reactions involved. These rate-limiting considerations are typical of biochemical reactions that do not proceed spontaneously, but, rather, require catalysis and/or energy to be effective physiological processes. For example, the breakdown of the energy source ATP (c.f., Chapter 4) into ADP and inorganic phosphate is an exergonic (energy-releasing) reaction that cannot occur in the absence of an active form of the enzyme ATP-ase. Neither can active transport proceed without activated and available carrier molecules. Indeed, transport processes that involve carrier-mediated mechanisms bring into the picture a whole new set of additional transport variables, to wit:

1) *How many* (concentration) carrier-molecules are available to help transport?

2) What is the carrier-substrate *affinity* (geometrical compatibility, stereo-specificity, chemical activity or electrostatic attraction) of these molecules for one another? A case in point: Hemoglobin has a 230 times *greater* affinity for Carbon Monoxide that it does for Oxygen. Thus, in competing reactions for binding sites on the hemoglobin molecule, carbon monoxide will almost always beat out oxygen. Furthermore, the binding of carbon monoxide with hemoglobin is virtually irreversible, so that, once bound with carbon monoxide, the hemoglobin molecules are rendered totally incapacitated for further transport. The result? When

carbon monoxide is present, blood loses its oxygen-carrying capacity and the individual involved becomes asphyxiated (i.e., suffocates from a deprivation of oxygen).

3) *How many binding sites* are available on these molecules to help carry species S? In the case of the sodium-potassium electrolyte pump, for example, there appear to be two binding sites for potassium and three for sodium on the carrier molecules involved.

4) To what degree are these binding sites already occupied? That is, how *saturated* are these carrier molecules?

5) How sensitive is the action or behavior of these carrier molecules to thermodynamic factors such as temperature and pressure? (e.g., see discussion in Chapter 4 of temperature effects in enzyme kinetics); Or to solution characteristics such as pH and tonicity? Or to the thermodynamic state of the solvent?

6) Are active-transport *facilitators* required? That is, must S in some sense be actively *placed* onto the carrier-molecule binding site (or, removed from same) via some related secondary chemical reactions at the inner or outer cell boundaries?

7) Do the carrier molecules, in some sense, have to be *activated* (perhaps by specific enzymes, hormones, or other chemical constituents) before they can assume their role in the transport process -- i.e., become geometrically compatible with the substrate so that it will "fit" into the binding sites? (See discussion in Concluding Remarks section of Chapter 4, for example, on the role of coenzymes in activating enzymes to catalyze reactions).

8) How strong is the *electrochemical gradient* driving these carrier molecules from one side of the membrane to the other, if, indeed, this is how they get there? In fact, how many of the transport mechanisms so far described act, as well, to drive the *carrier* molecules, themselves?

9) Do the carrier molecules affect transport through pores, or through the membrane material itself? If the former, how many of the considerations related to *that* type of transport (see earlier) regulate the action of carrier-molecules, as well, or their effectiveness in promoting

mass transport -- i.e., what is the pore resistance to transport? If carrier-mediated transport is through the membrane material, per se, then how many of the considerations related to membrane permeation act to control, as well, the action of the carrier-molecules, themselves -- i.e., what is the membrane resistance to transport?

10) Bearing in mind some of the earlier discussion of particle mobility, as it may relate to carrier molecules, how *mobile* are *these* molecules? For example, can they be *chelated*?

11) Is there some *pyramiding effect,* wherein carrier molecules carry carrier molecules, which carry carrier molecules, and so on? And if so, well, the reader can quickly see that the situation can become quite complex, indeed. However, since much of the theory of active transport and facilitated diffusion is still rather speculative and based on considerable circumstantial evidence, there would be nothing gained in going into much more detail on the subject at this particular time. The reader is referred to the literature for more information as we move on to say a few words about Pinocytosis and Phagocytosis.

Pinocytosis and Phagocytosis

There are, in the physiologic system, large molecular weight substances whose size is so large (8 nanometers or more) that their movement to desired locations on either side of a membrane barrier would be literally impossible if left to passive or even active processes of the type discussed until now. This includes some plasma proteins, lipoprotein molecules ("carrier molecules" for non-water-soluble fats), large polysaccharide (sugar) derivatives, and proteoglycan molecules, all of whose mobility across a membrane is essentially zero. For such particles, mother nature has provided still another means of transportation, called *Pinocytosis,* which means, "a condition (-osis) of cell (-kytos-) drinking (pinein-)."

The process of Pinocytosis is shown schematically in Figure 2-5. A particle approaches the membrane from side 1 and, coming in contact with the membrane surface, triggers a response that causes the membrane wall to ingest ("drink") a small amount of fluid (e.g., plasma), forming an invagination that surrounds and "sucks in" the particle. This invagination, shown in Figure 2-5B, gradually progresses to completely encapsulate the particle, forming a container or "vesicle", as shown in Figure 2-5C. The self-contained capsule now migrates (perhaps by passive diffusion, perhaps otherwise) in a few seconds from the outside cell surface 1 to the inside cell surface 2, carrying its cargo with it (Figure 2-5D). Having arrived at the inside cell surface (or, vice versa if the transport is in the other direction), the vesicle begins to open up, or evaginate to the interior of the cell, as illustrated in Figure 2-5E. This process continues until the particle is released into the cell, at which point the transport is complete and the membrane invagination disappears (Figure 2-5F).

Closely related to the process of Pinocytosis is that whereby phagocytes (white blood cells) ingest and subsequently digest bacteria and other foreign particles "tagged" by the body's immune system response (see earlier description). In this case, the foreign matter is encapsulated into a "vacuole" within the protoplasm of the leukocyte, following which digestive enzymes are secreted into the confined receptacle to destroy the material (literally, to "eat" it, from the Greek word element, "phago" which means "eating"). Its remnants are then disposed of by the body's sophisticated exhaust system. Phagocytes ("eating cells") are thus named because they accomplish this mission, but the ingestion and encapsulation of *any* large particles in the body has come to be called *Phagocytosis,* and is frequently used in transport jargon interchangeably with Pinocytosis. This distinction has to do with size -- Pinocytosis being reserved for more microscopic events, and Phagocytosis generally referring to more macroscopic events.

In either case, both of these are active processes requiring the expenditure of energy. The membrane invagination to form a capsule, the "revolving-door" type of movement of the capsule from one side of the membrane to the other, and the ultimate release of the transported particle, all require that work be done.

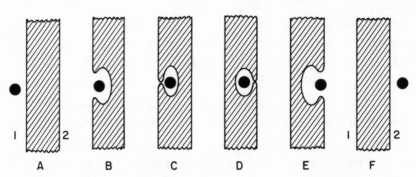

Figure 2-5 Pinocytotic transport resulting from the invagination, involution and closure of the boundary material, encapsulating the species to be transported and moving it in "revolving-door" fashion from one point to another.

Concluding Remarks

As if the reader has not already realized it by now, the transport of mass, energy, and momentum in physiologic systems involves extraordinarily complicated and diverse processes, most of which are *still* not clearly understood. Indeed, all of the mechanisms that govern such processes probably have not even been completely *identified* yet. And those that *have* are known to be complex, and to depend on a long list of variables -- all of which have to be functioning properly in order for the transport to proceed according to plan. The human body (as well as pharmaceutical companies) exploit these attributes as means for *controlling* and otherwise intervening in the physiological processes of life. Thus, there are innumerable control mechanisms that have evolved to limit the transport of various chemical constituents (and other quantities S) through semipermeable membranes to some *optimal* rates consistent with the overall well-being of the organism. The body can control the presence or absence of enzymes required to catalyze certain reactions, and, their degree of activation. It can control the permeability of membranes by manipulating any of the variables known to affect same. It can control the carrier-molecules required for active or facilitated transport. It can control the energy sources necessary for those processes that have to be driven. In short, the human organism has at its disposal an entire hierarchy of controllable parameters that give it ample flexibility in governing physiologic function.

But, as is to be expected, with so many controllable variables there are also so many potential break-downs in the system. Thus, there are bound to be a corresponding myriad of possibilities for failure, such that the malfunction of these limiting control mechanisms could produce a whole variety of disease states. No wonder, then, that these breakdowns manifest themselves as "diseases" of what appear to be an almost infinite variety.

The need to study further all of the transport processes in physiologic systems thus has a three-fold purpose: First, an obvious academic interest just to "know how we work" for whatever that may be worth; Second, to understand the probable relationship between certain disease states and a breakdown in metabolic transport pathways; and Third, in an even more fundamental sense, for the potential light that can be shed on the very process of evolution. Let us therefore back up now to examine in somewhat more detail many of the transport processes the essential features of which have been only briefly outlined in this chapter. Our intent is to analyze some of the specifics as they relate not only to the mathematical *description* of some of these processes -- as developed in the generic transport equation [2-4] -- but, also, to mechanisms by which these processes can be *controlled.*

We begin, in Chapter 3, with basic *analytical* techniques that can be used to describe the passive and active transport of *mass* in physiologic systems. The transport of energy and momentum are then addressed in Chapters 4 and 5, respectively, while Chapter 6 concerns itself with control and the "transport" of *information.*

Chapter 3

Basic Principles of Passive, Facilitated and Active Mass Transport Mechanisms in Physiologic Systems

Introduction

In Chapter 2, we observed that one can completely describe any transport process by defining for it a corresponding transport coefficient, α, intensive potential $\vec{\Omega}$ (or ϕ for conservative force systems), mass density, ρ, and respective constant of proportionality, $A_{i\,or\,j}$, and characteristic exponent, n ($n = 0$ for energy, $n = \frac{1}{2}$ for momentum, and $n = 1$ for mass) -- c.f., equations [2-3], [2-4], and associated discussions.

We also went on to say that kinematic transport coefficients and intensive "charge" potentials can generally be obtained by any or all of three basic methods: (i) purely analytic deductive techniques based on mathematical derivation using fundamental physical principles; (ii) purely experimental or empirical inductive techniques based on the generation of an arsenal of data; and (iii) techniques of nondimensional analysis based on an appropriate scaling of the physics involved in any transport process. In this and the chapters that follow, our intent is to illustrate each of these methods as they may be utilized to study mass, energy, and momentum transport in physiologic systems.

We begin here by addressing the deductive analysis of passive, facilitated and active mass transport, starting with basic principles of ordinary diffusion at the microscopic level.

Ordinary Diffusion

Consider the situation depicted in Figure 3-1A, where solute S, represented by large black dots is assumed to be suspended in a solvent represented by the smaller dots. In particular, at the instant shown, the large black dots representing S are distributed nonuniformly in the region occupied by the solute/solvent mixture. Let us examine the consequences of this situation as it relates to a fictitious "control volume", dV, contained within the arbitrary control surface, dA defining a "piece" of the solution. The control volume dV has been chosen to encompass the region containing most of S at the instant shown (see Figure 3-1B) and it is hypothetical in the sense that both solute and solvent molecules can cross dA freely with absolutely no problem. In other words, dV and dA are simply portions (or "boundaries", if you will) of the solution on which we have chosen to focus our attention.

Within dV, there are obviously very many large black dots per unit volume of solution, compared to those per unit volume contained outside of dV. Thus, one concludes that the concentration of S *within dV* (measured by the number of particles per unit volume in the region labelled 2 in Figure 3-1B, and designated $[S]_2$) is much higher than is the concentration of S outside of dV (measured by the number of particles per unit volume in the region labelled 1 in Figure 3-1B, and designated $[S]_1$). This leads to a spatial gradient in concentration of S, given by:

$$\text{grad}\,[S] = \nabla[S] = \frac{\partial[S]}{\partial x}\,\hat{i} + \frac{\partial[S]}{\partial y}\,\hat{j} + \frac{\partial[S]}{\partial z}\,\hat{k} \qquad \text{[3-1]}$$

where the brackets around S indicate that its concentration is being expressed as moles per liter of solution (recall that one mole of a substance whose molecular formula is known contains the Avogadro number, 6.02×10^{23}, of molecules); and where $\hat{i}$, $\hat{j}$, and $\hat{k}$ represent unit vectors along the x, y, and z directions in space, respectively, of an orthogonal, right-handed, Newtonian, cartesian coordinate system, as shown in Figure 3-1B. Note that the tiny dots representing the solvent molecules are

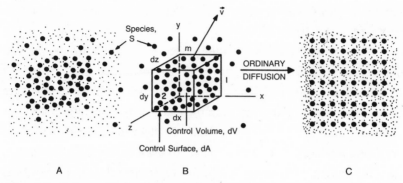

Figure 3-1 Ordinary diffusion that results from a concentration gradient coupled with the random Brownian motion of particles in solution.

omitted in this Figure to avoid clutter, but they are there just as in Figures 3-1A and 3-1C.

Now, all molecules in solution undergo random Brownian motion due to thermal energy (see page 177). Thus, there is constant movement of molecules in and out of control volume dV, which offers no resistance to such movement. However, since there are many more large black dots (high concentration of species S) *inside* dV than there are *outside*, there are correspondingly more dots available to *leave* the control volume than there are to return. In other words, from a statistical point of view, it is quite probable that, because of their nonuniform distribution in space, for every two or more molecules of large black dot material that leave the control volume, only one (on the average) will return. Over a long enough period of time, a mixing or redistribution of solute and solvent would thus take place randomly until the molecular distribution becomes homogeneous, or uniform, as shown in Figure 3-1C. At that point, statistically, as many black dots would be leaving the control volume as would be returning, and so the concentration of such dots would approach some steady-state uniform value throughout the region occupied by the solute/solvent mixture. This process of molecular rearrangement due to concentration differences, and driven by thermal Brownian motion forces, is called *ordinary diffusion*.

It is possible, in a deductive sense, to derive a means for measuring Brownian motion forces, and thus, to quantify ordinary diffusion in terms of α and ϕ, using the techniques of statistical mechanics. Consider, for example, the behavior of a *single* black dot of mass m, travelling within dV at some random Brownian velocity designated by:

$$\vec{v} = v_x \hat{i} + v_y \hat{j} + v_z \hat{k} \qquad [3\text{-}2]$$

as shown in Figure 3-1B. To get some "feel" for the *tendency* that m has to leave dV as a result of $\vec{v}$, let us "pretend" that the walls of dV are, in fact, fixed, rigid, and totally impermeable to m. Consider further, then, what happens as this particle *collides* with the inside face (say, the area bounded by sides dy and dz at location $x = dx$) of the volume dV within which it is contained. If $dy\,dz$ is immovable, and impermeable to m, then m will simply collide with the surface and bounce back. The *normal* force exerted by $dy\,dz$ on m during this collision prevents m from leaving dV, and so, in some sense is a measure of the tendency of m to *want* to leave dV (i.e., the energy inherent in thermal Brownian motion).

According to Newton's Third Law of Motion, the force exerted by $dy\,dz$ on m is equal and opposite to that exerted by m on $dy\,dz$. According to Newton's Second Law of Motion, the normal force exerted by m on the area $dy\,dz$ will be equal to the time rate of change of linear momentum of m in the direction normal to the surface $dy\,dz$, this being the x-direction for the case in point. Thus,

$$F_x = \frac{d(mv_x)}{dt} \qquad \text{[3-3]}$$

Now, if we assume the collision of m with $dy\,dz$ to be perfectly elastic, then this collision will cause m to rebound with an x-component of velocity exactly equal to $-v_x$, so that the total change in velocity of m in the x-direction will be $(v_x)_{\text{initial}} - (v_x)_{\text{final}} = v_x - (-v_x) = 2v_x$ per collision with $dy\,dz$, per molecule.

Going one step further, let us determine the time interval over which such velocity change takes place. That is to say, m in Figure 3-1B must travel an x-distance equal to $2dx$ between collisions with area $dy\,dz$ located at $x = dx$ -- i.e., a distance dx perpendicular to $dy\,dz$, along the x-direction, *away* from $dy\,dz$, in order to hit the "back wall" (at $x = 0$) of dV, and then another distance dx perpendicular to $dy\,dz$, along the x-direction, on the "rebound" from the back wall in order to hit $dy\,dz$ at $x = dx$ again. Assuming the distances involved are infinitesimally small, as are the time scales for the motion, m may be assumed to be travelling at an essentially constant x-velocity, v_x, between collisions, especially if they are assumed to be perfectly elastic. Thus, m will cover the distance $2dx$ in a time interval $t = \frac{2dx}{v_x}$. We conclude, therefore, that in its collisions with the area $dy\, dz$ located at $x = dx$, m has its velocity change by an amount $2v_x$ in the x-direction every $\dfrac{2dx}{v_x}$ time increments. With m, also assumed to be constant, then, we can rewrite equation [3-3] as:

$$F_x = m\frac{dv_x}{dt} = m\frac{2v_x}{\dfrac{2dx}{v_x}} = \frac{mv_x^2}{dx}, \qquad \text{[3-4]}$$

on the average.

Now define a corresponding thermodynamic "pressure" on this face of the parallelepiped as F_x divided by (or distributed over the surface) area $dy\, dz$. Then,

$$\begin{array}{c} p_{dy\,dz} \\ \text{at } x = dx \end{array} = \frac{F_x}{\text{Area of Face }(dy\,dz)} = \frac{mv_x^2}{dx\,dy\,dz} = \frac{mv_x^2}{dV}, \qquad \text{[3-5]}$$

where dV is the total volume of the container, which is equal to $dx\,dy\,dz$. Equation [3-5] defines the partial pressure exerted by a single molecule of mass m on face $dy\,dz$ of dV due to perfectly elastic collisions between $dy\,dz$ and m that result from the random thermal motion of the latter.

If we now consider that there are N^* particles of mass m contained in dV, and, further, that they interact with each other only elastically, then, by an extension of the reasoning presented above, one may logically conclude that:

$$\begin{array}{c} p_{dy\,dz} \\ \text{at } x = dx \end{array} = \frac{m_1 v_{1x}^2 + m_2 v_{2x}^2 + m_3 v_{3x}^2 + \cdots + m_{N^*} v_{N^*x}^2}{dV} \qquad \text{[3-6]}$$

For simplicity and convenience right now, let $m_1 = m_2 = m_3 = \cdots = m_{N^*}$ (i.e., let the big black dots in Figure 3-1 represent molecules of the *same* species S). Then equation [3-6] reduces to:

$$\begin{array}{c} p_{dy\,dz} \\ \text{at } x = dx \end{array} = \frac{m}{dV} \sum_{i=1}^{N^*} v_{ix}^2 = m\frac{N^*}{dV} \sum_{i=1}^{N^*} \frac{v_{ix}^2}{N^*} = m\frac{N^*}{dV}\,\overline{v_x^2}, \qquad \text{[3-7]}$$

where $\overline{v_x^2} = \sum\limits_{i=1}^{N^*} \dfrac{v_{ix}^2}{N^*}$ is the mean of the x-velocity-squared of all the molecules of mass m contained in dV (i.e., the "mean-square" x-velocity component of S). Now, if the motion of the N^* particles in dV is random in the true sense of the word, then it is reasonable to expect the distribution of the mean-square velocity components to be uniform in all directions -- or, at the very least, in three mutually perpendicular directions -- so that one may properly surmise that $\overline{v_x^2} = \overline{v_y^2} = \overline{v_z^2}$. In other words, it is of equal probability in a purely random motion that whatever the mean square velocity of m is in one direction, it will have the same value in any other direction because in a statistical sense, m has no preference for one direction over another. Thus, since the magnitude of $\vec{v}$ in equation [3-2] is equal to the square-root of the sum of the squares of its components, one may write further: $\overline{v^2} = \overline{v_x^2} + \overline{v_y^2} + \overline{v_z^2} = 3\overline{v_x^2}$, following the reasoning from above. Substituting this result into equation [3-7], one may conclude that the pressure developed on *any* face of dV can be written, in general, as:

$$p = m \frac{N^*}{dV} \overline{v_i^2} = \frac{m}{3} \frac{N^*}{dV} \overline{v^2} \qquad [3\text{-}8]$$

where $i = x, y,$ or z. Moreover, noting that mN^* is equal to the *total* mass, M, of S contained in the volume dV, equation [3-8] can be simplified to yield: $p = \left(\dfrac{2}{3} \right) \dfrac{1}{2dV} M\overline{v^2}$, or, more generally,

$$p = \frac{k_a}{V} \frac{1}{2} M\overline{v^2} \qquad [3\text{-}9]$$

Observe, from the analysis above, that thermodynamic pressure, p, is proportional to the translational molecular kinetic energy, $\frac{1}{2} M\overline{v^2}$, per unit volume, V, of any substance S. The particular substance involved is defined more specifically by the coefficient k_a which, for the *ideal* case we examined above (i.e., perfectly elastic collisions, no inter-molecular interference, dilute solution, and so on) is equal to $\left(\dfrac{2}{3} \right)$. For *real* materials, k_a would take on some other (lower) value, but the *functional* form of equation [3-9] would remain essentially the same. Observe further, since $\frac{M}{V}$ = mass density, ρ, that equation [3-9] may be rearranged to yield a characteristic transport velocity, v = the root-mean-square (RMS) value given by (c.f., Chapter 2, pgs. 110-111):

$$v = \sqrt{\overline{v^2}} = \sqrt{\frac{2p}{\rho k_a}}$$ [3-10]

The product of v and the mean free path, λ, of molecules of S contained in dV, yields a quantity having the dimensions $\left(\frac{L^2}{t}\right)$, which is a characteristic kinematic transport coefficient, α_S. This transport coefficient is called the *Diffusion Coefficient* for S, which accounts for mass transport due solely to the existence of a concentration difference such as is shown in Figure 3-1. Moreover, using the Van't Hoff analogy between dilute solutions of S and the perfect gas law (see later discussion of osmosis), we may write in equation [3-10], $p = \frac{\rho RT}{MW}$, so that,

$$\alpha_S = \lambda v = \lambda \sqrt{\frac{2RT}{k_a(MW)}}$$ [3-11]

The diffusion coefficient, otherwise known as the molecular or mass diffusivity of S is a measure of the "ease" with which a black dot (or molecule of species S) can travel by virtue of its own Brownian motion from region 2 towards region 1 (see Figure 3-1B) through the medium of "kinematic viscosity" α_S (i.e., the solvent or membrane material) in which it finds itself. This parameter quantifies how much effective resistance the membrane or solvent offers to having the solute pass through it ("mass conduction", if you will), and it is thus a lumped-parameter that accounts for the combined effect of many of the factors discussed qualitatively in Chapter 2.

For solutes with molecular weights, MW, up to a few hundred (i.e., most substances that are classified as being *crystalloid*), the diffusion coefficient is proportional to the reciprocal of the square-root of the molecular weight, i.e., $\alpha_S \sqrt{MW}$ = constant for isothermal physiologic systems, as suggested by equation [3-11]. In cases where the solute size or radius is large relative to the average distance between the solvent molecules (i.e., for substances classified as being *colloid*), α_S is inversely proportional to the cube-root of MW, i.e., $\alpha_S(MW)^{1/3} = $ constant. For large, spherical solute particles, the Stokes-Einstein equation may be applied:

$$\alpha_S = \frac{k_b T}{6\pi \mu_f r_o},$$

[3-12]

where k_b is Boltzmann's constant (1.38×10^{-16} erg/degree), T is temperature, μ_f is the coefficient of viscosity of the solvent in which S finds itself, and r_o is the average radius of the solute particle of mass m. For more concentrated solutions, one may further introduce an activity factor, α^*, as discussed previously in Chapter 2, to account for interactions *between* black dots and/or other dissolved species.

Before moving on, note in equation [3-11] that $v = 0$ when absolute temperature $T = 0$. In other words, absolute zero defines the state of affairs that exists when there is absolutely no molecular motion at all in the Brownian sense. For any finite temperature above $T = 0$, all molecules possess some inherent translational kinetic energy and thus the *tendency* for Brownian-motion-induced ordinary diffusion will always be there when there exists a difference in concentration of S within some region of space. That being the case, one may view *molecular kinetic energy* at the *microscopic* level as a form of *potential pressure energy* (c.f., equation [3-10]) in the more macroscopic continuum sense, or, equivalently, *concentration potential energy* which drives mass diffusion processes. The latter follows directly from the perfect gas law, $pV = NRT$, and from the definition of concentration, $[S] = \frac{N}{V}$ (in moles per liter), such that:

$$p = RT[S]$$

[3-13]

Energy is considered to be *potential* in the sense that it can be converted into useful work. The idea, then, is to see how much *work p* can do, and to derive from this concept a scalar intensive potential energy function, the gradient of which actually *drives* the mass diffusion process. Let us therefore *relax* the previously imposed assumption of total rigidity of the walls of dV (but *keep* the assumption of total impermeability to S), and let dV change by some amount δV as a direct result of the action of p. In a thermodynamic sense, the work done, δW, by p in changing dV by some amount δV is given by: $\delta W = p\,\delta V$. Since $V = \dfrac{N}{[S]}$, $\delta V = -\dfrac{\delta[S]N}{[S]^2}$; and, since $p = RT[S]$ (equation [3-13]), the corresponding work done can be written as:

$$\delta W = p\,\delta V = RT[S](-1)\frac{N\delta[S]}{[S]^2} \qquad \text{[3-14]}$$

If we expand equation [3-14] in accordance with the definition of a total differential, we get:

$$\delta W = \frac{RTN}{[S]}\left\{ \frac{\partial[S]}{\partial x}\delta x + \frac{\partial[S]}{\partial y}\delta y + \frac{\partial[S]}{\partial z}\delta z \right\}(-1), \qquad \text{[3-15]}$$

or, since $V = N/[S]$ and $\rho = M/V$, equation [3-15] becomes, per unit mass:

$$\frac{\delta W}{M} = -\frac{RT}{\rho}\left\{ \frac{\partial[S]}{\partial x}\delta x + \frac{\partial[S]}{\partial y}\delta y + \frac{\partial[S]}{\partial z}\delta z \right\}, \qquad \text{[3-16]}$$

which reduces to:

$$\frac{\delta W}{M} = -\frac{RT}{\rho}\left(\nabla[S]\cdot d\vec{r} \right), \qquad \text{[3-17]}$$

where: $\nabla = \dfrac{\partial}{\partial x}\hat{i} + \dfrac{\partial}{\partial y}\hat{j} + \dfrac{\partial}{\partial z}\hat{k}$ is the spatial gradient operator, and, $d\vec{r} = \delta x\,\hat{i} + \delta y\,\hat{j} + \delta z\,\hat{k}$ is a differential spatial displacement in the dimensions of dV due to the deformation induced by p.

Now, mechanical work is defined to be the dot product between a driving force and the resulting displacement. Clearly, then, from equation [3-17], we see that the work done per unit mass, $\delta W/M$, is equal to the dot product between some "concentration driving force per unit mass, $-(RT/\rho)\nabla[S]$", and the resulting displacement, $d\vec{r}$. For an isothermal process involving a material of constant density, this diffusional driving force per unit mass can be written as the negative gradient of the corresponding intensive scalar potential:

$$\phi = \frac{RT[S]}{\rho} = \frac{p}{\rho}, \qquad [3\text{-}18]$$

and it has a magnitude, a (characteristic transport acceleration), given by:

$$a = \left| \nabla\left(\frac{RT[S]}{\rho} \right) \right|. \qquad [3\text{-}19]$$

The negative sign in equation [3-17] indicates that the diffusional force acts opposite to $\nabla[S]$. That is, for the situation depicted in Figure 3-1, for example, $[S]_{\text{outside}} < [S]_{\text{inside}}$ of dV so the concentration of S is *decreasing* in the conventional *positive* direction, taken to be along the *outwardly* directed normal to dA. Thus, the gradient of $[S]$ is negative in this direction. The *actual* transport of mass, however, will tend to occur *from* inside dV *to* the outside of dV, i.e., *in* the *positive* spatial direction. One sees, therefore, that the *flux* of material (driven by $\nabla\phi$) is always *opposite* in direction to the gradient of $[S]$, so that material flows "down" the concentration gradient in a passive sense.

Fick's Laws of Diffusion

Let us summarize what we have derived thus far to describe the process of ordinary diffusion, and translate these results into what are known as Fick's first and second laws of diffusion. We observed that particles of a species S existing at some absolute temperature above zero degrees possess inherent translational kinetic energy that allows them to undergo random Brownian motion. This motion is associated with some characteristic transport speed, $v = \sqrt{\dfrac{2RT}{k_a(MW)}}$. The random movement of particles of S causes them to collide with one another, the mean distance travelled between collisions being designated the mean-free-path, λ. The rate at which an average particle suffers collisions is thus given by v/λ, which may be on the order of one billion (10^9) or more collisions per second.

Random movement of particles of S which exist at some concentration $[S]$ gives them a concentration potential energy, which has associated with it an intensive scalar potential function, $\phi = \dfrac{RT[S]}{\rho}$. Gradients in the latter may cause a redistribution of mass by a process called ordinary diffusion, driven by a characteristic force per unit mass whose magnitude is $a = \left| \nabla\left(\dfrac{RT[S]}{\rho} \right) \right|$, and associated with a corresponding kinematic transport coefficient, or mass diffusivity, defined by $\alpha_S = \lambda v$ (equation [3-11]).

Substituting the above results into equation [2-4], page 103 of Chapter 2 (i.e., the generic transport equation), with $n = 1$ for mass and $i = S$, we have:

$$\Phi \text{ (mass flux)} = -\frac{A_i \rho \alpha (\nabla \phi)}{(2\phi)^n}$$

$$= -\frac{A_S \rho \lambda}{2} \sqrt{\frac{2RT}{k_a(MW)}} \, \nabla\left\{ \frac{RT[S]}{\rho} \right\} \frac{\rho}{RT[S]}.$$

For constant ρ, R, T, and $(\rho/[S]) = MW$, this equation reduces to:

$$\Phi = -A_S \lambda \sqrt{\frac{RT(MW)}{2k_a}} \, \nabla[S] \qquad \text{[3-20]}$$

The coefficient A_S is related to both the type and permeability of the transport pathway involved. For intramural transport transport within a bounded region, or for transport across a biological membrane that does not distinguish between solute and solvent (i.e., is totally permeable to both in either direction through the membrane material itself, or through smooth, straight, unobstructed pores), A_S is generally equal to unity. For this situation, equilibration will eventually result in the steady-state, and there will be no subsequent flux of material once the concentration gradient has been dissipated.

For the case of a membrane or pathway that selects absolutely, such that only water may pass through but not any species of solute, S (i.e., in the case of total impermeability), $A_S = 0$. For this situation, equilibration will eventually result in the steady state due to *osmotic* pressure gradients generated by the concentration gradient in S (see later), but there will be no *net* movement of S across the membrane. Thus, $0 \leqq A_l \leqq 1$ measures the *semipermeability* of a membrane and the resistance of the transport pathway. Related to this also is the reflection coefficient or Staverman Factor (see discussion in Chapter 2, pages 104 and 131), $1 - A_S$; the Activity Factor (for more concentrated solutions, as also discussed in Chapter 2, pages 104, 124 and 153); and, in the case of transport through "tortuous" pores, the ratio (A_p/ξ), (c.f., Chapter 2, page 142).

If we now define a Binary Effective Diffusion Coefficient, or Fick's Mass Diffusivity, as:

$$D_S = A_S \lambda \sqrt{\frac{RT(MW)}{2k_a}} = A_S \frac{MW}{2} \alpha_S \qquad \text{[3-21]}$$

then equation [3-20] becomes:

$$\Phi = - D_S \nabla [S] \qquad \text{[3-22]}$$

Equation [3-22] is the well-known Fick's First Law of Diffusion for the unobstructed (i.e., ordinary, or free) conduction of mass in the presence of a concentration gradient such as is shown in Figure 3-1. It basically defines mass flux in terms of an effective kinematic transport coefficient, D_S, and the gradient in concentration of S that is driving the process.

In order to carry this analysis still one step further, let us now *relax* the assumption of impermeability of the boundaries of dV to S, and let the concentration potential actually result in some flux of S from region 2 to region 1 (see Figure 3-1B). This will cause an *efflux* of mass through the control surface, dA. Noting that $\vec{v} \cdot \hat{n}$ yields the component of $\vec{v}$ normal to dA, where $\hat{n}$ represents a unit vector in that direction, the efflux of S from dV may be written as the surface integral,

$$Efflux = \int_{dA} \rho \vec{v} \cdot \hat{n} \, \delta A \qquad \text{[3-23]}$$

where efflux has the dimensions of mass per unit time.

The efflux of S from dV will cause its concentration within dV to decrease with time. Such decrease can be calculated by observing the rate at which mass within dV is disappearing. The latter can be written as the volume integral,

$$"Fluxion" = -\frac{d}{dt} \int_{dV} \rho \, \delta V \qquad \text{[3-24]}$$

The word, "Fluxion" was coined by Newton to describe the rate of change or derivative of a varying function (it is more commonly used to define the time-rate-of-change of charge). The negative sign in equation [3-24] connotes the fact that the quantity of S in dV is *decreasing* with time. Now, in the absence of mass sources or sinks, conservation of mass requires that the fluxion of S within dV be exactly equal to the efflux of S from dV. Thus, from equations [3-23] and [3-24], we conclude that:

$$\int_{dA} \rho \vec{v} \cdot \hat{n} \, \delta A + \frac{d}{dt} \int_{dV} \rho \, \delta V = 0 \qquad \text{[3-25]}$$

Now, applying the Divergence Theorem, we note that $\int_{dA} \rho \vec{v} \cdot \hat{n} \, \delta A$ is equal to $\int_{dV} \nabla \cdot (\rho \vec{v}) \delta V$. Furthermore, for a continuous function, we can reverse the order of differentiation and integration, so that $\frac{d}{dt} \int_{dV} \rho \delta V = \int_{dV} \frac{\partial \rho}{\partial t} \delta V$. Thus, equation [3-25] now becomes:

$$\int \left\{ \frac{\partial \rho}{\partial t} + \nabla \cdot (\rho \vec{v}) \right\} \delta V = 0 \qquad \text{[3-26]}$$

with the integration being performed over the volume dV. The key here is that equation [3-26] should hold for *any* arbitrary control volume, dV. That is to say, the integration should not depend on dV. In order to guarantee that that should be the case, we can set the *integrand* identically equal to zero, all by itself, i.e., let:

$$\frac{\partial \rho}{\partial t} = -\nabla \cdot (\rho \vec{v}) \qquad \text{[3-27]}$$

But, recall from Chapter 2 (equation [2-1]) that $\rho \vec{v}$ actually is the directional mass flux, Φ, which we have just derived as Fick's First Law of Diffusion. Thus, substituting equation [3-22] into equation [3-27], we arrive at the conclusion that:

$$\frac{\partial \rho}{\partial t} = -\nabla \cdot \Phi = -\nabla \cdot (-D_S \nabla [S]) \qquad \text{[3-28]}$$

Finally, with $\rho = (MW)[S]$, $\nabla \cdot \nabla = \nabla^2$ (the LaPlacian operator), $D_S/MW = (\alpha_S A_S)/2$ (equation [3-21]) -- which we can arbitrarily set equal to some new constant, b^2 -- and, the two minus signs cancelling, equation [3-28] becomes Fick's Second Law of Diffusion:

$$\frac{\partial [S]}{\partial t} = b^2 \nabla^2 [S] \qquad \text{[3-29]}$$

The choice of b^2 for the constant in equation [3-29] simply puts it in the standard form of a time-dependent, second-order diffusion equation, where the kinematic transport coefficient governs the rate of the process. In general, S can actually be *any* quantity -- mass, energy, or momentum -- that is undergoing diffusion due to a gradient in charge as discussed in Chapter 2. The governing differential equation will always take the form of [3-29] as long as there are no sources or sinks. Bearing this in mind, then, let us develop the analysis still further as we examine the *energy* associated with taking a particle of *S* from region 2 to region 1.

Gibb's Energy Equation

Fick's Laws of Diffusion essentially address conservation of mass and momentum. In fact, equation [3-29] *is* a momentum equation that *includes* conservation of mass. But now that we have given *S* the freedom to move through the boundaries of dV, we have provided one mechanism for changing $[S]$, and we would like to address at this point the *energy* associated with such change. To do so, we go back to equation [3-14], which defines the energy required to change the concentration of *N* moles of species *S* from $[S]$ by an amount $\delta[S]$. In order to calculate the *total* energy involved in changing the concentration from $[S]_2$ inside dV to $[S]_1$ outside of dV, on a per mole basis, we can integrate equation [3-14] to get:

$$W \text{ (per mole)} = \int \frac{\partial W}{N} = \int_{[S]_2}^{[S]_1} - RT \frac{\delta[S]}{[S]}$$

$$= - RT \{\ell n\,[S]_1 - \ell n\,[S]_2\},$$

or,

$$W = RT \ell n \frac{[S]_2}{[S]_1} \text{ per mole of } S \qquad \text{[3-30]}$$

Equation [3-30] is called Gibbs Energy Equation and it defines the work done by concentration-gradient forces to carry one mole of *S* (i.e., an amount of *S* whose weight in grams numerically equals its formula weight) from an area of high concentration (region 2 located inside dV in Figure 3-1B) to an area of low concentration (region 1 located outside of dV in Figure 3-1B), *along* a naturally-occurring concentration gradient -- i.e., spontaneously, or passively. Thus, if we wanted to carry *S* the other way, *solely against* the concentration gradient which tends to drive the black dots in the opposite direction, then *we* would have to do work W *actively* to accomplish that task. Or, stated another way, if we want to maintain the control volume at dV, and to *maintain* the concentration difference $[S]_2 - [S]_1$, then *we* must do work $- W$. The minus sign in this case indicates that the particles of S are being crowded together *against* their natural tendency to expand or spread apart as a result of the $\Delta[S]$ that is shown in Figure 3-1.

The change in energy, $RT\{\ell n\,[S]_2 - \ell n\,[S]_1\}$ is called the Gibbs Free Energy change for this process, and it represents the difference between the work done to compress one mole of species S from an infinite volume (i.e., zero concentration) to concentration $[S]_2$, and the work done to compress one mole of S from an infinite volume to concentration $[S]_1$. For a spontaneous process at constant temperature and pressure, the Gibbs Free Energy decreases, from some larger initial value, $g_{initial}$, to some smaller final value, g_{final}. Since the corresponding free-energy change is traditionally defined to be $g_{final} - g_{initial}$, we see that a *negative* free-energy change so defined constitutes a passive, or spontaneously occurring process. Conversely, a *positive* free-energy change corresponds to an energy-consuming or *active* process. Note further from equation [3-30] that if $[S]_2 = [S]_1$, then $W = \Delta g = 0$, and the system is in equilibrium. Thus, for an equilibrium situation, the Gibbs Free Energy change in the system is zero.

The Gibbs Free Energy of a species S is called its chemical potential, a subject about which we shall have much more to say in Chapter 4. Suffice it to say here that chemical reaction rates can be calculated by applying the equations of wave mechanics to colliding particles; and that chemical reactions that occur between colliding particles may be quantified in terms of differences that appear between the concentration of the reactants and the concentration of the products of the reaction. Using equation [3-30], then, one may quantify chemical kinetics in terms of Gibbs Free Energy changes that occur between reactants and products.

And finally, observe further that not only is $\Delta g = 0$ at equilibrium, but so, also, is $\partial[S]/\partial t$, since there is no mass flux. Thus, at equilibrium, we have, from equation [3-29], the additional requirement that $\nabla^2[S] = 0$.

Diffusion Through a Membrane

If control surface dA actually represents in real life a thin membrane of thickness δ, separating two solutions having different concentrations of black dots, then the characteristic transport velocity, v, across the membrane is called its *permeability coefficient,* designated P_S, for species S. The membrane permeability coefficient represents the number of moles of S crossing a unit area of dA per unit time, when a unit concentration difference is applied across the membrane, i.e., when $[S]_2 - [S]_1 = 1$ mole per liter of solution, for example. Thus, P_S is a measure of how *fast* molecules can penetrate the membrane barrier, and it is often determined experimentally for a given membrane.

The permeability coefficient for a membrane quantifies many of the extramural transport considerations discussed qualitatively on pages 125 to 157 and 159 to 164 of Chapter 2, such as (i) the geometry of the pore for transport via that route; (ii) the characteristics of the material that fills the pore; (iii) factors specific to permeation through the membrane material, itself, for transport via *that* route; and (iv) certain of the variables involved in carrier-mediated transport. In terms of P_S, one can write the flux of species S, in moles per unit area, per unit time, net, as $P_S(\Delta[S])$, where $\Delta[S]$ is the *actual* difference in concentration of species S between the outside of the membrane (region 1 in Figure 3-1) and the inside (region 2).

If we divide and multiply the flux equation above by δ, we get flux $= P_S\delta \dfrac{\Delta[S]}{\delta}$, which, in finite-difference form, is analogous to equation [3-22], i.e., Fick's First Law of Diffusion. We can thus let $P_S\delta$ represent the effective diffusion coefficient for the membrane, i.e.,

$$D_{Sm} = P_S\delta \qquad\qquad [3\text{-}31]$$

For biological membranes, δ is typically very small, on the order of 50 to 100 angstrom units for a bimolecular lipid cell membrane (see Figure 2-1), which is around 8 nanometers.

It should be re-emphasized that Φ represents the *net* mass transfer across a membrane, per unit time, per unit area. That is, it is the difference between efflux and influx as discussed in the introductory comments made at the beginning of this section on ordinary diffusion. Furthermore, the general conservation principle requires that:

$$\text{Fluxion} + \text{Effluxes} - \text{Influxes} = 0$$

And, finally, the transport described usually assumes that the only resistance encountered by species S in moving from one side of the membrane to the other is the membrane itself. Thus, "correction" factors, such as the nondimensional "Activity Factor", α^*, must be introduced in real life, where the above assumptions do not hold.

Osmosis

Water molecules in the liquid state undergo the same random thermal motion that is characteristic of all particles in solution. Thus, if a membrane permeable to *water only* separates two aqueous solutions of different solute concentrations, $[S]_{outside}$, and $[S]_{inside}$, then *water* will move across the membrane in an effort to equalize the concentration of solute on both sides. In effect, the flow of water across the membrane attempts to dilute the more concentrated side of the membrane, and to concentrate the more dilute side. Experiments have revealed, however, that such water flow, which is called *Osmosis*, cannot be regarded as just simple diffusion of H_2O molecules across the membrane. That is, a difference in water activity across a membrane leads to a *greater* water flux than can be explained by simple diffusion principles alone, such as those discussed earlier.

This observation can probably be explained in part by the fact that most diffusion considerations ostensibly neglect any particle-to-particle interactions, and inherently assume very dilute solutions of S, on the order of 0.1 molar or less. In an aqueous solution, however, water obviously cannot be considered "dilute", since *most* of the solution is water! Thus, whereas particles attempting to "diffuse" or be conducted *through* water have their motion *impeded* by collisions with water molecules, the movement of water *itself* (a large body of mass compared with individual solute molecules) is actually aided or *facilitated* by particle-to-particle momentum transfers and other inertial considerations. Inertia and bulk fluid properties, then, at least partially explain why total mass fluid movement in osmosis exceeds predictions that are based on simple, virtually mass-less, single-molecule-at-a-time diffusion considerations. This suggests, then, that mathematical descriptions of osmotic-type mass transfers should be based more on continuum concepts than on statistical or molecular theories, and, indeed, they are.

The continuum description of the flow of a fluid ascribes one cause of such flow to a gradient in hydrostatic pressure between two points (say, the opposite sides, 1 and 2, of a membrane permeable to water). It is thus convenient to quantify flow due to osmosis by measuring the hydrostatic pressure difference necessary to *prevent* such flow and to *maintain* an unequal concentration of total dissolved particles in two solutions on either side of a membrane. Such a hydrostatic pressure difference may be calculated from thermodynamic principles elucidated in 1885 by a scientist named Van't Hoff. He showed that there is a certain similarity between dilute aqueous solutions and the behavior of perfect gases. The formal proof of this analogy is available in many advanced text books on Thermodynamics, but what it amounts to is that: "The osmotic pressure of a dilute solution of N moles of solute S is the same as the thermodynamic pressure of a perfect gas, which would be measured if N moles of the gas exerted pressure on the walls of a vessel of total volume V, equal to that of the solvent plus the solute." This is known as Van't Hoff's Proposition -- which was alluded to in the earlier development of equation [3-11] -- and it translates mathematically into *Van't Hoff's Law*, which is a direct analog of the perfect gas law, i.e.:

$$\Pi V = NRT \qquad \qquad [3\text{-}32]$$

where Π now represents *osmotic* pressure (defined below), and where, since $(N/V) = [S]$, we further' conclude that: $\Pi = RT[S]$, as in equation [3-13]. The *net* osmotic pressure developed across a membrane is the *difference* between the osmotic pressure on each side (see, for example, Figure 3-1):

$$\Pi_{net} = \Pi_1 - \Pi_2 = RT\left(\Sigma[S_i]_1 - \Sigma[S_i]_2\right) \qquad [3\text{-}33]$$

The terms on the right side of equation [3-33] represent the difference between the summed molar concentrations of all species S_i having differing concentrations on either side (1 and 2) of the membrane. Those species which have the *same* concentration on both sides of the membrane do not contribute to the generation of any osmotic pressure gradient. Note that for $R = 8.21 \times 10^{-2}$ liter-atmospheres per mole, per degrees Kelvin, $T = 310°K\,(37°C)$, and $\Sigma[S_i]_1 - \Sigma[S_i]_2 = 1$ mole per liter, equation [3-33] yields the result that $\Pi_{net} = 25.451$ atmospheres (1 atmosphere $= 760$ mm Hg $= 14.696$ pounds per square inch, psi).

The hydrostatic pressure necessary to *keep* a net concentration difference of one mole per liter across a membrane permeable only to water is defined to be *one osmol* of *osmotic pressure*. That is to say, 1 osmol represents the osmotic pressure associated with the physical situation wherein 6.02×10^{23} *more* particles of S per liter of solution are maintained on one side of the membrane than there are on the other side. Under these circumstances, water would tend to want to flow across the membrane to "neutralize" this concentration gradient, since the membrane is considered to be impermeable to S, itself. The *hydrostatic pressure* that must be applied to *prevent* water flow from *its* more concentrated side of the membrane (i.e., the *dilute* side) to *its* less concentrated side (i.e., the *concentrated* solute side) in this case turns out to be 25.451 atmospheres, as calculated above, so one concludes that: 1 osmol of osmotic pressure $= 25.451$ atmospheres of hydrostatic pressure.

Note that 1 osmol is a rather high unit of pressure -- about 374 pounds per square inch -- and, as such, is somewhat unrealistic from a physiologic point of view. Thus, the *milliosmol* (10^{-3} osmols) is a more convenient unit of measure. The milliosmol is equal to 0.025451 atmospheres, or, equivalently, some $19\frac{1}{3}$ millimeters of mercury, and it corresponds to a concentration difference across a membrane which is maintained at 1 millimole per liter (1 millimolar).

Note further that it is the *total number of particles* on either side of the membrane that determines the osmotic pressure gradient across the membrane. Thus, the total osmolarity of a solution is the *summed* molar concentrations of all species in the solution. For example, a dilute one millimolar solution of Calcium Chloride, $CaCl_2$, existing on one side of a membrane would generate an osmotic pressure (or have an "osmolarity") of 3 milliosmols, because such a solution would dissociate into one millimole per liter of calcium and two millimoles per liter of chlorine, yielding a total of three millimoles per liter of *actual particles other than water* on that side of the membrane relative to the other side.

If the solutions involved should become so concentrated that they could no longer be considered dilute, then particle-to-particle interactions, again, would have to be taken into account. As is typically the case in such situations, "correction factors" are introduced to compensate for these interactions. Thus, the osmolarity of a concentrated solution may be multiplied by an *osmotic coefficient*, or *factor*, which is always less than unity because the osmotic effect of concentrated solutions is less than that of dilute solutions. This is because particle-to-particle interactions slow down the rate of transport, making it "appear" as if the actual concentration difference across the membrane is less than it really is.

The term "tonicity" is used to describe the osmolarity of one solution relative to another. Thus, a solution is said to be *hypertonic, isotonic,* or *hypotonic* with respect to another solution, depending upon whether it contains more, the same number of, or fewer dissolved particles per liter, respectively, than does the solution to which it is being compared. If the solution on one side of a membrane is *hypertonic* to that on the other side, then water will tend to flow osmotically *toward* the hypertonic side; for example, a hypertonic cell -- i.e., a cell placed in a hypotonic *solution* -- will tend to swell. The exact opposite will occur if the solution on one side of a membrane is *hypotonic* to that on the other side; for example, a hypotonic cell -- i.e., a cell placed in a hypertonic solution -- will tend to shrink. And *no* osmotic mass flux will transpire across a membrane separating two solutions that are *isotonic* relative to one another. Mammalian serum, for example, contains approximately 310 milliosmols of osmotic pressure (see later) per liter, and is frequently considered to be a reference solution for determining the tonicity of other substances in the body. Thus, a 155 millimolar solution of Sodium Chloride, NaCl, is isotonic to mammalian serum.

The situation depicted in Figure 3-1 represents a cell containing a solution that is hypertonic relative to its environment (or, equivalently, the cell is considered to exist in a hypotonic environment or solution). If dA, or the cell membrane is permeable to S, then *ordinary diffusion* would ultimately cause the situation to become isotonic in accordance with the mass flux considerations already described. On the other hand, if the cell membrane is *impermeable* to S, but totally permeable to H_2O, then *osmosis* would ultimately bring about the same result (i.e., isotonicity) by allowing water to flow *into* the cell (at the expense of causing the cell to swell, however). Cells placed in hypotonic solutions will thus tend to swell (and possibly burst), or become "cytoedemic"; whereas cells placed in hypertonic solution will tend to shrink (or become "cytolytic") in a process called "plasmolysis" (or, in the case of red blood cells, "crenation" or "hemolysis"), which may lead to cellular rupture, as well.

For the process of osmosis, one may define a characteristic length scale, which is the ratio of membrane surface area, A, (or, if more appropriate, the total cross-sectional area of all *pores* in the membrane, c.f., equation [2-14]), to membrane thickness, δ; a characteristic transport speed, which is the *osmotic permeability coefficient*, P_{H_2O} that represents the phenomenological coefficient relating the molar flux of water to its conjugate driving force per unit area, Π_{net}; an *osmotic pressure-filtration kinematic transport coefficient*, $\alpha = D_{H_2Om} = P_{H_2O} \dfrac{A}{\delta}$ (c.f., equation [3-31]); a characteristic intensive scalar potential, $\phi = \dfrac{\Pi}{\rho} = \dfrac{\Sigma_i[S_i]RT}{\rho}$, (c.f., equations [3-18] and [3-32]); and a characteristic transport driving force per unit mass, which is an acceleration equal to the magnitude of $\nabla\phi$. Thus, with $n = 1$ for mass, equation [2-4] yields, for osmosis:

$$\Phi = - \frac{A_{H_2O}\rho P_{H_2O}A}{2\Pi\delta} \nabla\Pi \qquad [3\text{-}34]$$

where A_{H_2O} and P_{H_2O} are usually determined for a given membrane in experiments wherein the membrane permeable *only* to water, and nothing else, is used.

Thus, we see that osmosis proceeds in response to a gradient in osmotic pressure, as defined by equation [3-34], and that this gradient, like the diffusional force per unit volume (or per unit mass) represents an osmotic force per unit volume (or mass). Moreover, osmotic pressure, too, is a form of fluid potential energy defined in terms of molecular kinetic energy (as measured by T) per unit volume (V).

In real life, if the membrane "leaks", i.e., if it will not prevent entirely solute transport across it, then the movement of the solute must be taken into account by once more introducing a "correction factor". In this case, the appropriate quantity is the so-called *Staverman Factor*, or, *reflection coefficient*, $\beta = A_{H_2O}$. The role of the Staverman Factor in defining membrane semipermeability has already been discussed (see Chapter 2, pg 104 at the end of the section entitled, *Solution of the Generalized Transport Equation;* the discussion on page 131 surrounding equation [2-12] in the section entitled, *Extramural Transport Through Pores;* and in this chapter, pages 180-181 in the discussion following equation [3-20] before equation [3-21]. Suffice it to say again here that the reflection coefficient has values between 0 and 1.

While we are on the subject of the transport of water across a membrane by osmosis, it is convenient to recognize also that H_2O has a second way of passing from one side of a permeable barrier to the other. It can be physically "pushed" through the membrane by ordinary mechanical forces that are defined in terms of the negative mean of the three normal stress components that act on a differential element of fluid, i.e., the so-called *hydrostatic* fluid pressure, p. Rather than being generated by a concentration difference in S, as is *osmotic* pressure, hydrostatic pressure is generated mechanically by, for example, the action of a pump such as the human heart. The discussion surrounding equations [2-8] to [2-10] and, in Chapter 2, the dynamic state of the material which fills the pore in the case of extramural transport by this mechanism, is particularly relevant here. That is to say, water flux due to a hydrostatic pressure gradient can be written by a direct analog of equation [3-34] as:

$$\Phi = -\frac{A^*_{H_2O}\rho P^*_{H_2O}A}{2p\delta}\nabla p \qquad [3\text{-}34a]$$

where, for this case, $\phi = p/\rho$, $\alpha = P^*_{H_2O}A/\delta$ is called the *hydrostatic pressure-filtration kinematic transport coefficient* (a characteristic transport coefficient for transudation); $P^*_{H_2O}$ is the hydrostatic permeability coefficient of the membrane (a characteristic transport speed given, for example, by equation [2-8], $A^*_{H_2O}$ is a corresponding scaling factor for $P^*_{H_2O}$; A/δ is a characteristic length for the transport (i.e., the ratio of membrane surface area to thickness), and, for a thin membrane, $|\nabla p| = \dfrac{p_2 - p_1}{\delta}$ is a characteristic transport acceleration multiplied by mass density.

In its simplest form, the hydrostatic pressure-filtration kinematic transport coefficient can be calculated by assuming that the water flow takes place through circular cylindrical pores oriented at right angles to the inside and outside surfaces of the membrane. Then, $\alpha = \dfrac{A}{\delta}P^*_{H_2O} = \dfrac{\mu_f}{\rho_f}$, as given by equation [2-9], and $P^*_{H_2O}$ corresponds to the amount of water crossing a unit area of membrane, per unit time, when a unit pressure gradient is applied across a given pore. Finally, if there are n^* pores per unit surface area of membrane, each having a mean pore-diameter d, then (c.f., equation [2-10]):

$$\Phi = \frac{\rho_f}{\mu_f} \frac{d^2}{32} \left(\frac{p_2 - p_1}{\delta} \right) n^* \qquad [3\text{-}35]$$

for each unit area of membrane; or, written in terms of volumetric flow rate, one obtains Poiseuille's Equation,

$$\begin{array}{l} \text{Volumetric Flow Per Unit} \\ \text{Time, Per Unit Area Of} \\ \text{Membrane} \end{array} = \frac{\pi d^4}{128} \frac{n^* \rho_f}{\delta \mu_f} (p_2 - p_1) \qquad [3\text{-}36]$$

When water leaves one side of a membrane -- irrespective of whether or not it carries solute particles with it -- the concentration of solute particles on that side of the membrane increases, and that on the other side decreases. Thus, the movement of water may establish concentration gradients among transmembrane species within which such a gradient might not have been otherwise present. There is thus a tendency for these particles to move in the same direction as the net flux of water -- not because the water is dragging them along (as in solvent drag), but because an established concentration gradient is causing them to move by ordinary diffusion. The *sum* of solvent drag *plus* concentration effects thus generated may be so great that it actually dominates the movement of a given species of dissolved particles. Such total particle movement is called *bulk flow,* and is due to net water flux that results from a combination of both osmotic and hydrostatic pressure gradients. Bulk flow is characteristic of extremely permeable membranes, where the interaction forces between particles and membrane (τ_S) are small, e.g., in the systemic capillaries and the renal glomerulus. This concept shall be addressed further in the next section, where we return again to phenomena associated with Gibbs-Donnan equilibrium effects.

Gibbs-Donnan Equilibrium and Capillary Mass Transport

There is a special case of mass transport which has particular physiologic relevance in capillaries. This involves a biological membrane which has semipermeable characteristics such that one of the nonpermeating parti-cles happens to be an electrically charged ion. The condition of electroneutrality across the membrane requires that for every charged macromolecule that cannot penetrate the membrane, an equivalent oppositely charged particle must stay behind with it, effectively becoming immobilized even though the latter particle, itself, might otherwise have no trouble at all getting across. Thus, the concentration of oppositely charged particles on that side of the membrane is higher than it would be normally and the equilibrium concentrations that are eventually ap-proached differ from those that would be predicted from purely passive electrochemical considerations.

As mentioned in Chapter 2 (pgs. 153-154), these phenomena in-volving fixed ions were investigated experimentally by F. G. Donnan and analytically by W. J. Gibbs; hence, they have come to be known as Gibbs-Donnan equilibrium effects. Let us examine the particular case involving Sodium ions, Chlorine ions (both of which pass easily across cell membranes) and *large* (molecular weight greater than 7000) plasma protein ions, which carry a predominantly negative charge and which, because of their large size and lipid-insolubility, normally do not pene-trate the walls of mammalian capillaries.

From thermodynamic principles, Gibbs showed that, if the osmotic pressure gradient caused by the impenetrable plasma proteins is balanced exactly by an externally applied fluid dynamic hydrostatic pressure, then Sodium and Chlorine ions will equilibrate across the membrane in such a way that the change in free energy across the membrane will be equal to zero (see earlier discussion of Gibb's Energy Equation). Utilizing equation [3-30], it can be shown that this last statement translates into the thermodynamic equation:

$$RT \ell n \, \frac{[Na^+]_2}{[Na^+]_1} + RT \ell n \, \frac{[Cl^-]_2}{[Cl^-]_1} = 0 \qquad \text{[3-37]}$$

where the subscript 2 designates "inside" the membrane (or on the plasma side of a capillary) and the subscript 1 designates "outside" the membrane (or on the interstitial side of a capillary). Thus, based on ordinary diffusion considerations, equation [3-37] yields the result that:

$$[Na^+]_2[Cl^-]_2 = [Na^+]_1[Cl^-]_1 \qquad \text{[3-38]}$$

where *only* Sodium and Chlorine are assumed to be capable of permeating the capillary wall. Now, if contained within the capillary is a large plasma protein having concentration $[S]_2$ and negative valence $-Z$, then charge neutrality on either side of the membrane imposes the further restrictions that:

$$[Na^+]_2 = [Cl^-]_2 + Z[S^{-Z}]_2, \text{ and,}$$

$$\text{[3-39]}$$

$$[Na^+]_1 = [Cl^-]_1$$

Substituting the relations [3-39] into equation [3-38] leads to the following quadratic equation for the concentration of Chlorine:

$$[Cl^-]_2{}^2 + Z[S^{-Z}]_2[Cl^-]_2 = [Cl^-]_1{}^2 \qquad \text{[3-40]}$$

Typically, the concentration of chlorine in interstitial fluid is about 155 millimoles per liter. That of plasma proteins is around 1 millimole per liter, and the latter generally have a valence of some -18. Thus, equation [3-40] becomes:

$$[Cl^-]_2{}^2 + 18[Cl^-]_2 - (155)^2 = 0, \text{ from which,}$$

$$[Cl^-]_2 = \frac{-18 \pm \sqrt{(18)^2 + 4(155)^2}}{2} = 146.26 \text{ or } -164.26$$

Since negative concentrations have no physical meaning, we conclude that:

$$[Cl^-]_1 = 155 \frac{\text{Millimoles}}{\text{liter}} \qquad [Na^+]_1 = 155 \frac{\text{Millimoles}}{\text{liter}}$$

$$[Cl^-]_2 = 146.26 \text{ mM}/\ell \qquad [Na^+]_2 = 164.26 \text{ mM}/\ell$$

The "Donnan Concentration Ratio" for Sodium is thus:

$$\frac{[Na^+]_{\text{outside}}}{[Na^+]_{\text{inside}}} = \frac{155.00}{164.26} = 0.94$$

Observe that the concentration of Sodium inside the capillary is higher than that outside, and higher than it *would* be were it not for the negatively charged plasma proteins that are holding it back. This is the Gibbs-Donnan phenomenon, confirmed experimentally by Donnan. It also illustrates a point made several times in Chapter 2 relative to particle mobility -- i.e., that it is affected by whether or not the solute molecules are "bound" in solution. In this case, the binding is caused by electrostatic considerations.

Taking this analysis one step further, recall that the *total* osmotic pressure is proportional to the difference between the total particle concentrations on the two sides of a membrane (equation [3-33]). On the *plasma* side 2 of the capillary, the total particle concentration is $[Cl^-]_2 + [Na^+]_2 + [S^{-z}]_2 = 146.26 + 164.26 + 1 = 311.52$ millimoles per liter. On the *interstitial* side 1 of the capillary, the total particle concentration is $[Cl^-]_1 + [Na^+]_1 = 155 + 155 = 310$ millimoles per liter. The difference is $311.52 - 310 = 1.52$ millimoles per liter, or, about 50% more than that due to the nondiffusible anion, S^-, alone -- whose concentration difference across the capillary membrane is only 1 millimole per liter.

We see then, that the Gibbs-Donnan phenomenon contributes also to establishing a higher osmotic pressure than would be the case for a simple *uncharged* nonpermeable species S. The difference is caused by totally *diffusible* ions which are electrostatically *trapped* on the wrong side of the membrane by the presence of the charged nondiffusible particles. The *osmotic* pressure in a capillary is thus about 50% greater than the value of $19\frac{1}{3}$ millimeters of mercury calculated earlier (page 190) for a nonionized concentration difference which is maintained at 1 millimole/liter across a membrane. In fact, the total osmotic pressure tending to drive fluid *into* the capillary (since the inside is *more* concentrated than the outside) is actually $(19.34276)(1.52) = 29.4$ mm Hg, and this is counterbalanced by a capillary *hydrostatic* pressure (or capillary blood pressure) of equal amount in order to prevent capillary swelling due to hypertonicity.

The total osmotic pressure generated by the impermeable plasma proteins is called the *colloid osmotic pressure* (COP), or *oncotic* ("tending to cause swelling") *pressure* of the system. Added to the small interstitial fluid pressure of some 2-5 mm Hg, the oncotic pressure constitutes most of the force opposing filtration from capillaries to surrounding tissue. This opposing force of 30-35 mm Hg is counteracted by the actual fluid dynamic pressure of blood in the capillaries. At the arterial end (i.e., entering the capillary) the latter is measured to be on the order of 40-45 mm Hg, while at the venous end (i.e., leaving the capillary) this pressure is around 10-15 mm Hg. The frictional pressure loss in a capillary is thus some 25-35 mm Hg, and the mean pressure generally sits at about 27 or 28 mm Hg. More importantly, however, is the fact that at the arteriolar end (inlet) a *net* effective *filtration pressure* of 5-15 mm Hg drives material *out* of a capillary *into* the interstitial fluid by bulk convection, while at the venule end (outlet) a net pressure difference of 15-25 mm Hg drives material *back into* the capillary by the same mechanism.

Indeed, the *Starling Hypothesis* of fluid distribution between blood plasma and interstitial compartments postulates that fluid filters out at the arteriolar end of the capillary, circulates through the tissue interstices, and is returned to the capillary at the venular end in a "sub-microscopic recirculation pattern." This, of course, is a somewhat oversimplified description of what really happens (which involves, as well, factors related to capillary permeability, tissue distensibility, vasomotion and lymphatic drainage), but it is self-consistent with mass transport considerations and Gibbs-Donnan phenomena discussed thus far.

It also further illustrates LaBarbera and Vogel's (1982) contention (see Chapter 2, pg. 116) that, "while diffusion is always used for short-distance transport, it is augmented by bulk flow ... for any long-distance transport." One might add a corollary to this "first principle" of transport, i.e., that the "augmentation" is accomplished by energy-consuming *active* (rather than spontaneous passive) processes -- in this case, the blood pressure generated by contractions of the heart, which also maintain the *mean* capillary hydrostatic pressure at values just sufficient to prevent net swelling due to the hypertonicity of the inside relative to the outside.

Other than blood pressure, per se, several other factors can alter capillary permeability. These include both chemical and mechanical agents or events, such as:

1) A fall in plasma protein concentration below 5 gram-percent (5 grams per 100 ml of blood). This will cause increased capillary leakage by lowering the vascular osmotic pressure and thus causing water and electrolytes to flow from the blood to the tissues. Such "Vascular Edema" sometimes is caused by a protein-deficient diet, or, by illnesses that cause plasma protein deficits -- most especially in the *ratios* between *different* protein fractions, since each has a different colloid osmotic pressure.

2) The activity of *pre-and-post capillary sphincters*. These are circular rings of muscular fibers that surround the entrance (pre-) and/or exit (post-) of a vascular capillary, so that, by contracting, they can constrict the respective inlet and outlet orifices of the blood vessel. Downstream constriction of post-capillary sphincters increases capillary hydrostatic back-pressure, thus enhancing permeability. Upstream constriction of pre-capillary sphincters tends to do just the opposite, while also restricting flow into the capillary involved. These sphincters are significantly controlled by certain chemical substances, among them:

a) *Kinins,* or other substances of a protein or polypeptide nature. These small polypeptides that are split off from an α-globulin in
plasma or tissue fluids by proteolytic enzymes are very powerful smooth-
muscle relaxants. Thus, they can induce vasodilation (an increase in
blood vessel diameter), hypotension (reduced blood pressure), hyperemia
(increased blood flow), enhanced capillary permeability, interstitial edema
-- and, sometimes, the pain that accompanies the latter. One of these
kinins, Bradykinin, is believed to be present in exocrine glands, but its
exact physiologic significance is still a controversial issue.

b) *Histamine,* a chemical substance (amine) derived from the
amino acid histidine and bound to cellular and extracellular proteins.
This amine, when released by injured tissue, induces a powerful three-fold
effect: (i) it is a powerful arteriolar vasoconstrictor (i.e., it induces
smooth muscle contraction in these small blood vessels); (ii) it is a powerful venodilator; and, (iii) it dramatically increases capillary permeability
by allowing these tiny blood vessels to dilate and become very porous.
Among the substances that can cause histamine release are Curare,
Morphine, Atrophine, and a host of allergens.

c) Certain groups of *nucleotides* and *nucleosides* derived from
the breakdown of nucleoproteins. A nucleoside is a glycoside formed by
the union of a purine or pyrimidine base with a sugar (usually pentose).
It becomes a nucleotide when it combines further with phosphoric acid
to constitute the structural unit of nucleic acids. We shall get back to this
in greater detail in Chapter 5. Suffice it to say here that these chemical
substances can produce inflammatory reactions that result from increased
capillary permeability.

3) Certain *Enzymes* (e.g., Hyaluronidase and other so-called
Permeases). These have the ability to depolymerize the gel-state of the
mucopolysaccharide (many-sugared) matrix (including Hyaluronic Acid)
that gives tissue fluid certain glue-like properties. Disolution of such
binding and intercellular protective agents *increases* vascular permeability
by essentially getting rid of the substances (sugar "glues") that hold
endothelial cells together.

4) Certain *Hormones* (e.g., Glucocorticoids). These have the ability to stabilize or otherwise mediate the action of potentially destructive enzymes, or, to affect electrolyte and salt balance across a membrane, thereby acting mainly to *reduce* capillary permeability. This is why they are used a great deal for the treatment of allergic reactions, especially those that are associated with Histamine release (see paragraph (2)b above).

5) *Blockage of Lymphatic Vessels* increases downstream peripheral resistance and thus leads to increased capillary permeability.

6) Temperature, pH, Carbon Dioxide concentration, Oxygen concentration, endothelial cell injury, and even some Vitamins can influence capillary permeability; not to mention all of the factors discussed in Chapter 2. These contribute to the redistribution of body fluids in both the extracellular and intracellular spaces.

Electromotive Diffusion and the Nernst Equation

While we are on the subject of situations where the transported species S may have electrical properties, let us consider further the case where transport results not from a nonuniform distribution of *mass* (as in *ordinary* Brownian motion diffusion), but from a nonuniform distribution of *electric charge* across a region in space. In this case, electric field forces will come into play in the over-all mass transport process. The effect of these field forces may be quantified by going through a formal analysis similar to the one developed earlier for the case of ordinary diffusion. Now, however, an *ionic flux* of material is considered to result from a difference in *electrical potential energy* (rather than concentration potential energy), this difference being defined from a consideration of *Coulomb's Law* (rather than the perfect gas law or Van't Hoff's law). The analysis as developed below assumes that:

1) The *only* forces acting on the transported charged particles are due to the electrical properties of the situation that prevails, i.e., the

transport is driven entirely by gradients in *electric charge* distribution, not mass distribution (concentration).

2) The individual ions move through space independently of one another, i.e., again, no particle-to-particle interactions are to be considered just yet.

3) The system under consideration is otherwise in osmotic equilibrium and no significant gradients in mass concentration act to modulate the transport of ions; i.e., the transport is not under the influence of any bulk-flow considerations.

With the above in mind, Charles Coulomb formulated, in 1787, a relationship that defines the force of attraction or repulsion experienced by a "point charge", q_1, when it enters the "sphere of influence" of another "point charge", q_2. The "sphere of influence" of q_2 (or q_1, for that matter) is defined to be the region around an electric charge where its influence may be detected, this region being measured radially outward from the charge and encompassing a control volume $V = \frac{4}{3}\pi r^3$, bounded by a control surface $A = 4\pi r^2$. In more technical terms, the sphere of influence around an electric charge is called its *electric field*. The field is assumed to be generated by a "point charge" in the sense that its largest dimensions (i.e., the physical dimensions of the charge, itself) are orders of magnitude smaller than the distances r at which its influence can still be measured; so that q_1 or q_2 may effectively be considered to be concentrated at a single point in space. If this is the case, then for a homogeneous, isotropic medium, Coulomb showed experimentally that:

$$F = \frac{q_1 q_2}{r^2} k \qquad\qquad [3\text{-}41]$$

where F is the electromotive force of attraction (if q_1 and q_2 are of opposite charge) or repulsion (if q_1 and q_2 are both positive or both negative) between the charges, separated by some distance, r, in space; and k is a constant of proportionality (called the electric Coulomb constant) having the value 9×10^9 joule-meter/coulomb^2 when charge is expressed in coulombs (a coulomb is the charge associated with 6.25×10^{18} electrons, or, the charge of one electron is 1.6×10^{-19} coulombs), force is expressed in Newtons (positive for repulsion and negative for attraction), and r is expressed in meters (positive in a direction measured radially outward from q).

From *Coulomb's Law* (i.e., equation [3-41]), and, following a reasoning similar to that used in defining *thermodynamic* pressure earlier, one may define a corresponding *electric* pressure, p, as electromotive force, F, divided by the spherical surface area, A, over which it is spread at a distance r from q. That is to say, let:

$$\text{"Electric Pressure"}, \; p = \frac{F}{A} = k\,\frac{q_1 q_2}{r^2}\,\frac{1}{4\pi r^2} \qquad \text{[3-42]}$$

Now we further define the *Electric Field Intensity* at r as the electrical force per unit charge, and designate it E_e. In other words, E_e is the force that would be exerted on a charge of one coulomb ($q_1 = 1$, for example) placed a distance r away from a charge q_2 of q coulombs, and, from equation [3-41], we see that the magnitude of the electric field intensity is given by:

$$E_e = k\,\frac{q}{r^2} \qquad \text{[3-43]}$$

Substituting equation [3-43] into equation [3-42], we may write:

$$p = E_{e1} E_{e2}\,\frac{1}{4\pi k} \qquad \text{[3-44]}$$

where E_{e1} is the electric field intensity at q_2 due to q_1 and E_{e2} is the electric field intensity at q_1 due to q_2. Equation [3-43] may be generalized in the sense that $p = E_e^2/4\pi k$ at a distance r from q. Moreover, with r very small, i.e., for two point charges separated in space by a very short distance (such as a physiological membrane of microscopic thickness, δ), one may assume that p varies approximately linearly in the region between

$r = 0$ and $r = r$, so that the *average* electrical pressure at any point *between* the two charges is given by $\frac{1}{2} E_e{}^2 \frac{1}{4\pi k}$. Finally, if we divide p by density, ρ, we obtain an *intensive scalar potential*, ϕ (c.f., equation [3-18]) for *electromotive diffusion*:

$$\phi = \frac{1}{8\pi} E_e^2 \frac{1}{\rho k} \qquad \qquad [3\text{-}45]$$

From equation [3-45], one may now define a characteristic transport velocity, $v = \sqrt{\phi}$, transport acceleration, $a = -\nabla\phi$, and, with a characteristic length, $\ell = r$, a corresponding kinematic transport coefficient, $\alpha = vr$ for electromotive diffusion. Substituting these into equation [2-4], one obtains the electromotive mass flux equation:

$$\Phi = -\frac{A_i \rho r}{\sqrt{8\pi\rho k}} \nabla E_e, \qquad \qquad [3\text{-}46]$$

which shows that a gradient in electric field intensity causes the movement of *charged* particles, just as a gradient in concentration (equation [3-20]) causes the net movement of *uncharged* particles. Again, the minus sign appears because mass flux is in a direction opposite to the electric field intensity gradient, i.e., mass flows "down" a gradient in field intensity.

Recall for the case of the Brownian motion field, that we introduced a correction factor, α^*, called the "activity factor" to take into account mass transport variations resulting from particle interference in other-than-dilute solutions. In the present case, we can introduce a similar correction factor, called the *dielectric constant*, κ^*, which reduces the actual electromotive forces by the factor $\left(\frac{1}{\kappa^*}\right) = A_i$. For a perfect vacuum, the dielectric constant is exactly equal to unity; $\kappa^* = 1.00059$ for air at standard temperature and pressure, and 80.0 for liquid water at 20°C. Typically, dielectric constants for physiologic cell membranes range in value from 5 to 10, depending on a number of factors, and they can be found tabulated in a number of physiologic references.

We can express Φ (mass per unit time, per unit area) on the left side of equation [3-46] in terms of the flow of electric charge by introducing the following definitions: *electric current* is defined to be the charge movement per unit time, i.e., $I = \dfrac{q}{t}$ (or, in differential form, $\delta q/\delta t$); and *current density* is defined to be the charge per unit time per unit area, i.e., $J = \dfrac{I}{A}$. Recall that the charge on one electron is 1.6×10^{-19} coulombs, and that one mole of a substance contains 6.02×10^{23} particles. Thus, one mole of *univalent* ($Z = 1$) ions carries a charge of $(1.6)(6.02)(10^4) = 96,320$ coulombs. This conversion factor is called the Faraday Constant, $\overline{F}$, more precisely equal to 96,500 coulombs per mole of monovalent ions when the charge on an electron and the Avogadro number are carried out to a greater number of decimal places.

The valence, Z, of an ion represents the number of "free" electrons that are responsible for the reacting capacity and mobility characteristics of the species. If a particular species S has a valence equal to Z, then the *total charge* carried by *one mole* of the ionized species would be equal to $\overline{F}Z$ coulombs per mole of S, and if S has a molecular weight MW (in grams per mole), then one *gram* of S carries a charge of $\dfrac{\overline{F}Z}{MW}$ coulombs per gram. Note that the total charge carried by one *ion* of S would be equal to $1.6 \times 10^{-19}Z$. With these considerations in mind, we can now write the mass flux due to electromotive forces as:

$$\Phi = \frac{J(MW)}{\overline{F}Z} = \frac{\text{coulombs}}{(\text{time})\text{area}} \times \frac{\text{moles of } S}{\text{coulombs}} \times \frac{\text{grams of } S}{\text{mole of } S} = \frac{\text{grams of } S}{(\text{time})\text{area}} \; ,$$

or, using equation [3-46],

$$I = -\frac{\overline{F}Z}{MW} 4\pi r^2 \frac{A_i \rho r}{\sqrt{8\pi\rho k}} \nabla E_e \qquad \text{[3-47]}$$

Voltage Energy Equation

As we did in the derivation of the Gibb's Energy Equation, let us now formulate a means for quantifying the *work* the electromotive forces can do (or which *we* must do *against* them) in *moving* an electric charge from one point to another while it is within the electric field of a nearby charge. With F given by equation [3-41] and $q_1 = 1$ (i.e., on a *per unit charge* basis), let $q_2 = q$. Then, if *we* do work to move q_1 some differential distance δr within the field of q_2, this work will be equal to (force × distance):

$$\delta W \text{ (per unit charge)} = -F\,\delta r = -k\frac{q}{r^2}\,\delta r \qquad \text{[3-48]}$$

Again, the negative sign takes into account the fact that the charge q_1 is being moved *towards* q_2 in the direction *opposite* to the outer normal in the positive direction of r, so that *we* are doing the work in the case of q_1 and q_2 *both* positive or negative (i.e., when they repel each other and we move them closer together), and q_2 is doing the work when the charges are opposite and tend to attract one another. In the latter case, *we* do the work if we attempt to keep the charges apart from one another. In any event, the *total* work done in moving q_1 from a position r_1 to a new position r_2 is:

$$W \text{ (per unit charge)} = \int \delta W = \int_{r_1}^{r_2} -k\frac{q}{r^2}\,\delta r$$

$$W = kq\left(\frac{1}{r_2} - \frac{1}{r_1}\right) \qquad \text{[3-49]}$$

Now, if r_1 is infinity, then equation [3-49] gives the net work done to bring a unit charge *in* from a very large distance away to some location r_2 relative to the location of a charge $q_2 = q$. This work is defined to be the *electrical potential* or *voltage* existing at the point r_2 in the field of q, so that, in general,

$$\text{Voltage, } B = \frac{kq}{r} \text{ (Joules/coulomb)} \qquad \text{[3-50]}$$

A comparison of equation [3-50] with equation [3-43] reveals that the latter is the negative derivative of the former, i.e.,

$$-\frac{\delta B}{\delta r} = E_e$$

In a more generalized sense, then, one may conclude that the electrical force per unit charge, or electric field intensity, E_e is the negative gradient of the electrical potential or voltage existing at some point in an electric field, i.e., $\vec{E}_e = -\nabla B$. A *voltage gradient,* then, is responsible for the forces causing movement in an electric field, just as a *concentration gradient* is responsible for the forces causing movement or net mass transfer in a Brownian motion field, as mentioned earlier in the discussion of mass flux.

Equation [3-49] thus defines the potential difference, or "voltage drop" between two points as the amount of work done against electrical forces to carry a *unit* charge of 1 coulomb from one point r_1 to another, r_2. It is the difference between the work done to bring a unit charge in from $r =$ infinity to $r = r_2$, and that done to bring the same charge in from $r =$ infinity to $r = r_1$. Moreover, if the *total* charge being moved is $\bar{F}Z$ per mole of S, then the corresponding work done per mole of S moved is:

$$W \text{ (per mole of } S) = \bar{F}Zkq\left(\frac{1}{r_2} - \frac{1}{r_1}\right) \qquad \text{[3-51]}$$

where W is measured in Joules per coulomb (which is the definition of a *volt*) in equation [3-49], or Joules per mole of Z-valent ions in equation [3-51]. Note the similarity between these energy considerations related to a *voltage* difference in an *electric* field, and the energy considerations related to a *concentration* potential difference in an unevenly distributed *mass* field, i.e., equation [3-30]. Indeed, we may develop this similarity even further as follows.

The Nernst Equation

Using the concepts of concentration potential energy (equation [3-30]) and electromotive potential energy (equation [3-51]) per mole of S, one may combine ordinary mass diffusion with electromotive ionic diffusion by defining an *electrochemical potential difference,*

$$\Delta\mu_c = \overline{F}Zkq\left(\frac{1}{r_2} - \frac{1}{r_1}\right) + RT\ell n \frac{[S]_2}{[S]_1} \text{ per mole of } S \qquad [3\text{-}52]$$

Equation [3-52] represents the total potential energy difference, per mole of S, driving transport processes that result from *both* gradients in concentration and in electric charge distribution. There is a special case of this which has particular physiologic relevance, and this involves a combination of mass diffusion in one direction due strictly to chemical concentration gradients, which is balanced *exactly* by diffusion in the *opposite* direction due solely to electrical voltage gradients. What we have here is zero *net* mass flux because Φ in one direction due to $[S]_2 - [S]_1$ is exactly cancelled or counterbalanced by Φ in the opposite direction due to $B_2 - B_1$.

If we go back to Figure 3-1 for the moment, and assume that control surface dA represents a thin membrane separating regions 1 and 2 in space, then instead of considering dA to be impermeable to S -- which we did in order to develop the theory for ordinary diffusion -- we can postulate based on the discussion above that we apply a *voltage* to dA that keeps $[S]_1$ and $[S]_2$ constant by *repelling* electrically the ionic flux of S across dA. This voltage difference, $B_2 - B_1$ thus creates an equilibrium situation that is defined by setting $\Delta\mu_c$ in equation [3-52] equal to zero. That is to say, there is no net flux of S because there is no net electrochemical potential difference driving the process. Then, from equation [3-52], we conclude that:

$$\overline{F}Z(B_1 - B_2) = RT\ell n \frac{[S]_2}{[S]_1} \qquad [3\text{-}53]$$

The voltage difference, $B_2 - B_1$ is called the *membrane equilibrium potential* (a shortened version of membrane equilibrium potential difference), designated ΔB_m, for a species of ion, S. If the steady state membrane potential of an actual cell membrane happens to equal the equilibrium potential of an ion species, S, that ion is said to be distributed *passively* with respect to that membrane in the sense that the *net* flux of S across dA is zero because there are no electrochemical forces acting to cause movement of that ion across the membrane in any particular preferential direction. From equation [3-53], the membrane equilibrium potential is calculated to be:

$$\Delta B_m = \frac{RT}{FZ} \, \ell n \, \frac{[S]_{outside}}{[S]_{inside}} \qquad \text{[3-54]}$$

Equation [3-54] is known as the *Nernst Equation,* and it would be wise at this point to review the assumptions that are inherent in this equation:

1) The *only* forces acting on the ions involved are due to concentration gradients driving particles in one direction, which are *exactly* balanced by voltage gradients driving particles in the opposite direction;

2) The individual ions move through the membrane independently of one another (i.e., ionic concentrations are low), and the membrane is considered to be a thin plane oriented perpendicular to the flux direction;

3) The system is in osmotic equilibrium, i.e., there is no transport of water across the membrane;

4) the electric field is zero (or at least uniform) everywhere in each solution *except* in the immediate neighborhood of the membrane, where a potential difference exists;

5) Concentrations are the same everywhere in each solution but may differ between solutions on either side of the membrane;

6) The concentration and electric fields that exist *only* in and very near the membrane itself create diffusional and voltage forces that are in equilibrium, i.e., there is *no net mass flux.*

Reiteration of these basic assumptions should give the reader no excuse for attempting to use equation [3-54] in situations for which it does not apply! That is, in most human cells, most ions are not equilibrated and the net passive fluxes of permeating ion species, S, are not zero. One must therefore be very careful to examine separately mass fluxes due to concentration gradients, and those due to voltage gradients, and to add these vectorially to get the net mass flux into or out of the cell for each ion species.

Note further in equation [3-54] that ΔB_m is the voltage drop $B_{inside} - B_{outside}$, so that, by convention, the electrical potential of the inside of a cell is always expressed relative to that on the outside of the cell, and not vice versa. For $R = 8.31$ Joules per mole per degree Kelvin, $T = 310$ degrees Kelvin, $\overline{F} = 96,500$ coulombs per mole of monovalent ions, and $Z = +1$, i.e., for a monovalent cation, the Nernst Equation can be written with respect to the common logarithmic base as:

$$\Delta B_m \text{ (in millivolts)} = 61 \log \frac{[S]_{outside}}{[S]_{inside}} \qquad [3\text{-}55]$$

As an example, recall (pg. 197) that the Donnan concentration ratio for sodium across the capillary wall is: $[Na^+]_{outside} \div [Na^+]_{inside} = 0.94$. Thus, we conclude that the equilibrium potential for sodium is 61 log (0.94) $= -1.64$ millivolts across the capillary wall, with the plasma side (inside) negative relative to the interstitial fluid side (outside). And finally, from the definition of the equilibrium potential, observe that equation [3-52] can be written in the equivalent form:

$$\Delta\mu_c = \overline{F}Z(\Delta B - \Delta B_m) \text{ per mole of } S \qquad [3\text{-}56]$$

The mechanism used to *generate* ΔB's in physiologic systems involves active transport processes such as those that were discussed qualitatively in Chapter 2, and will be addressed more quantitatively later in this chapter. These potential differences are called *transmembrane voltages* or *potentials* (inside, or intracellular voltage, minus outside, or interstitial voltage). The equilibrium potential, ΔB_m is called the *resting membrane potential* in the case of *excitable* tissue (see later), and it depends upon the resistance-capacitance properties of the tissue, which are discussed further below. Generally speaking, however, the resting membrane potential is on the order of 0.10 volts (100 millivolts) or less for most physiologic membranes.

Electromotive Diffusion Equation, Resistance, Capacitance

With coordinate r functioning as an independent variable, we can write, from equation [3-50],

$$\left\{\frac{\partial B}{\partial t}\right\}_{\text{with } r \text{ constant}} = \frac{k}{r}\frac{\partial q}{\partial t} = \frac{k}{r}I \quad (I = \text{current}) \qquad [3\text{-}57]$$

Substituting equation [3-57] for I into equation [3-47], together with the result obtained earlier that $\vec{E}_e = -\nabla B$, and, letting $b^{*2} = \dfrac{\bar{F}Zr^2}{\kappa^*MW}\sqrt{2\pi\rho k}$, we obtain the electromotive diffusion equation,

$$\frac{\partial B}{\partial t} = b^{*2}\nabla^2 B \qquad [3\text{-}58]$$

Comparing equation [3-58] with equation [3-29] again reveals the analogy between B for electromotive diffusion and $[S]$ for ordinary diffusion. We therefore conclude that the transport coefficient b^{*2} (which has dimensions L^2/t) is in some sense analogous to the transport coefficient b^2 for ordinary diffusion, and may thus be referred to as an electrical or electromotive "diffusivity". If $r = \delta$, this becomes the *membrane electromotive diffusivity*, and, correspondingly, the quantity $\dfrac{\bar{F}Z\delta}{\kappa^*MW}\sqrt{2\pi\rho k}$, having velocity dimensions L/t, becomes the *electrical permeability coefficient*, P^*_S of the membrane to ionic species S. The membrane electromotive diffusivity is a measure of the ionic flux crossing a membrane of thickness δ when a unit voltage difference is applied across the membrane. It is therefore, in some sense, a measure of the *resistance* offered by that membrane to the flow of ions across it. We can quantify this relationship using *Ohm's Law*:

$$\Delta B = IR^* \qquad [3\text{-}59]$$

where R^*, having dimensions ML^2/tq^2, is the electrical resistance of the membrane, expressed in ohms $\left(\dfrac{\text{Joules-seconds}}{\text{coulombs}^2} = \text{volts/ampere} \right)$.

For $\Delta B = 1$ (i.e., for a unit voltage difference), we see that current flow (as measured by electromotive diffusivity) is inversely proportional to membrane resistance (as measured by R^*). In other words, the higher the resistance of the conducting material to the passage of charged particles through it, the lower the resulting ionic flux across the membrane. This suggests that R^* and b^{*2} are somehow inversely related, and, indeed they are. Observe that if we divide the dimensions of R^* (ML^2/tq^2) by those of the product $k\delta$ (ML^4/t^2q^2), we get exactly the dimensions (t^2/L), which are the inverse of the dimensions of b^{*2}. Thus, we may define the *membrane resistance*, R^*_m, by:

$$R^*_m = \frac{k\delta}{b^{*2}} = \frac{MW\kappa^*\sqrt{k}}{\overline{F}Z\delta\sqrt{2\pi\rho}} \qquad [3\text{-}60]$$

There is a huge variation in cell membrane resistances. The values range from about 5 ohms per cm² of membrane area for glial cells (the non-nervous or supporting structures of the central nervous system) and red blood cells, to larger than 100,000 ohms per square centimeter of surface area for some eggs. More typical values are of the order of 1000 ohms/cm² of membrane surface area.

Electrical resistance decreases as the area through which the charge must move increases, or the path which it must negotiate decreases. This can be expressed mathematically by writing $R^*_m = k^* \dfrac{\delta}{A}$, where the constant of proportionality, k^* is called the *specific resistivity* of the conducting material. Typically, mammalian extracellular fluids have resistivities on the order of 60 ohm-centimeters (the unit, "ohm" is equal to volts per coulombs per second, or volts per ampere, where an ampere is a coulomb per second); while that of axoplasm (the cytoplasm of an axon) is around 200 ohm-cm. In general, concentrated ionic solutions containing species having big valence numbers have correspondingly low resistivities if the ions are relatively mobile.

The reader may often find in the literature a further quantity called the *conductivity*, σ_e, of a material. Electrical conductivity is simply the reciprocal of the specific resistivity, k^*, so that it is closely related to the electromotive diffusivity. In fact, $b^{*2} = \sigma_e A k$. the conductivity of a physiologic cell membrane is the sum of the conductivities due to each ionic species. And finally, the quantity $(1/k^*\delta) = (1/R^*_m A)$ is called the *specific membrane conductance* for the particular ionic species S.

The discussion above has addressed specifically the *flow* of electrical current, or, equivalently, the *flux* of ionized material through a medium offering resistance R^* -- such flux being driven by a voltage difference ΔB. We turn now to a discussion of how well physiologic membranes can *maintain* the voltage difference ΔB, i.e., the ability that such biological structures have to keep charges separated from one another. This ability is called the *Capacitance* of the membrane.

Capacitance is defined to be the ratio of the quantity of charge q being kept from flowing, to the potential difference ΔB attempting to move q. Mathematically, we may write:

$$C \equiv \frac{q}{\Delta B} \qquad \text{[3-61]}$$

and, from equation [3-50], we see that C is in some sense related to the ratio (r/k), having dimensions $\dfrac{q^2 t^2}{ML^2}$. If q is expressed in coulombs and B in volts, then C is expressed in farads, which are relatively large units for biological membranes. More commonly, the microfarad (10^{-6} farads) is used to describe the capacitance of a physiologic membrane. The latter is equal to the material dielectric constant, κ^*, times the material surface area, A (in cm^2), divided by the charge spacing or membrane thickness, δ (in cm), and the product $36\pi \times 10^5$. Note that the ratio $\left(\dfrac{A}{\delta} \right)$, which has length units, gives the characteristic $"r"$ for capacitance, and the factor $\kappa^*/36\pi \times 10^5$ provides the appropriate corrected $"\dfrac{1}{k}"$ value for capacitance to give the results in microfarads. Letting C^*_m, then, represent the capacitance of a biological membrane, we can write:

$$C^*_m = \kappa^* \left(\frac{A}{\delta} \right) \frac{1}{36\pi \times 10^5} \propto \frac{r}{k} \qquad [3\text{-}62]$$

Cell membranes have a capacitance on the order of one microfarad (10^{-6} coulombs of charge separation per unit of transmembrane voltage generated) per square centimeter, i.e., $(C^*_m/A) = 0(1)$, which is about what is predicted from equation [3-62] for a 50 angstrom units ($\delta = 50 \times 10^{-8}$ cm) thick layer of substance having a dielectric constant κ^* of 5. The proportion of the *total* number of ions in the intracellular fluid that is required to charge a cell membrane to its equilibrium potential will vary with the size of the cell. For a 20-micron-diameter cell, for example, it will require only about one thirty-thousandth (3.33×10^{-5}) of the negative ions inside to charge the inside of the membrane to its equilibrium potential relative to the outside.

In summary, the typical cell membrane illustrated in Figure 2-1 can be represented by an electrical analog which contains a resistance, R^*_m (equation [3-60]) of order 1000 ohms/cm^2 in parallel with a capacitance C^*_m (equation [3-62]) of order 1 microfarad/cm^2. Moreover, the product, $R^*_m C^*_m$ has units of time, and represents a characteristic time-constant for the parallel resistance-capacitance membrane network. This characteristic time-constant, $\tau^*_m = R^*_m C^*_m$ is of order $(1000)(10^{-6}) = 10^{-3}$ sec, or 1 millisecond per cm^2 of typical physiologic membrane. We shall have more to say about this in Chapters 5 and 6 when we address action potentials and information transport in physiologic systems. For now, we move ahead to examine active transport mechanisms, which, in addition to providing a very effective means for moving mass (as discussed in Chapter 2), are also the means by which excitable tissues can generate voltage differences across membranes.

Carrier-Mediated Active Transport

As mentioned in Chapter 2, pgs. 154-166, when species S cannot get across a membrane passively by any of the mechanisms discussed -- or, when S is to be delivered to a location that involves moving it against an electrochemical gradient that would normally drive it back the other way -- the species can be physically carried across by active processes that involve the expenditure of energy. A mechanism that has been proposed to describe how active transport is accomplished across a physiologic cell membrane is shown schematically in Figure 3-2.

Suppose species S_1 is to be carried *actively* from the cell interior to the cell exterior, against an electrochemical gradient that would actually drive it from outside the cell in. Suppose further that species S_2, like species S_1 is exposed to an electrochemical gradient that tends to drive *it* from the cell exterior to the cell interior, as well -- but, that this would normally take place very slowly for a variety of reasons (two of which being that S_2 may have a very low solubility in the cell membrane material, and it may have no alternate access to the cell interior through a pore). Finally, suppose that the desired situation is to have S_1 outside and S_2 inside. Then consider the following sequence of events, starting with a certain *carrier molecule, Y*, in the membrane boundary at its interface with the inside cell surface:

1) S_1 has a great affinity for Y, such that when it collides with the inner membrane surface during its random wanderings on the inside of the cell, it tends to form the chemical complex $S_1 Y$ according to the activated-enzyme ($E_1 - P_i$) catalyzed reaction:

$$S_1 + Y \quad \underset{\substack{\varepsilon_1;\, E_1 - P_i \\ \text{catalyzed} \\ \text{association}}}{\overset{\substack{\text{spontaneous} \\ \text{dissociation} \\ \varepsilon_2}}{\rightleftharpoons}} \quad S_1 Y \qquad \text{[3-63]}$$

where ε_1 (in liters per mole, per unit time) is the rate constant controlling the *association reaction* forming the quantity $S_1 Y$ actively, and ε_2 (in units of reciprocal time) is the rate constant controlling the *dissociation reaction* forming the species S_1 and Y passively. At the input side of the mem-

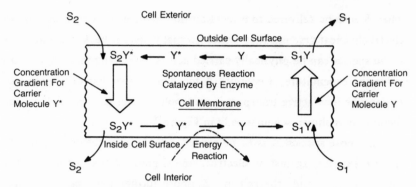

Figure 3-2 Schematic representation of mechanism by which active transport may be accomplished across a cell membrane.

brane, activated enzyme catalysis together with the great affinity that S_1 has for Y causes the rate of association to exceed greatly the rate of dissociation of the quantity $S_1 Y$, so equation [3-63] tends to be driven to the right. Note that a similar equation could be written for active transport by directional sites (c.f., Chapter 2, pg. 161), where Y would now represent a binding site along the pore, rather than a carrier molecule; and the rate constant for the formation of $S_1 Y$ at the binding site would again be much higher than the rate constant for the dissociation of $S_1 Y$, so the reaction would go essentially to the right at each binding site.

2) The catalyzing enzyme, $E_1 - P_i$ is believed to be "activated" by an energy-consuming phosphorylation reaction that is driven by the breakdown of the high-energy compound Adenosine Tri-Phosphate (ATP), which is considered further in Chapter 5. Thus,

$$E_1 + ATP \quad \xrightleftharpoons{\text{Electrolytes}} \quad E_1 - P_i + ADP^- \qquad [3\text{-}64]$$

where ADP^- represents Adenosine Di-Phosphate, the dephosphorylated ion of ATP, P_i represents inorganic phosphate, and, as indicated, the reaction appears to be dependent upon the presence of certain key electrolytes (see later).

3) Within the cell membrane itself, several things may now happen, most of which are still conjectural to the extent that the *exact* events taking place are still uncertain and subject to debate (Kaplan, 1985). Nevertheless, $S_1 Y$ *somehow* gets from the inside cell surface to the outside. Figure 3-2 illustrates one proposed mechanism, wherein a concentration gradient exists for carrier molecule Y (and for $S_1 Y$ as well), such that the movement of $S_1 Y$ is thus driven from the inside cell surface of the membrane to the outside cell surface of the membrane. All of the earlier discussions related to electrochemical diffusion and mass transport would then apply as well to the movement of species $S_1 Y$ within the cell membrane, if this, indeed is how it gets transported; and the appropriate dif-

fusion and mass transport coefficients could be derived to quantify the flux of $S_1 Y$ to the outside cell surface.

Another proposed mechanism for the transport of $S_1 Y$ is similar to a "revolving door" or "candy-ball machine." Here, $S_1 Y$ is assumed to be facing "in" to the cell interior while reactions [3-63] and [3-64] are taking place, and then the complex somehow "pivots" (like turning the crank on a candy-ball machine or rotating a revolving door), transforming its orientation from facing "in" to facing "out" to the cell exterior. Other mechanisms have been proposed and are still under investigation, but the end result is that the movement of $S_1 Y$ terminates at the lining of the outside cell surface.

4) Reaching the outside cell surface, the species $S_1 Y$ spontaneously dissociates according to the unidirectional reaction:

$$S_1 Y \xrightarrow[\text{spontaneous}]{\varepsilon_3} S_1 + Y \qquad [3\text{-}65]$$

where ε_3 (in units of reciprocal time) is the rate constant controlling this dissociation. The reaction at the outside cell surface tends to be unidirectional to the right for at least two plausible reasons: (i) presumably, all the molecules of Y reaching this point are *already* carrying S_1 with them (i.e., there are no, or few "free" carrier molecules available for an association reaction to occur along the outside cell surface); and (ii) once the dissociation has occurred at the outside cell surface, carrier molecule Y undergoes an activated-enzyme-catalyzed reorganization to form a new species, Y^*, which, unlike Y, has virtually no affinity at all for S_1. This "reorganization reaction" takes place at a rate much greater than the association rate constant for $S_1 Y$, so at the outside cell surface $S_1 Y$ is basically only dissociating.

5) Further assisting the dissociation of $S_1 Y$ at the outer cell surface is the fact that the *enzyme*, $E_1 - P_i$ that catalyzes the association reaction is *also* gone. This is because at this location, $E_1 - P_i$ undergoes, itself, an electrolyte-dependent conversion to a *new* enzyme, $E_2 - P_i$, which, in fact,

turns out to be the very activated-enzyme that drives the reaction converting Y to Y^*. Thus, at the outer cell surface, we have:

$$E_1 - P_i \xrightarrow[\quad Y \quad]{\text{Electrolytes}} E_2 - P_t \xrightarrow{\quad\quad} Y^* \qquad \text{[3-66]}$$

To summarize, S_1Y releases S_1 to the exterior of the cell and leaves behind a new carrier molecule Y^*. The facts that S_1Y is saturated when it arrives at this point, that $E_1 - P_i$ catalyzing the formation of S_1Y disappears here, and that the conversion of Y to Y^* (where the latter has virtually *no* affinity for S_1) occurs much faster than the formation of S_1Y, all act to drive reaction [3-65] virtually entirely to the right. We shall get back to Y^* later when we discuss facilitated carrier-mediated transport.

Michaelis-Menten Kinetics

From the above considerations, one can write an equation for the rate of accumulation of species S_1 in the cell membrane, i.e.,

$$\frac{\delta[S_1Y]}{\delta t} = \varepsilon_1[S_1][Y] - (\varepsilon_2 + \varepsilon_3)[S_1Y]$$

$$= \begin{array}{c}\text{Rate of} \\ \text{Formation}\end{array} - \begin{array}{c}\text{Rate of Dissociation at} \\ \text{inside and outside surfaces}\end{array} \qquad \text{[3-67]}$$

In the steady state, this rate will be equal to zero, so that:

$$\varepsilon_1[S_1][Y] = (\varepsilon_2 + \varepsilon_3)[S_1Y] \qquad \text{[3-68]}$$

If $[Y_m]$ designates the *total* concentration of carrier molecules in the membrane, then, at any given time, those available to undergo the association reaction at the inside surface will be equal to:

$$[Y] = [Y_m] - [S_1Y] \qquad \text{[3-69]}$$

Substituting equation [3-69] into equation [3-68], and rearranging, one obtains:

$$[S_1 Y] = \frac{[S_1][Y_m]}{K_m + [S_1]} \text{, moles per liter,} \qquad \text{[3-70]}$$

where:

$$K_m = \frac{\varepsilon_2 + \varepsilon_3}{\varepsilon_1},$$

having units of moles per liter, is called the dissociation constant, or Michaelis constant, or "Affinity Factor" of the active transport process because, for a given solute, the affinity of carrier molecule Y for species S_1 is *inversely* proportional to K_m. The affinity factor thus depends upon the chemical properties of the carrier and solute.

The rate of change of the concentration of species S_1 in the cell interior due to active transport *out* of the cell is now given by $\varepsilon_3[S_1 Y]$ (c.f., equation [3-65]), so, from equation [3-70], we have:

$$-\frac{\delta[S_1]_2}{\delta t} = \varepsilon_3[S_1 Y] = \frac{\varepsilon_3[S_1]_2[Y_m]}{K_m + [S_1]_2} = \frac{\text{moles per liter}}{\text{per unit time}} \qquad \text{[3-71]}$$

Multiplying equation [3-71] by the molecular weight, MW, of S_1, dividing by its density, ρ_S, and introducing again the characteristic length "r" $= \frac{A}{\delta}$, one may define an active transport velocity, or permeability coefficient as:

$$P_m = \frac{\varepsilon_3[S_1]_2[Y_m](MW)A}{(\rho_S \delta)(K_m + [S_1]_2)} \text{ cm/sec} \qquad \text{[3-72]}$$

Note from above that the *maximum* possible rate of transport occurs when K_m is zero -- all other things being the same -- for which the affinity of carrier molecule Y for species S_1 is infinite (i.e., $\varepsilon_2 + \varepsilon_3 = 0$ and $\varepsilon_1 \to \infty$), meaning that there is an infinite tendency for the *association* reaction to occur and no tendency for *dissociation* to transpire once S_1 and Y have "mated". That being the case, *all* of carrier molecules Y will soon become saturated, so one observes that the *maximum* rate of disappearance of S_1 from the cell interior is limited to $\varepsilon_3[Y_m]$, which corresponds to a maximum permeability coefficient $\varepsilon_3[Y_m](MW)A/(\rho_S\delta)$. Letting this maximum transport velocity be equal to P^*_m, we get, from equation [3-72], with subscript 2 implied:

$$\frac{P_m}{P^*_m} = \frac{[S_1]}{K_m + [S_1]} \qquad [3\text{-}73]$$

Note that when the concentration of S_1 inside the cell is equal to K_m, $P_m = \left(\frac{1}{2}\right) P^*_m$, or the *actual* velocity of transport is one-half the maximum. Thus, when $K_m = 0$, the rate of transport is maximum, when K_m is equal to the concentration of S_1 inside the cell, the rate of transport is one-half maximum, and when K_m is infinite (equation [3-73]), the rate of transport is zero -- verifying that the affinity of carrier molecule Y for species S_1 is inversely proportional to K_m. Upon rearranging equation [3-73], one gets equation [3-74] or equation [3-75]:

$$K_m = \frac{\varepsilon_2 + \varepsilon_3}{\varepsilon_1} = [S_1]\left\{\frac{P^*_m}{P_m} - 1\right\} \qquad [3\text{-}74]$$

$$\frac{[S_1]}{K_m} = \frac{\left[\dfrac{P_m}{P^*_m}\right]}{1 - \left[\dfrac{P_m}{P^*_m}\right]} \qquad [3\text{-}75]$$

Equation [3-75] is called the Michaelis-Menten Equation for active transport, and it is plotted in Figure 3-3. Observe, when there is no substrate ($[S_1] = 0$), there is no reaction ($P_m = 0$). When there is a paucity of substrate ($[S_1]$ very small compared with K_m), solving the Michaelis-Menten Equation for P_m, i.e., from equation [3-73] we have, for very small $[S]_1$ relative to K_m:

$$P_m \simeq \frac{P^*_m}{K_m}[S_1], \qquad [3\text{-}76]$$

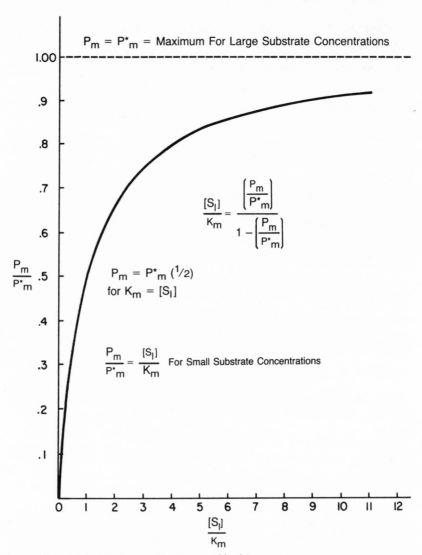

Figure 3-3 Michaelis-Menten active transport kinetics.

such that the steady-state velocity of the reaction increases approximately linearly with the concentration of the substrate at a rate determined by the value of $(\varepsilon_3[Y_m]/K_m)$. The more carrier molecules that are present $([Y_m])$, the greater their affinity for S_1 (i.e., for large $\dfrac{1}{K_m}$), and the more spontaneous the dissociation reaction at the outer membrane surface (ε_3), the greater the *rate* at which P_m will increase with $[S_1]$. On the other hand, when the substrate concentration is so large compared with K_m that K_m can effectively be neglected in equation [3-73], then $P_m \simeq P^*_m$, the velocity of the reaction will be approximately at its maximum, and independent of the concentration of $[S_1]$. It will still, however, depend linearly on the concentration of carrier molecules, at a rate determined by ε_3, since:

$$P^*_m = \varepsilon_3 [Y_m] \frac{(MW)A}{(\rho_S \delta)} \qquad [3\text{-}77]$$

Again, the more carrier molecules that are present, the greater the *maximum velocity* at which the transport will progress at high substrate concentrations. Moreover, for $[S_1]$ large, the limiting factor will be $[Y_m]$ because essentially all of the carrier molecules present will be in their bound form, $S_1 Y$.

If we assume that S_1 is disappearing (i.e., its concentration is *decreasing*) due to active transport out of region 2 in Figure 3-1, then we may write, from equations [3-71] and [3-72]:

$$-\frac{\delta[S_1]_2}{\delta t} = \varepsilon_3 [S_1 Y] = \frac{P_m(\rho_S \delta)}{(MW)A} \qquad [3\text{-}78]$$

where the minus sign has been inserted to show that the concentration $[S_1]_2$ is decreasing due to a positive efflux of S_1 (i.e., $+P_m$) out of the interior of the cell. Based on the preceding discussion, let us examine three special cases of equation [3-78]:

Case 1: Large substrate concentration:

$$P_m = P^*{}_m = \varepsilon_3 [Y_m] \frac{(MW)A}{(\rho_S \delta)}.$$

Then:

$$-\frac{\delta [S_1]_2}{\delta t} = \varepsilon_3 [Y_m] = \text{a constant, } a_m \qquad \text{[3-79]}$$

This is called a *zero'th order reaction*, in that, for a given concentration of carrier molecules, the rate of disappearance of $[S_1]_2$ is constant -- independent of the amount $[S_1]_2$ present at any given time (since there is so *much* of it); but still dependent upon the concentration of carrier molecules present and the rate of dissociation of the carrier-molecule-substrate complex, $S_1 Y$ at the outer surface of the membrane. One integration of equation [3-79] yields:

$$\Delta [S_1]_2 = a_m \Delta t \qquad \text{[3-80]}$$

where $\Delta [S_1]_2 = [S_1]_{2,\text{time } t_1} - [S_1]_{2,\text{time } t_2}$ is the amount of $[S_1]_2$ (in moles per liter) transformed into the complex $[S_1 Y]$ during the time interval $\Delta t = t_2 - t_1$; and we note that this amount varies *linearly* with time in region 2 of Figure 3-1, at a rate set by a_m.

Case 2: Small substrate concentration, $P_m = \dfrac{P^*{}_m}{K_m} [S_1]_2$. Then:

$$-\frac{\delta [S_1]_2}{\delta t} = \varepsilon_3 [Y_m] \frac{[S_1]_2}{K_m} = \frac{a_m}{K_m} [S_1]_2 = b_m [S_1]_2 \qquad \text{[3-81]}$$

This is called a *first order reaction*, in that, for a given concentration of carrier molecules, the rate of disappearance of $[S_1]_2$ is linearly proportional to the amount of substrate present at any given time. Note that b_m, like ε_2 and ε_3, has dimensions of $\dfrac{1}{t}$, and represents a rate constant for equation [3-81], i.e.,

$$b_m = \frac{a_m}{K_m} = \frac{\varepsilon_1 \varepsilon_3}{\varepsilon_2 + \varepsilon_3} [Y_m], \qquad \text{[3-82]}$$

so that in this case, the rate of disappearance of S_1 depends directly on the amount of carrier molecules present, and the rates of association and dissociation of the complex $S_1 Y$. One integration of equation [3-81] yields:

$$-\int_{[S_1]_{2,t_1}}^{[S_1]_{2,t_2}} \frac{\delta[S_1]_2}{[S_1]_2} = \ell n \; \frac{[S_1]_{2,t_1}}{[S_1]_{2,t_2}} = \int_{t_1}^{t_2} b_m \delta t = b_m (t_2 - t_1),$$

or,

$$b_m = \frac{1}{\Delta t} \; \ell n \; \frac{[S_1]_{2,t_1}}{[S_1]_{2,t_2}} = \frac{\varepsilon_1 \varepsilon_3}{\varepsilon_2 + \varepsilon_3} [Y_m] \qquad [3\text{-}83]$$

where $[S_1]_{2,t_i}$ $(i = 1,2)$ is the concentration of substrate S_1 in region 2 of Figure 3-1 at time t_i, starting with a concentration $[S_1]_{2,t_1}$ at time t_1 and ending with a concentration $[S_1]_{2,t_2}$ at time t_2.

Case 3: For *arbitrary* intermediate substrate concentrations, equation [3-71] can be integrated directly to yield:

$$K_m \; \ell n \; \frac{[S_1]_{2,t_1}}{[S_1]_{2,t_2}} + \Delta[S_1]_2 = \varepsilon_3 [Y_m] \Delta t \qquad [3\text{-}84]$$

This is a more complicated relationship involving *both* first-order and zero-order reaction terms.

Using equation [3-72], one can define an "Active Transport Mass Diffusivity," $\alpha = P_m \delta$ (c.f., equation [3-31]. Then,

$$\alpha = \frac{\varepsilon_3 [S_1][Y_m](MW)A}{\rho_S (K_m + [S_1])} \qquad [3\text{-}85]$$

Moreover, also from equation [3-72], one may further define an active transport "flux" as the product $P_m \rho_S$, so that:

$$\Phi = P^*_m \; \rho_S \frac{[S_1]_{\text{inside}}}{K_m + [S_1]_{\text{inside}}} \qquad [3\text{-}86]$$

And finally, since species S_1 is being transported out of the cell (i.e., *from* region 2) *against* an assumed electrochemical gradient that would tend to drive it back the other way, work must be done to achieve this result. Thus, a continuous supply of energy is required for active transport, this energy being associated with the phosphorylation reaction [3-64] that makes available an "activated" enzyme to catalyze reaction [3-63] removing S_1 from region 2; and, with a subsequent hydrolysis reaction (see below) that provides a continuous supply of carrier molecules Y to affect the transport.

The amount of energy required, per mole of species transported, per unit area of membrane, per second, can be determined from the concept of electrochemical potential difference defined earlier (equation [3-52]) and from equation [3-86] for the active transport flux of S_1. That is, if the total transmembrane potential energy difference, per mole of ions, tending to drive S_1 *into* the cell as the result of concentration gradient potentials and transmembrane electrical potential differences is $+ \Delta\mu_c$ as given by equation [3-52], and, if the active ion efflux is $\dfrac{\Phi}{MW}$ moles per unit area, per unit time as defined from equation [3-86], then the *active transport power* required to affect such transport actively is given by:

$$\begin{array}{l}\text{Active Transport}\\ \text{Power}\end{array} = -\Delta\mu_c\left(\frac{\Phi}{MW}\right) = \begin{array}{l}\text{Transport work per mole}\\ \text{of ions times moles of}\\ \text{ion transported per}\\ \text{unit area, per unit time.}\end{array} \qquad \text{[3-87]}$$

We shall get back to the work done by active transport in a later section (pg. 239). But first, let us continue to discuss the events depicted in Figure 3-2. To this point, carrier molecule Y has picked up a particle (or more) of species S_1 at the inside surface of the cell membrane, carried it across to the outside surface of the membrane, released it to the cell exterior, and, has been converted to the new carrier molecule Y^* (equation [3-66]). Meanwhile, enzyme E_1, utilizing the energy released from the breakdown of ATP, has undergone conversions that result in the phosphorylated activated complex $E_2 - P_i$, and a negatively charged ion of ADP, i.e., ADP^-.

Facilitated Diffusion

There now follows a hydrolysis reaction wherein, in the presence of certain electrolytes (see later), $E_2 - P_i$ and ADP^- react with H_2O to form the de-phosphorylated enzyme, E_2, plus phosphoric acid, H_3PO_4, plus Adenosine Di-Phosphate, ADP:

$$E_2 - P_i + ADP^- + H_2O \xrightarrow{\text{Electrolytes}} E_2 + H_3PO_4 + ADP \qquad \text{[3-88]}$$

It turns out that the enzyme E_2, in turn, catalyzes a subsequent reaction wherein carrier molecule Y^* -- which, incidentally, happens to have a great affinity for species S_2 -- combines with S_2 to form the complex $S_2 - Y^*$, i.e., at the outside surface of the membrane:

$$S_2 + Y^* \underset{\substack{\varepsilon_5 \\ \text{spontaneous} \\ \text{dissociation}}}{\overset{\substack{\text{catalyzed} \\ \text{association} \\ \varepsilon_4;\ E_2}}{\rightleftarrows}} S_2 Y^* \qquad \text{[3-89]}$$

where, again, ε_4 and ε_5 are rate constants controlling the respective association and dissociation reactions. Now, recall that S_2, driven by a favorable electrochemical gradient, is trying to get *into* the cell *anyway* (but perhaps more slowly than metabolic requirements would prefer). Thus, carrier molecule Y^* comes along to *assist* (not *affect*) this transport, which would take place anyway, on its own, but more slowly. The enzyme catalysis of reaction [3-89], the affinity of Y^* for S_2, and the favorable electrochemical gradient, all act quite strongly to drive the reaction almost entirely to the right.

This process, where a carrier molecule is also involved, but where the ion being transported is also moving in the direction of a favorable electrochemical driving force (rather than opposite to it as in active transport) is called *facilitated* diffusion. Thus, carrier molecule Y^* *facilitates* the transport of species S_2 by combining with it at the outside cell surface boundary, diffusing down a concentration gradient -- or, by whatever other means, getting to the inside cell surface boundary, and then spontaneously releasing S_2 into the cell interior:

$$S_2 Y^* \xrightarrow[\text{spontaneous}]{\varepsilon_6} S_2 + Y^* \qquad \text{[3-90]}$$

Having accomplished this mission, Y^* finally undergoes a transformation back to Y to recycle the entire sequence of events which actively carries another particle (or more) of S_1 out of the cell. This last transformation of Y^* into Y is an enzyme catalyzed endergonic (energy-absorbing) reaction that utilizes some of the energy released in reaction [3-64], assuming only one ATP molecule is used up for every "turn" of the active transport pump. There is still considerable controversy on this issue, some investigators believing that a second ATP-reaction drives the conversion of Y^* to Y, so that two molecules of ATP are required per turn of the pump. In any case, the enzyme that drives the transformation is E_1, so to complete the process, we have a spontaneous conversion of E_2 back to E_1, so that:

$$E_2 \xrightarrow[Y^*]{\text{spontaneous}} \quad \xrightarrow[]{E_1} \quad Y \qquad \text{[3-91]}$$

In summary then, equations [3-63, 64, 65, 66, 88, 89, 90, and 91] describe the cycle of events that is currently believed to represent a typical active transport process across a cell membrane. The entire cycle ultimately loses energy, which must be replaced by the nutrients delivered to the cell (see Chapters 4 and 5).

Facilitated Kinetics

In facilitated diffusion, $K^*_m = \dfrac{\varepsilon_5 + \varepsilon_6}{\varepsilon_4}$ is the same for transport in either direction across the membrane, and the *net* migration rate from side 1 to side 2 of the membrane is given by the mass flux relationship (see equation [3-86]):

$$\Phi = P^*_f \rho_S \left\{ \frac{[S_2]_{outside}}{K^*_m + [S_2]_{outside}} - \frac{[S_2]_{inside}}{K^*_m + [S_2]_{inside}} \right\}, \quad [3\text{-}92]$$

where P^*_f is the maximum facilitated transport permeability coefficient, and K^*_m is as defined above. Equation [3-92] can be manipulated into the form:

$$\Phi = P^*_f \rho_S K^*_m \left\{ \frac{[S_2]_{outside} - [S_2]_{inside}}{(K^*_m + [S_2]_{outside})(K^*_m + [S_2]_{inside})} \right\} \quad [3\text{-}93]$$

In equation [3-93], if K^*_m is large compared with the concentrations of species S_2 inside or outside of the cell (i.e., if the carrier molecule Y^* is far from being saturated, and/or, if the carrier molecule has a relatively low affinity for S_2), then the equation approaches the result:

$$\Phi = \frac{P^*_f \rho_S}{K^*_m} \left([S_2]_{outside} - [S_2]_{inside} \right) \quad [3\text{-}94]$$

Dividing and multiplying equation [3-94] by membrane thickness, δ, we note that:

$$\frac{[S_2]_{outside} - [S_2]_{inside}}{\delta} = - \frac{([S_2]_2 - [S_2]_1)}{\delta} = - \frac{\Delta[S_2]}{\delta},$$

which, in the arbitrary three-dimensional case, can be generalized to $- \nabla[S_2]$. Then, defining a "facilitated transport binary effective diffusion coefficient," or, a facilitated diffusivity,

$$D^*_S = \frac{P^*_f \rho_S \delta}{K^*_m},$$

equation [3-94] reduces to:

$$\Phi = - D^*{}_S \nabla [S_2] \qquad\qquad [3\text{-}95]$$

Observe the similarity in form between equation [3-95] and equation [3-22] for ordinary diffusion. In fact, since facilitated diffusion operates in the same direction as ordinary diffusion, one may combine the two equations to yield the *total* flux when the latter is being "assisted" by the former. In other words, when ordinary diffusion is being assisted by facilitated diffusion, then:

$$\Phi_{\text{total}} = - (D_S + D^*{}_S)\nabla[S], \text{ or,}$$

$$\Phi_{\text{total}} = - D_S\left(1 + \frac{D^*{}_S}{D_S}\right)\nabla[S] \qquad [3\text{-}96]$$

The ratio,

$$\frac{D^*{}_S}{D_S} = \frac{\varepsilon_4\varepsilon_6[Y^*{}_m]A\sqrt{2k_a(MW)}}{(\varepsilon_5 + \varepsilon_6)A_S\lambda\sqrt{RT}}$$

is a dimensionless "facilitation factor" which is a measure of by how much ordinary diffusion is "amplified" as a result of facilitation assistance. As was indicated in Chapter 2, pgs. 157-159, (in the section entitled, "Intramural Facilitated Transport"), this amplification factor is around 67 for the oxyhemoglobin system, where Y^* is the hemoglobin carrier molecule and S_2 is oxygen. The facilitation factor can be as high as 150 when the concentrations of S_2 are large compared with $K^*{}_m$ and/or if the carrier molecule has a great affinity for S_2. In the latter case, equation [3-93] approaches the result:

$$\Phi = P^*{}_f \rho_S K^*{}_m\left\{ \frac{1}{[S_2]_{\text{inside}}} - \frac{1}{[S_2]_{\text{outside}}} \right\} \qquad [3\text{-}97]$$

By contrast to equation [3-94], where, at low degrees of carrier saturation, solutes with the *greatest* affinity (smallest $K^*{}_m$) will be transported most rapidly, we see in the case of equation [3-97] that at high degrees of carrier saturation, solutes with the *lowest* affinities (larger K^*_m's) will be transported most rapidly. Moreover, the form of equation [3-96] suggests that, in laboratory experiments, if one measures diffusion-type kinetics, this does not necessarily exclude the possibility of carrier transport or chemical mediation of transport in the system.

Passive Gradients Produced by Active Transport

Several times thus far we have cited examples where the "spin-off" of one type of transport process will lead to other types of mass transport that did not exist previously. Still another example of this involves the consequences of active transport. For instance, suppose the species being transported from one side of a membrane to the other by active processes happens to be positively charged (such as a sodium ion). Then there will soon develop a situation where there is an unequal distribution of electrical charge across the membrane. Consequently, this will produce an electromotive force tending to drive negatively charged ions (e.g., chlorine) passively across the membrane to achieve electroneutrality. Thus we have the result that *active* transport of sodium carries along with it *passive* transport of chlorine. And, going one step further, the transport of *both* of these species across the membrane generates an osmotic pressure gradient that may not have existed previously; so water, too, now moves across the membrane from one side to the other. These considerations have particular relevance in terms of the generation of action potentials in the physiologic system (c.f., Chapters 5 and 6), and in terms of the establishment of a resting membrane potential in excitable tissue.

Generation of a Resting Membrane Potential in Excitable Tissue

The Sodium-Potassium Electrogenic Pump

One prime example of active transport in the human physiologic system is the so-called Sodium-Potassium Pump, wherein sodium ions (three, to be exact) are "pumped" by a mechanism similar (if not identical) to that described by equations [3-63] to [3-87] *from* inside the cell to the outside, and potassium ions (two, to be exact) are "pumped" by a mechanism similar (if not identical) to that described by equations [3-88] to [3-97] from outside the cell *to* the inside, with each "turn of the crank". This pumping action may be for the purpose of polarizing the cell membrane, as is the case for nerve cells and muscle cells; or, it may be for the purpose of establishing an osmotic pressure gradient to keep the extracellular fluid from penetrating the cell and causing it to swell (and, ultimately, to burst), as is the case for red blood cells; or, it may be for the purpose of enhancing passive transport mechanisms, as is the case in the kidneys and the intestines.

In the case of the Sodium-Potassium electrogenic pump (hereinafter referred to simply as the "Sodium Pump"), S_1 in equation [3-63] represents three sodium ions and Y a globulin carrier molecule having a molecular weight of around 95,000. In equation [3-64], E_1 represents a glycoprotein (the amino sugar *hexosamine* combined with a globular protein) enzyme of molecular weight 55,000, and the electrolytes involved are sodium (Na^+) and magnesium (Mg^{++}). The inorganic phosphate, P_i, which "activates" E_1 is actually split off from the ATP complex by the enzyme action of $Y + E_1$, acting as a *Sodium-Potassium-ATPase* in the presence of Na^+ and Mg^{++}. The magnesium is believed to facilitate the process by combining with an intermediate form of the ATP complex during the dephosphorylation steps of the overall reaction. In equation [3-65] (indeed, for *all* of the dissociation reactions), note that the sodium release is spontaneous, it requires no energy, and, to the contrary, energy is *released* during these and the subsequent potassium-binding processes.

In equation [3-66], the magnesium electrolyte catalyzes the conversion of $E_1 - P_i$ to the alternate configuration $E_2 - P_i$; and in equation [3-88], the electrolytes involved in the hydrolysis reaction are magnesium again, but this time, potassium instead of sodium. Thus, Na^+ catalyzes its own efflux reactions, which are associated with the phosphorylation reactions [3-63] - [3-66], and potassium catalyzes its own influx reactions, which are associated with the hydrolysis reactions [3-88] - [3-90]. This idea that the actual substrate involved in a chemical reaction catalyzes its own "fate", if you will, is addressed further in Chapter 5, and underlies a basic principle of metabolism. This principle essentially asserts that the enzymes necessary to catalyze any given biochemical reaction are manufactured as needed only when the substrates involved in that reaction are present. This allows the organism to economize on the necessity for storing the incredibly large number of enzymes that are required to catalyze all of the reactions associated with metabolism. Instead, the physiologic system only stores the smaller number of "ingredients" necessary to *make* the enzymes, and the "recipes" (i.e., DNA-formula) for each one. It then synthesizes them, as necessary, depending on what substrates are present. More on this in Chapters 4 and 5.

In equation [3-89], S_2 represents two potassium ions, and Y^* a converted form of the same carrier molecule Y. Similarly, E_2 is a converted form of E_1, and, again, the dissociation reaction [3-90] is spontaneous and exergonic (energy-releasing). Thus, all of the energy that drives the Sodium pump is needed for reactions that take place at the *inside* of the cell membrane, i.e., reactions [3-63, 64 and 91 that converts Y^* back to Y]; and therefore, this energy is derived from *intracellular* sources of ATP. Indeed, when the Sodium pump of a cell is *blocked* by a variety of mechanisms, its oxygen consumption (used primarily to manufacture ATP -- see Chapter 5) correspondingly drops as much as 30 to 40% (Hall and Baker, 1977), suggesting that a considerable portion of a cell's metabolism is devoted to driving the Sodium pump (see later).

One of the additional peculiarities of this active transport mechanism is the fact that it normally transports three sodium ions to the outside of the membrane for every two potassium ions that it transports to the inside (Guyton, 1981). Exactly *how* it does this is still not entirely clear, but speculation has it that since sodium is smaller than potassium (see Table 3-1) -- having, in fact, a little less than $\frac{2}{3}$ of its atomic weight and just $\frac{3}{4}$ of its mean ionic radius -- carrier molecules involved in the active transport process can bind, or have binding sites that can geometrically accommodate 3 sodium ions for every 2 potassium ions. This gives a net transfer of 50% more positive charges outward through the membrane than inward, and, therefore, creates an *electrical potential* across the membrane, as well as a chemical concentration gradient. For this reason, the Sodium-Potassium-Electrolytic-Pump is also called an *Electrogenic* pump because, whenever it pumps, it creates negativity in the Intracellular Fluid (ICF) and positivity in the Extracellular Fluid (ECF).

Resting Membrane-Potential Equilibration

As mentioned in Chapter 1, when properly stimulated, "excitable tissue" is characterized as having the ability to generate action potentials, i.e., waves of membrane depolarizations. *How* this is accomplished is addressed further in Chapter 5, but the process *starts* from an equilibrium configuration involving a resting polarized membrane. The polarization, in turn, is achieved by a balance of passive and active transport mechanisms as follows:

1) Membranes of excitable tissue (particularly nerves and muscles) typically contain 100-200 Sodium-Potassium pumps per square micrometer of membrane surface. These are activated by increases in intracellular sodium concentration -- such as those that result from Gibbs-Donnan equilibrium effects discussed earlier, or from membrane permeability disturbances due to excitation (to be discussed later). In fact, studies show that pump activity increases as the *cube* of ICF $[Na^+]$, so that even a *slight* excess buildup of sodium ions inside the cell, such as might be caused by Gibbs-Donnan phenomena involving impermeable ICF proteins, serves to activate the pump very strongly (Guyton, 1981). Again, we see that the substrate, itself, determines its own subsequent disposition.

2) When the cell's sodium pumps are activated, they begin to lower the sodium concentration in the interior of the cell, while increasing the potassium concentration inside the cell. As this is happening, of course, the exact reverse is occurring in the extracellular fluid surrounding the cell membrane. Thus, the electrolytic pump *establishes,* actively, a concen-

Table 3-1

Properties Of Key Ions Involved In

The Establishment Of A Resting Membrane Potential

In Excitable Tissue

Species [S]	Atomic Weight	ECF $[S]_1$ millimoles liter	ICF $[S]_2$ millimoles liter	Mean Ionic Radius Angstroms	Resting Membrane Permeability Relative To Potassium	Valence
Na^+	23	145 – 160	7 – 12	0.95 – 0.98	0.02	+1
K^+	39	4	150 – 155	1.33	1.00	+1
Ca^{++}	40	1.2	0.0001 – 0.2	0.94 – 0.99	N/A	+2
Mg^{++}	24.3	0.75 – 1.25	25 – 26	0.65	N/A	+2
Cl^-	35.5	120 – 155	3.8	1.81	2.00	–1

tration gradient across the cell membrane, such that Na^+ *wants* to move *into* the cell and K^+ *wants* to move out of the cell. Although neither potassium nor sodium ions can move easily through the cell membrane material, itself, each is small enough to pass with ease through a membrane pore, as indicated by the data given in Table 3-I. Recall, however, that large protein molecules, having a predominantly negative valence, *cannot* pass easily through the cell membrane by *any* means, i.e., they are effectively "trapped" inside the cell.

3) At a certain point in this process, the build-up of potassium in the ICF (and depletion in the ECF) generates such a strong concentration gradient that, despite the action of the pump, K^+ starts to diffuse slowly out of the cell through the membrane pores. As it does so, it leaves behind still more ICF negativity. This strong negativity causes K^+ to be electrostatically "stuck" to the opposite (outside) side of the membrane. Sodium also clings to this side of the membrane as a result of the concentration gradient generated by the pump; and large protein anions accumulate on the inside of the membrane, held there by both a concentration gradient for S^{-z} and the electrostatic "pull" of K^+ and Na^+. Because of certain selective membrane permeability characteristics, Na^+ does not "leak" quite as easily back into the cell as does K^+ out of the cell at this time. The net result, then, is that the membrane capacity gets "charged up", with the outside more positive and the inside more negative.

4) This whole charging process will not continue indefinitely, because the accumulating positivity on the outside and negativity on the inside will soon begin to impede (by electrostatic repulsion) the further movement of potassium ions. That is, the *electrical* force generated by charging up the membrane, and the action of the *sodium-potassium pump*, will counter the *concentration* gradient forces generated by the pump, itself, and, eventually, the whole process will move asymptotically towards a self-limiting equilibration point. At that point, there will be no further *net* motion of potassium, although specific ions will continue to move randomly in both directions across the membrane. At equilibrium, how-

ever, the flux *in* (electrolytic pump plus electrostatic repulsion) will be exactly balanced by the flux *out* (concentration gradient), so that the *net* flux will be zero.

What we have described here is an example of electrochemical diffusion such as is governed by the Nernst equation [3-54] derived earlier. However, the actual physiologic situation is much more complicated because there, we are dealing with *more* than just one ionic species, such as potassium. Indeed, we are dealing with several (including magnesium and calcium), but the three of immediate interest are sodium, potassium, and chlorine. In order to deal with these in a quantitative sense, we must modify the Nernst equation as described below.

The Goldman Equation

Mechanisms associated with active transport and selective membrane permeability eventually lead to an equilibration of ionic concentrations at the values shown in Table 3-I, columns 3 and 4 for an excitable nerve or muscle membrane at rest (i.e., undisturbed). These ionic distributions charge up the cell membrane, generating a transmembrane potential difference or voltage called the *resting potential*, ΔB_m, or membrane equilibrium potential of the excitable cell. Now, the cell membrane is freely permeable to sodium, potassium, and chlorine, so it would seem reasonable to expect the *actual* resting membrane potential to be somewhere between the extremes of all equilibrium potentials predicted by the Nernst Equation [3-54] applied separately to each of these three ionic species.

In 1943, Goldman developed flux equations that lead to a reasonable estimate of the actual resting membrane potential if two basic assumptions are imposed: (a) that the total current due to passive flux of all ions is zero; and (b) that the electric field across the membrane is constant (i.e., that all transport follows the principles of *constant field theory*). Although neither one of these assumptions is actually valid, they are a close enough approximation to allow the so-called *Goldman*

Equation to work *most* of the time. Thus, Goldman's modification of the Nernst equation for more than one ionic species takes the form:

$$\Delta B_m = \frac{RT}{\overline{F}Z} \ \ell n \ \frac{P_K[K^+]_1 + P_{Na}[Na^+]_1 + P_{Cl}[Cl^-]_2}{P_K[K^+]_2 + P_{Na}[Na^+]_2 + P_{Cl}[Cl^-]_1} \quad [3\text{-}98]$$

where: P_K, P_{Na}, and P_{Cl} are, respectively, the permeability coefficients of the membrane to potassium, sodium and chlorine; $[S]_j$ are the concentrations (or, more accurately, the activities) of species S outside $(j = 1)$ and inside $(j = 2)$ the cell, respectively; and the other quantities are as previously defined.

The Goldman equation [3-98] suggests that the transmembrane potential of the cell is a function of the concentrations of the various ions, and of the permeability of the membrane to these ions (note from Table 3-I that potassium permeability is about 50 times that of sodium, so that potassium permeability tends to dominate the establishment of a resting membrane potential). Note, too, that chlorine permeability is the highest of all, so that it tends to equilibrate at the values shown because of its affinity for sodium, and due to the passive gradients generated by the sodium-potassium active transport pump (see earlier discussion). From the means of the data of Table 3-I, and with $R = 8.31$, $T = 310°K$, $\overline{F} = 96,500$, and $Z = +1$ (since the sodium pump establishes a *net* transfer of $+1$, exchanging 3 sodium ions for every 2 potassium ions), the Goldman equation yields:

$$\Delta B_m = \frac{(8.31)(310)}{96,500} \ \ell n \ \frac{P_K[4] + 0.02P_K[157.5] + 2P_K[3.8]}{P_K[152.5] + 0.02P_K[9.5] + 2P_K[137.5]}$$

$$[3\text{-}99]$$

$$\Delta B_m = - \ 0.08989 \ \text{volts, or, about} - 90 \ \text{millivolts}$$

This value for the resting membrane potential is not only in line with, but virtually exactly equal to those values actually measured in excitable tissue, and it predicts a negative ΔB_m by convention. That is, as already discussed (see equation [3-55] and corresponding text), it is a convention to refer to the electrical potential of the *inside* of the cell with respect to the outside (by letting $Z = +1$ rather than -1 above), and not vice versa. Thus, if the *inside* of a cell contains more negative charge than does the *outside* of the cell, the *transmembrane potential* is negative, and vice versa. The *more* negative is the *resting potential*, the more *hyperpolarized* is the cell membrane said to be. The *less* negative (towards zero) is the resting potential, the more *hypopolarized* is the cell membrane. Resting membrane potentials are always negative in both plant and animal cells, and can range from -10 to -100 millivolts, with the latter being about the value for some skeletal muscle tissue. Cardiac muscle cells have resting potentials in the range -70 to -90 millivolts, but certain kinds of heart, or cardiac tissue -- most notably the sino-atrial (SA) and Atrio-Ventricular (AV) nodes, have resting membrane potentials down in the -40 to -60 millivolt range.

Active Transport Energy Required to Maintain a Resting Potential

We can use equations [3-30] and [3-54] to get an idea of the energy required to maintain ΔB_m. That is, From [3-54]:

$$\overline{F}Z\Delta B_m = RT\,\ell n \,\frac{[S]_{outside}}{[S]_{inside}} \qquad \text{[3-100]}$$

and, from equation [3-30],

$$RT\,\ell n\,\frac{[S]_{outside}}{[S]_{inside}} = -W, \text{per mole} \qquad \text{[3-101]}$$

Thus, substituting equation [3-101] into equation [3-100], and, with $\Delta B_m = -0.090$, $\overline{F} = 96,500$, and $Z = +1$, we have,

$$W = -\,(96,500)\,(+1)\,(-0.090)\,= 8685 \text{ Joules per mole,}$$

where the positive sign reflects the fact that the sodium-potassium pump is doing work *on* the system, i.e., the ions are moving "uphill" in the same direction as that in which the mechanism is pumping. Using the conversion factor Joules $\times$ 0.239 = Calories, we see that 8685 Joules per mole = 2075.7 Calories per mole.

Now, the hydrolysis of a mole of ATP releases some 7800 ± 500 calories of usable "standard" free energy (see Chapters 4 and 5). Although it is still not clear whether each turn of the sodium-potassium pump utilizes one, or two molecules of ATP, let us assume for the time being that the pump exchanges two potassium ions for three sodium ions per single molecule of ATP hydrolyzed. Then there will be a *net* exchange of one mole of univalent ions for every mole of ATP hydrolyzed, suggesting that the efficiency of the electrogenic pump is somewhere around: $\frac{2075.7}{7800} = 0.266$, or, on the order of 26.6%. This value seems reasonable when compared with efficiencies of man-made engines, in general, and most other physical and physiological processes. For example, the thermal efficiency of steam turbines is around 17%, that of Otto (internal combustion) engines is between 20 and 27%, and the efficiency of Diesel engines rises to somewhere between 30 and 35%; while muscular performance under favorable circumstances achieves a mechanical efficiency approaching 20-25%.

Various estimates suggest that as much as 20% of resting oxygen consumption in mammalian muscle tissue may go entirely for the purpose of actively transporting sodium ions via the sodium pump (Lenchene, 1988). Moreover, the pump is powerful enough to transport sodium ions against concentration gradients as great as 20 to 1, and potassium ions against concentration gradients as great as 30 to 1 (Guyton, 1981). It is also worth noting that there have been *no* observed physiological conditions where the pump, per se, is *extrinsically* controlled by mechanisms affecting the various reactions or enzymes. Direct control of the pump, itself, is strictly intrinsic and limited to intracellular events -- particularly as they relate to sodium concentration. There may, however, exist extrinsic control in the form of factors that affect membrane *permeability* to $[S_l]$, temperature, pH, ATP synthesis, and so on. It is worth examining a few of these in particular, as we address some aspects of physiologic control of transport.

Physiologic Control of Transport

Substrate Control

We have already mentioned substrate control of active transport (see discussion of Michaelis-Menten equation [3-75] and $[Na^+]$ control of the sodium pump above), and will get back (in Chapters 4 and 5) to the role of the substrate, itself, in both the synthesis and activation of the enzymes that are required to catalyze its very disposition. In a more passive sense, however, substrates are also directly involved in the establishment of the concentration gradients that *drive* transport processes (equations [3-22], [3-29], [3-86], [3-95], and [3-96]). To illustrate this, consider the *Oxygen-Reserve Mechanism* in the cardiovascular system.

Recall (Chapter 2, pgs. 157-159) that normally, each 100 cc of arterial blood entering a capillary network contains approximately 19.6 cc of oxygen. Of these, 5 cc or so are removed on the average during routine metabolism (Chapter 4). Thus, the *arterio-venous O_2 difference*, $AV - \Delta O_2$ is equal to $\dfrac{5 \; m\ell \; O_2}{100 \; m\ell \; \text{Blood}} = 5$ vol. %.

During exertion, tissues use up oxygen at a higher rate. Since this lowers $[S]_2$, the concentration gradient driving O_2 *from* the blood *to* the cells increases, allowing the tissues to extract *more* oxygen from the blood, and leading to a widened arterio-venous-oxygen difference. Moreover, blood coming back *to* the lungs with less oxygen, by the same mechanism, is able to extract more oxygen from the alveoli, so the concentration $[S]_1$ of O_2 in the blood itself also tends to go up. The net result, known as the oxygen-reserve mechanism and based on the intrinsic metabolic rate of tissues, is that during exertion, *both* the concentration of oxygen in blood, $[O_2]_1$, increases, while the concentration of oxygen in tissues, $[O_2]_2$, decreases, so that the concentration gradient for O_2

transport to the tissues is greatly enhanced. Obviously, the reverse trend is manifest during periods of idleness.

The oxygen-reserve mechanism is particularly important in the coronary circulation which feeds the musculature of the heart. As little as a 10% increase in the metabolic activity of this tissue can produce as much as a 68% increase in the $AV - \Delta O_2$.

Membrane Permeability Control

Having a significant concentration gradient driving a mass transport process is, of course, meaningless if the membrane involved is impermeable to the species $[S]$ trying to get across. Thus, control of membrane permeability gives the physiologic organism a great deal of flexibility in affecting transport processes. In addition to having at its disposal all of the mechanisms described earlier with regard to capillary filtration, and those discussed in Chapter 2 with respect to carrier-facilitated transport, the cell can also manufacture specific permeability enzymes, called *permeases.*

Permeases respond to environmental and genetic as well as metabolic stimulants to adjust the cell membrane permeability to a particular substrate. The cell can therefore arrange to have $[S]$ present, or absent, to any desirable level of concentration. Indeed, as we already suggested earlier (Chapter 2, pgs. 91-93), one of the corollaries of evolution theory hypothesizes that the "fittest" were those animal species which were able to develop semipermeable membranes that could, with the least expenditure of energy, concentrate their desired chemical nutrients *from* the environment and excrete *to* the environment their undesired end products of metabolism.

In a sense, although not manufactured within the cell, per se, Insulin is a permease that enhances cellular uptake of blood glucose by affecting the membrane permeability to same. In the absence of Insulin, cells cannot remove glucose effectively from the blood; blood sugar levels therefore rise; ICF concentrations of glucose fall; there is a paucity of available substrate for the manufacture of ATP -- and, consequently, cellular metabolism is significantly impaired. The same can happen if certain permeases required for the active absorption of iron in the duodenum are absent. In this case, however, the effect is to impair the formation of hemoglobin, which hinders the oxygen-carrying capability of red blood cells, which produces a hypoxemia condition, which blocks the oxidation reactions required to convert glucose into ATP (see Chapter 5), which, finally, depletes the cell of its energy source for metabolism. And, since *extraction* of that energy requires an ATP-ase catalyzed hydrolysis, anything inhibiting this enzyme -- such as certain cardiac glycosides (most especially, Ouabain) -- will depress both cellular metabolism and the activity of the sodium-potassium electrolytic pump.

Control of Active Transport

As an integral part of membrane permeability, active transport mechanisms provide the cell with still further opportunities for control. In particular, recall that substrate control $([Na^+], [K^+], [Mg^{++}])$ is especially effective in regulating pump activity, as are those factors (glucose, oxygen, ATP-ase, permeases) that govern the energy sources that drive the transport. Under the heading above we can now add to these *additional* constraints that are peculiar to carrier-facilitated transport, i.e., (i) those factors that affect P_m: specifically, carrier molecule concentration, $[Y_m]$; carrier-substrate affinity, K_m; and dissociation constant ε_3 (c.f., equations [3-72] and [3-85]); (ii) the number of binding sites on Y or Y^* for S_1 or S_2 (e.g., three for sodium on Y and 2 for potassium on Y^*); (iii) the degree of carrier-molecule saturation; (iv) the activity of carrier-molecule "activators" and active transport "facilitators" $(E_1 - P_i, E_2 - P_i, E_1$ and $E_2)$; (v) concentration gradients driving the transport of Y and Y^*, together with the associated mobility of same; and (vi) thermodynamic and solution characteristics such as temperature, pressure, and pH -- to the extent that they influence D_S (equation [3-21]) for Y and Y^*. Several of these have been addressed in this and the last chapter. Others we will get back to in the next two chapters. For now, we will close this discussion of passive, facilitated and active mass transport by examining a useful continuum approach to the study of physiologic mass distributions and fluid balance. The approach involves the concept of *compartmental kinetics,* as developed below.

Compartmental Kinetics Analysis of Physiologic Transport

Recall from Chapter 1 that the *total* amount of fluid contained *within all* of the estimated 10^{14} cells of the human body is referred to *collectively* as the Intracellular Fluid, ICF -- or, the Intracellular Fluid *Compartment* of the body. Likewise, *all* of the remaining fluid in the body that lies *outside* of the cells is called, in a collective sense, the Extracellular Fluid, ECF -- or, the Extracellular Fluid *Compartment* of the body. This conceptualization of regions of the body that share, in a global sense, a common set of anatomic and/or physiologic state of affairs, as constituting a single hypothetical domain for the purposes of analytical treatment, is called *compartmentalization*; and the corresponding analytical treatment of the interaction of such domains is called *Compartmental Analysis.*

Note, especially, that the *contents* of physiologic "compartments" such as ICF and ECF do not necessarily have to occupy a common *physical* proximity to one another, in the sense of occupying a common geometric or "geographical" region in *space.* They are lumped together not by *location,* but based on specific *characteristics* that they share in common (such as all fluid lying *inside* of cells, or *outside* of cells, and so on) -- and, on the similarity of the mechanisms by which they interact with the environment or other regions of the body (such as transcellular fluids, all of which must pass through a continuous layer of epithelial cells, i.e., *two* cell membranes, en route from one portion of the ECF to another). Other than for transcellular fluids, the common mechanism in the case of ICF/ECF interactions always involves transport across a cell membrane, by processes already described.

Of greatest interest and concern in the compartmental analysis of ICF/ECF interactions is the transport of fluid from the intravascular ECF space (i.e., the blood plasma) across a capillary wall (via pores) to the extravascular ECF space (i.e., the interstitial fluid), and then across the cell membrane (via pores) to the ICF compartment, and then back again the other way. Such transport constitutes the physiologic process of *fluid balance* that is critical to life. That is, should this balance be upset, then the proper gradients (i.e., transport driving forces) necessary for getting things into and out of the physiologic system would likewise be upset, and a critical state of affairs leading ultimately to death would ensue. We have already examined transport as it occurs at the capillary level between plasma and interstitial fluid. Let us therefore address transport as it occurs between the ECF and the ICF using the methods of compartmental kinetics analysis.

Consider the situation depicted in Figure 3-4. The arrow labelled "External Input" accounts for all of the fluid that enters the body daily -- both directly as drinking water (usually around 1500 mℓ), and mixed in with the food we eat (usually around 700 mℓ for a well-balanced 2600-Calorie diet). Under normal circumstances, this water is absorbed from the gastrointestinal tract by *passive* mechanisms (i.e., secondary to and as a consequence of solute absorption), but in clinical situations it may also enter through intravenous infusion, blood transfusions, or any of a number of other invasive mechanisms. In any case, the external input generally is to the intravascular component of the extracellullar fluid compartment (see Table 1-I).

To the amount of water coming in to the ECF from *external* sources must be added that formed as the product of the metabolic oxidation of nutrients (see Chapters 4 and 5), and that secreted *internally* by various glands and organs. In order to prevent cellular swelling, with the ultimate possibility of bursting, water manufactured as the result of metabolic processes must leave the ICF and enter the interstitial fluid compartment of the ECF, to be eventually excreted. Similarly, to prevent cellular shrinking, with the ultimate possibility of a degenerative plasmolysis, ex-

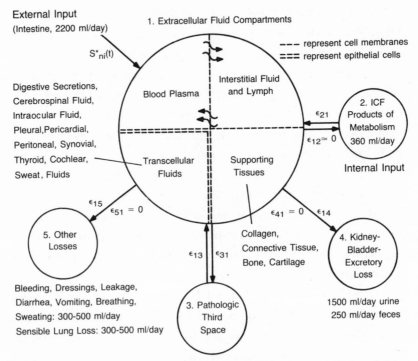

Figure 3-4 Compartmental kinetics representation of fluid transport mechanisms between functional extracellular fluid spaces and various input-output pathways.

cessive quantities of water must not leave the ICF as a result of osmotic pressure gradients. In other words, the ICF fluid compartment must maintain a relatively constant volume, and it is not as flexible in this respect as is the ECF fluid compartment, which has much more compliant boundaries.

Now, the daily caloric requirement for an "average" 70 kilogram individual is some 2600 *dietary* calories (2,600,000 *thermodynamic* calories), as discussed in the chapter which follows. Of these, in a *balanced* diet, 1400 should come from carbohydrates, 900 from fats, and 300 from proteins. The physiologic oxidation of carbohydrates has been found to yield about 4100 thermodynamic calories per gram; so 1,400,000 calories corresponds to some 341.46 grams, or, for glucose, $C_6(H_2O)_6$, of molecular weight 180, about 1.897 moles. Going one step further, the oxidation of 1 mole of glucose generates 6 moles of water, or, at 18 grams per mole, 108 grams total. One gram of water corresponds very nearly to 1 cc or 1 mℓ of water at 37°C. Thus, the daily oxidation of carbohydrate (1.897 moles) forms somewhere in the neighborhood of $(1.897)(6)(18) = 205$ mℓ of water.

Similarly, the physiologic oxidation of 1 kilogram of neutral fat forms slightly over 1 liter of H_2O, and generates 9,300,000 thermodynamic calories. Thus, 900 dietary calories of fat oxidized daily yields about $\frac{900,000}{9,300,000} = 0.097$ liters of water, or, about another 100 mℓ per day. And finally, the physiologic oxidation of one kilogram of lean protein yields about 900 mℓ of sodium-free water -- 750 mℓ from protein or glycogen and 150 mℓ from organic material -- and generates some 4,100,000 thermodynamic calories, as well. Thus, 300 dietary calories of protein oxidized daily yields an additional $\frac{300,000}{4,100,000}(750) = 54.878$, or 55 m$\ell$ of water to the body, from the protein fraction itself. The remaining water fraction is recycled into cellular hydrolysis reactions such as those involving ATP.

Summing all of this up, we see that the products of the metabolic oxidation of nutrients, which *occurs* in the ICF, but which the ICF must ultimately discard into the interstitial fluid in order to keep its own volume constant, adds to the ECF compartment some 360 additional milliliters of water daily. This water, which eventually finds its way into the blood plasma, brings the total ECF input to 2560 cc. This is an *average,* plus or minus perhaps 100 cc (4-5%) for the typical individual.

The transcellular and supporting tissue fluid compartments shown in Figure 3-4 tend more to be *recirculating* in terms of their daily activity, rather than mass sources or sinks. They are thus minimally involved in considerations of fluid balance to the extent that no *net* additions or subtractions usually originate here. For example, internal secretions into the alimentary canal may actually total as much as 8.2 liters daily (1500 mℓ of saliva, 2500 mℓ of gastric secretions, 500 mℓ bile, 700 mℓ of pancreatic juices, and 3000 mℓ of intestinal secretions), but all except a very small percentage (perhaps 2% or less) of these are reabsorbed during the digestive process. Similarly, Endocrine organ secretions are not normally intended to *leave* the body; nor are supporting tissue fluids. So the net result is that, in the steady state, the human body must deal with getting rid of around 2500-2600 mℓ of fluid per day, or, 3-4% of total average body weight.

Some of this fluid never actually makes it into the body, i.e., perhaps 200 mℓ or so to as much as 300 passes right on through the Gastrointestinal system and exits with the feces during the daily evacuation of the bowels. Insensible water diffusion through the skin (see Chapter 4) accounts for an additional fluid loss of anywhere from 350-500 mℓ per day, and insensible respiratory losses (ibid.) can total the same amount. These two thermoregulatory mechanisms thus account for an additional 700-1000 mℓ of fluid loss per day, averaging more nearly 800. The remainder (and *most* of it) leaves as urine from the kidney via the bladder. This is usually some 1000-1500 mℓ daily. If we total all the losses, we get 1900-2800 mℓ, which just balances the normal intake discussed earlier. This, of course is under *normal* conditions.

Under abnormal conditions, significantly *more* water can be lost through a variety of mechanisms. For example, the loss of water through copious sweating (sensible perspiration) may reach 2.0 liters per hour during strenuous exercise; and this may be further aggravated by respiratory sensible losses that accompany panting and hyperventilating during physical exertion. Similarly, significant fluid losses may occur during psychologically stressful situations that lead to hysterical crying, vomiting, and other gastrointestinal reactions, such as diarrhea.

The extracellular fluid volume is normally maintained by the kidney, which can regulate its 1500 mℓ per day output depending on whether there is an excess or deficiency of ECF. Stretch receptors ("strain gauges") located at strategic points in the physiologic system monitor total fluid volume, and relay information to the central nervous system, which, in turn, modulates renal function. This is accomplished via sympathetic and parasympathetic (vagal) nerve fibers emanating from both the renal plexuses and cranial nerves, which richly innervate the blood vessels of the kidney. If ECF drops, the volume receptors cause kidney urinary output to decrease accordingly. If ECF rises, the volume receptors cause kidney urinary output to increase a corresponding amount. This very neat feedback control system can backfire, however, if the signals received are misleading. Thus, for example, the pooling of blood in the core of the body, due to reduced flow to the periphery during cold stress, gives a *false* "ECF overload" alarm to the core stretch receptors, which triggers a tendency towards excessive urination. The latter, in turn, results in the danger of dehydration because the urination is not accompanied by a compensatory thirst mechanism. In other words, the body "thinks" it is expelling *excess* ECF, not fluid required for normal function, and so it makes no effort to replenish it by stimulating the desire to drink additional fluid (Schneck, D. J., 1983e).

A very similar problem occurs during exposure to a sub-gravity environment. Here, however, the pooling is in the *peripheral* blood vessels (mostly of the head and neck) due to a sub-gravity-induced impairment of venous return to the heart. In addition to possible dehydration that results from peripheral-receptor-induced urination, astronauts frequently suffer from related symptoms that include head "fullness", nasal stuffiness, facial plethora (distended blood vessels), distended neck, scalp, and peripheral veins, lowered heart rate, higher diastolic blood pressure, lower systolic pressure, lower pulse pressure, and other consequences of a peripheral shift in body fluids (NASA, CR-3487, 1981).

Other clinical situations in which fluid losses beyond normal may occur include bleeding, hemorrhaging, tube suctioning, fistula and canulla insertion, and so on.   In extreme cases, a so-called "Pathological Edematous Third Space," or "Third Non-Functional," Inflammatory ECF Compartment may actually be identified, in addition to the "Functional" ECF (i.e., blood plasma together with interstitial fluid and lymph compartments) and ICF compartments. This is common during surgery, and requires immediate fluid therapy in order to maintain the proper fluid balance between the ICF and ECF compartments.   Indeed, during the decade 1955-1965, a series of studies by Shires and his associates (1959, 1960, 1961, 1964) suggested that the extracellular fluid volume is the most labile, the most responsive, and the most frequently involved compartment of the body fluids in surgical illnesses. Primarily, this is due to the accumulation of water in soft tissues following surgery and acute trauma.

For the purpose of formulating some rational means for quantifying such fluid volume shifts under both normal and pathologic conditions, it is of interest to apply unsteady fluid dynamic principles to the compartmentalized global scheme illustrated in Figure 3-4.

Generic Compartmental Kinetics Equation

In Figure 3-4, a "black-box" type of input-output flow diagram is used to simulate mass transfer into and out of a "functional" Extracellular Fluid Space (ECF Compartment). The latter is intended to include essentially blood plasma (i.e., the *intravascular* compartment) and interstitial extracellular fluid and lymph (i.e., about one-half of the *extravascular* compartment, c.f., Table 1-I) -- all of which exchange mass at a rapid rate and thereby equilibrate (i.e., reach a steady-state distribution of mass) within 20 to 30 minutes following some step disturbance. By contrast, the functional ECF compartments exchange mass relatively slowly -- that is, with an equilibrium time of 7 hours or more -- with other extracellular spaces, such as transcellular fluids and supporting and connective tissue.

In Figure 3-4, let: $S_{nj}(t)$ = the total amount (not concentration) of material S_n in compartment j at time t ($n = 1, 2, 3, \ldots$; $j = 1 =$ extracellular fluid space, $j = 2 =$ intracellular fluid space, $j = 3 =$ pathologic third space penetration, $j = 4 =$ all excretory losses, $j = 5 =$ "other" losses, as shown). Note that the double-subscript notation in this case does *not* mean an implicit summation on the respective indices; it is simply being used as an identification scheme for species n and compartment j. Bearing this in mind, further let $S_{nji}(t)$ represent the total amount of S_n *per unit time* flowing *from* compartment j *to* compartment i at time t, where $i = 1, 2, 3, 4, 5$ as above for j. And finally, let

$$\varepsilon_{nji}(t) = \frac{S_{nji}(t)}{S_{nj}(t)} \, ,$$

which has units of reciprocal time, represent a rate constant governing the fractional amount of S_n flowing per unit time from the jth fluid space to the ith fluid space at time t.

The expressions above define basically the exchange of mass as it flows from compartment j *to* compartment i (i.e., the *input* to compartment i from the other fluid spaces). To these, we add an *external* input, $S^*_{ni}(t)$, which could represent all of the inputs discussed earlier, or, in a clinical situation, the *rate* of injection of some tracer substance S_n to compartment i (as a function of time), or the infusion rate of external fluids to bring into balance the ECF/ICF ratio. In a similar manner, we can define the *output* from compartment i by introducing the functions $S_{ni}(t)$, $S_{nij}(t)$, and $\varepsilon_{nij}(t)$. Then, applying the principle of conservation of mass to the quantity $S^*_{ni}(t)$ introduced into compartment i, one may show in a straightforward manner that the rate of change of the amount of S_n in compartment i at any time t may be expressed by the equation:

$$\frac{\delta S_{ni}(t)}{\delta t} = S^*_{ni}(t) + \varepsilon_{nji}(t)S_{nj}(t) - \varepsilon_{nij}(t)S_{ni}(t) \qquad \text{[3-102]}$$

where: $S^*_{ni}(t)$ is the *rate* of infusion of S_n into compartment i at time t; $\varepsilon_{nji}(t)S_{nj}(t)$ is the rate of influx of S_n into compartment i from compartment j; and, $\varepsilon_{nij}(t)S_{ni}(t)$ is the rate of efflux of S_n from compartment i into compartment j.

When tracer substances are used to estimate the total ECF or ICF volume, they are generally administered as a single dose of radioactive material at time $t = 0$; and urine and plasma concentrations of S_n are then measured during successive intervals of time to determine its ultimate distribution among the various body fluid compartments. In this case,

$$S^*_{ni}(t) = F(t)\delta(t). \tag{3-103}$$

The function $\delta(t)$ is the *Dirac Delta Function*, defined to be equal to zero for $t \neq 0$, and,

$$\int_{0^-}^{0^+} \delta(t)dt = 1 \tag{3-104}$$

for t in the neighborhood of zero. Thus,

$$\int F(t)\delta(t)dt = F(0), \tag{3-105}$$

where $F(0)$ is the quantity of tracer substance S_n administered at $t = 0$, and $F(t)$ otherwise has no value. Methods utilizing this approach are called *dye-dilution techniques* for the estimation of ECF or ICF. There are also *thermal dilution techniques* that operate under similar principles, but use *heat*, rather than *mass* as the tracer substance.

Equation [3-102] states simply that the rate of change of the amount of S_n remaining in compartment i at time t is the difference between what is being infused, $S^*_{ni}(t)$, and the *net* result of a series of exchanges between compartment i and other fluid compartments, j The generalization of equation [3-102] for a compartmental system with m compartments becomes (Sandberg, 1978a):

$$\frac{\delta S_{ni}(t)}{\delta t} = S^*_{ni}(t) + \sum_{j=1}^{m} \varepsilon_{nji}(t)S_{nj}(t) - \sum_{j=1}^{m} \varepsilon_{nij}(t)S_{ni}(t) - S^{*o}_{ni}(t)$$

$$t > 0; \quad S^o_{ni} = S_{ni}(O)$$

$$i = 1, 2, 3, \ldots, m$$

[3-106]

where: S_{ni} is the amount of material S_n in compartment i; S^o_{ni} is the *initial* value of $S_{ni}(t)$ for $t = 0$; $S^*_{ni}(t)$ and $\varepsilon_{nji}(t)S_{nj}(t)$ are the flow rates for the flows *into* compartment i from the environment and compartment j, respectively, for $j \neq i$; and, $S^{*o}_{ni}(t)$ and $\varepsilon_{nij}(t)S_{ni}(t)$, respectively, are the flow rates for the flows of S_n from compartment i to the environment and into compartment j, for $i \neq j$.

It is assumed that $S_{nii}(t) = 0$ for each i (no sources or sinks) and that, for each $i \neq j$ the flows whose rates are $S_{nij}(t)$ and $S_{nji}(t)$ occur by way of separate processes. The mathematical foundations for equation [3-106], including the uniqueness and stability of various solutions to this equation have been considered by Sandberg (1978a and 1978b) and Brown (1985), among others (see, for example, Llaurado, 1973, Cooney, 1976, Maeda, et al., 1976, and Jones and Godfrey, 1981). We shall not dwell on them further in the scope of this text, but the reader is referred to the literature for more specific details.

Concluding Remarks

In this chapter, we have addressed both micro- and macro-continuum approaches to the mathematical derivation of physiological mass transport parameters, such as characteristic velocities, kinematic transport coefficients, intensive scalar potentials, driving forces, and so on. In those situations where the appropriate transport coefficient, L^2/t, or energy function, L^2/t^2, cannot be so uniquely defined analytically for a specific situation from the properties of the system involved and/or related parameters of physics, an order of magnitude approach can yield a generalized transport coefficient and intensive energy function based on the proper scaling of any arbitrary process. That is, if one can at least identify an appropriate length scale, L, for the process, and a characteristic time scale, t, for the same process, then taking the ratio of $\left(\dfrac{L^2}{t}\right)$ and $\left(\dfrac{L^2}{t^2}\right)$ gives an order-of-magnitude measure of the transport quantities associated with that process.

As a brief illustration, consider flow through a capillary. The appropriate length scale in this case is the average length of a human capillary, which is on the order of 1 mm. The time-scale involved is the average amount of time that blood spends in a capillary, which is on the order of 2 to 2.5 sec. Thus, $\left(\dfrac{L^2}{t}\right) \triangleq \dfrac{(1.00)^2}{(2.25)} = 0.44$ mm^2/sec. Comparing this to the kinematic viscosity of blood, $\dfrac{\mu_f}{\rho_f} = 4.84$ mm^2/sec, we conclude that the transport of momentum by conduction (i.e., friction-driven viscous shear) in capillaries is an order of magnitude larger than is the transport of momentum by convection (i.e., pressure-driven flow). Dimensional analysis is a very powerful technique which has particular relevance in physiological research because of the complexity of the latter. This is illustrated in the following chapter.

Chapter 4

Basic Principles of Bioenergetics, Metabolism and Thermoregulation in Physiologic Systems

Introduction

The raw materials that are required to keep the normal "human machine" alive and functioning properly are air, food and water. Air supplies oxygen, which accounts for some 65% of all the elements in the body, and is an essential ingredient in the processes concerned with metabolism (see below). The gas is extracted from the atmosphere and mixed with blood in the "carburetor" of the machine, i.e., the lungs and respiratory system. Food supplies nutrients and other life-sustaining ingredients. These are absorbed and enter the blood and lymphatic streams in usable form as a result of having been processed in the "refinery" of the machine, which is the alimentary canal and gastrointestinal system. Water, as discussed at the end of Chapter 3, enters basically the same way, to supplement that which is already present in the organism.

These processed raw materials are delivered to the functional elements of the machine, i.e., *cells*, or, if you will, the 10^{14} "cylinders" of the "engine", by the river of life -- blood -- which, driven by the system's "fuel pump" -- which is the heart -- courses through thousands of miles of a complex vascular system. Having thus been delivered to the cells, oxygen, nutrients and water enter these tiny masses of protoplasm by mass transport processes that have already been described, and the waste products of metabolism leave accordingly. Once inside the cell, the raw materials or ingredients of life undergo a series of transformations that generate products designed to:

1) satisfy the immediate energy needs of the organism;

2) store energy in a retrievable form;

3) serve as physiologic structural components, biochemical catalysts, or regulatory hormones;

4) repair and maintain worn down engine parts; and/or,

5) provide repositories for the genetic information that characterizes the species.

The latter then serves as the "cookbook", containing as "recipes" for survival the sum total of mankind's evolutionary experience and heritage.

The above transformations, encompassing all of the physical and chemical changes that take place within the cells of an organism, together with all of the energy considerations involved, are collectively referred to as *metabolism*, from the Greek, "metaballein," meaning to turn about, alter, or undergo some specific transformation.

Metabolism actually involves two types of processes: *Anabolism*, wherein simple substances are converted into more complex macromolecules (chiefly proteins, nucleic acids, lipids and polysaccharides); and *Catabolism*, wherein complex macromolecules are converted into simpler ones. This breakdown is usually for the purpose of: (a) extracting the chemical bond energy inherent in the larger complexes; (b) reducing them to a size suitable for absorption and transport (as in digestion of proteins into amino acids, carbohydrates into simple sugars like glucose, and fats into free fatty acids); or (c) converting them from one complex form into another to satisfy whatever the metabolic needs of the organism are at any given time. Vital physiologic processes that require energy include, among others, synthetic reactions (e.g., protein synthesis and the synthesis of high-energy phosphates), muscular contraction, nerve conduction, active mass transport (c.f., Chapters 2 and 3) and reproduction.

In Anabolism, biopolymers are constructed from discrete monomer units, ingredients, or "building blocks", according to instructions or "recipes" coded into the DNA stores (see Chapter 5). For nucleic acids, the monomer units are *nucleotides*. These are compounds formed from phosphoric acid (H_3PO_4), a sugar (usually pentose, $C_5(H_2O)_5$, or deoxypentose, $C_5H_{10}O_4$) and a base (purine, $C_5H_4N_4$, or pyrimidine, $C_4H_4N_2$ in various states of oxygenation). For complex polysaccharides, the monomer units are sugar derivatives and simple carbohydrates in glycosidic linkage. Most frequently involved is the monosaccharide D-glucose, $C_6(H_2O)_6$, otherwise known as aldohexose. For proteins, the monomer units are *amino acids*. These consist of the combination of an amino group (NH_2), a carboxyl group (COOH), and any aliphatic (i.e., containing open or branched carbon chains) or aromatic (i.e., containing closed carbon rings such as benzene) radical. And, for lipids, the monomer units are *fatty acids*. These are hydrocarbons in which one of the hydrogen atoms has been replaced by a carboxyl group. Alternatively, lipids can be synthesized from simple sugars and starches.

Catabolism, on the other hand, results in end products that include Urea [$CO(NH_2)_2$], Carbon Dioxide (CO_2), Water (H_2O), and nonusable heat energy (entropy, s, per unit mass, per unit temperature). Urea is the refuse from the destruction of the ammonia portion of tissue proteins. Carbon dioxide and water are the end products from the oxidation of the non-nitrogenous part of protein molecules, from aldehydes (oxidation products of alcohols, having the characteristic group CHO) and carbon acids resulting from the breakdown of glucose and glycogen; and, from the oxidation of glycerol and fatty acids that accumulate from the catabolism of fats and lipids.

Since the body is an *electrochemical* engine, and not a *heat* engine, when the biochemical reactions comprising all of metabolism release energy (more than 75% of it appearing as heat) this nonusable form of energy must be handled in a very unique way. In order to understand the ultimate dissipation of heat by the organism, the basic bioenergetic principles involved in metabolic processes, and the role played by enzymes and catalysts in biochemical reactions, we first review some basic principles from thermodynamics and chemical kinetics.

Basic Thermodynamic Principles

Recall from Chapter 2 that to describe the configuration of a real (i.e., dissipative), homogeneous, thermodynamic system, one needs to define for it a (or sets of) characteristic length scale(s), L, intensive scalar potential energy function(s), ϕ, intensive dissipation function(s), ψ, and, if the system is not in equilibrium, corresponding flux(es), Φ, which depend on $\nabla\phi$ (c.f., equation [2-4]). These four quantities are in the nature of basic dimensions of state which constitute the basis of irreversible, nonequilibrium thermodynamics.

Network Thermodynamics

Historically, the notion of dimensions of state have been most exploited in electrical network theory. That is, early in the development of this theory it was recognized that the "whole" (i.e., a complete electrical system) could be effectively analyzed as a conglomeration of "parts", which include resistors (energy dissipators, ψ), capacitors (energy storers, ϕ), and inductors (inertial elements, Φ). More recently, this idea has carried over into thermodynamic modelling of physical systems, in general, and the term "Network Thermodynamics" has been coined (Mikulecky, 1983).

In network thermodynamics, a physical system is defined in terms of *thermodynamic* elements that store (ϕ), dissipate (ψ), or transport (Φ) energy. These are called *ports*, and, in physiologic systems, are identified as closely as possible from the anatomy and morphology of the structures involved (Mikulecky, 1984, Mikulecky and Thomas, 1979). The ports are then considered to be a representation of the whole as a result of the particular manner in which they are connected.

The functions ϕ and ψ (note that Φ is a function of ϕ) are then written in terms of *variables* (as opposed to *dimensions*) of state. For example, four such variables are pressure, p, specific volume, $v = \frac{1}{\rho}$, specific entropy, s, and temperature, T, in the sense that to describe the state of a simple, homogeneous thermodynamic system, one may define

$\phi = pv$ (c.f., equation [3-18]), and $\psi = sT$ (from the second law of thermodynamics). More generally, one may choose variables that are associated with, or in some sense describe, "effort", e (energy per unit mass), "flow", f^* (mass per unit time), generalized displacement, ℓ, and generalized momentum, α (in the form of a kinematic transport coefficient). The latter two are defined by the relationships: $\ell = \ell(0) + \int_0^t \dfrac{f^*(t)}{\rho A}\, dt$; $\alpha = \alpha(0) + \int_0^t e(t)dt$; and the former two are chosen so that the power consumed and/or stored by the system is given by $\Sigma_i e_i \dot{f}^*_i$.

Given the four basic state variables -- in this case, p, v, s, and T -- where p and v address the mechanical degrees of freedom of the system and s and T address the thermal degrees of freedom of the system -- there are four possible binary combinations that include one mechanical variable and one thermal variable in each: s, v; s, p; T, v; and T, p. Corresponding to each of these, one may define thermodynamic energy functions as follows:

$u(s, v) =$ Internal Energy $=$ The total energy per unit mass internal to a thermodynamic system. It includes energy associated with mechanical degrees of freedom (molecular translational kinetic energy, rotational kinetic energy, and vibrational energy), energy associated with excitation (electromagnetic and nuclear), and energy associated with the inherent atomic, molecular and other chemical bonds of the system. It accounts for both the potential energy by which the characteristics and/or composition of a given mass under observation may be changed, and the non-usable energy which may ultimately become manifest as heat. It does *not* account for energy that may be brought *into* the system from outside due to flow or other exchanges with the environment. Such energy is called "flow work."

$h_E(s, p) =$ Enthalpy $= u[s, v(p)] + pv(p) =$ Internal energy plus the energy brought into the system as a result of flow work. This thus represents the *total heat content* of an open thermodynamic system that can exchange both mass and energy with the environment.

$f_H(T, v)$ = Helmholtz Free Energy = $u[s(T), v] - Ts(T)$ = the maximum amount of *useful* work (internal energy minus entropy) that can be extracted in a theoretical sense (i.e., reversibly) from a non-flow, "closed thermodynamic system", capable of exchanging energy, but not mass with its environment (hence, constant mass system), when it is allowed to undergo a reversible, isothermal (no heat generated) change from an initial state with a free energy level, f_{H_1}, to a final state with a free energy level f_{H_2}. This can be seen by writing $\delta f_H = \delta u - T\delta s - s\delta T$, such that $\delta f_H = \delta u$ when $\delta s = 0$ (reversible, isentropic) and $\delta T = 0$ (isothermal).

$g(T, p)$ = Gibbs Free Energy (or "free enthalpy") = $u[s(T), v(p)]$ $- Ts(T) + pv(p) = f_H + pv = h_E - Ts$ = the maximum amount of *useful* work (enthalpy minus entropy) that can be extracted in a theoretical sense from a non-flow (constant mass), isothermal, isobaric system, or, from an open isothermal flow system, when either of these is allowed to change in a reversible (isentropic) fashion from an initial state with a free energy level g_1 to a final state with a free energy level g_2. Writing $\delta g = \delta u - T\delta s - s\delta T + p\delta v + v\delta p$, we see that $\delta g = \delta u$ for an isentropic ($\delta s = 0$), isothermal ($\delta T = 0$), isobaric ($\delta p = 0$), closed ($\delta v = 0$) system; $\delta g = \delta u + p\delta v$ for an isentropic, isothermal, isobaric, open system.

It should be mentioned that each of the above energy functions has the property of a thermodynamic potential *only* with respect to the variables that define it -- for example, T and p in the case of the Gibbs Free Energy function. The remaining two variables, s, v, are assumed not to be independent, and are called *conjugates* of the independent variables T, p that define g. For a different choice of variables, the Gibbs Free Energy *loses* (as do the other functions) its (their) property as a potential.

Constitutive Equations

Note in the above definitions for h_E, f_H and g, the dependence of v on p, and s on T. That is, the structure and physical properties of a system can impose certain relations among the state variables, so that they become dependent upon one another. Equations that describe this dependence are called *Constitutive Relations*. In general, for the development of a complete thermodynamic theory to describe any physical system, one normally needs to know at least four types of constitutive relations for the network branches of the system. These are: (i) Rheological stress-strain or stress-strain-rate equations that define the degree to which a network branch can *respond* to a *mechanical* disturbance (Silver, 1987, Desai and Siriwardane, 1984, Walburn and Schneck, 1976); (ii) Energetic descriptions of the internal energy characteristics, $u(s, v)$ of a network branch -- such as equations of state $v(p)$ in classical thermodynamics or "phenomenological" equations in nonequilibrium thermodynamics (Rice, 1967, Hill, 1960); (iii) Thermal heat transport equations for conductive, convective, or radiated flux (see later in this chapter); and (iv) Entropic relationships for $s(T)$ (Astarita and Marucci, 1974, Sommerfeld, 1956).

The ability of a system, in a constitutive sense, to *respond* to a disturbance is how we define many of its physical properties, such as moment of inertia, modulus of elasticity, thermal conductivity, and so on. Thus, most constitutive equations are determined experimentally and/or from other theoretical considerations such as those comprising the fields of statistical, quantum, or continuum mechanics. Since one may find very fine discussions of constitutive modelling in the literature (ibid.), it shall be considered -- you should excuse the expression -- "beyond the scope of this text", and so we shall move on with the realization that the development of constitutive models of *physical* systems is an integral part of the engineering analysis of *physiologic* systems.

The Laws of Thermodynamics

Choosing internal energy, $u(s, v)$ as a thermodynamic state potential, with entropy, s, and specific volume, v, considered to be independent variables, one can write:

$$\delta u = \left(\frac{\partial u}{\partial s} \right)_v \delta s + \left(\frac{\partial u}{\partial v} \right)_s \delta v \qquad \text{[4-1]}$$

Now, the existence of temperature as a property of a system, and the further requirement of equality of temperature as a condition of *thermal equilibrium* between two systems (or between two parts of a single system) is sometimes called the *Zero'th Law of Thermodynamics*. The zero'th law basically defines temperature as:

$$T = \left(\frac{\partial u}{\partial s} \right)_v \qquad \text{[4-2]}$$

Going one step further, we saw in Chapter 3 that thermodynamic pressure is basically a measure of molecular translational kinetic energy per unit volume (c.f., equation [3-9]). In a situation where there is no dissipation of energy (entropy, $s = $ constant), it is reasonable to expect internal energy per unit volume to be in a translational kinetic energy mode, since rotational kinetic energy and vibrational energy are both dissipative. In other words, we can surmise from our analyses thus far that p and $\left(\frac{\partial u}{\partial v} \right)_s$ are somehow related. In fact, since a system *loses* internal energy when it does work in a $p \, \delta v$ sense, we can write:

$$p = - \left(\frac{\partial u}{\partial v} \right)_s \qquad \text{[4-3]}$$

where the minus sign implies that conservative internal energy in forms other than rotational, vibrational, and other dissipative modes, *decreases* as it is converted reversibly (s constant) into translational kinetic energy per unit volume, which appears as pressure, p. In other words, the definition of pressure by equation [4-3] is essentially a conservation statement among the various forms of internal energy that are non-dissipative.

Substituting equations [4-2] and [4-3] into [4-1] yields the *First Law of Thermodynamics* in its traditional form:

$$\delta u = T \, \delta s - p \, \delta v \qquad \text{[4-4]}$$

where the independent state variables, s and v, have now been joined by their dependent conjugate pair, T and p, to complete the set of variables required to define the thermodynamic state of a system.

Having thus defined temperature, one may now go on to define *heat* as that form of energy which passes from one body to another (or between two contiguous regions in space) as the result of a temperature *gradient* between the two -- i.e., due to a lack of *thermal* equilibrium (c.f., Chapter 2).

It has always been observed that whereas it is possible to convert mechanical work completely into heat, it is not possible to convert heat completely into work; that is, the work-to-heat transition is not completely reversible. There always remains a quantity of heat energy which is unavailable to do useful work. It turns out that this unavailable energy is related to the disorderly activity (or, to the "state of disorder") of the molecules involved, in that, *useful* work can only be derived from ordered, directed molecular motion. Imagine, for example, the simple case of a tug-of-war between two individuals. With the two pulling in *opposite* directions in a disorderly fashion, their combined effort cannot be harnessed as effectively as if they were both pulling in the *same* direction with a co-ordinated level of activity. So it is that *random* molecular motion is "wasted" -- in the sense that, for every molecule moving one way, the probability is that another molecule is moving the opposite way, so that the two cancel one another in terms of any meaningful coordinated effort. On the other hand, *directed*, organized, coordinated molecular motion can result in a net useful production of some well-defined task.

Such disorder in a physical system is related to a thermodynamic system property called *entropy, s,* per unit mass. That is, entropy is a *measure* of the number of possible configurations that a system having a *given total energy* may assume, with that corresponding to the most disordered configuration (i.e., maximum entropy) representing a state of equilibrium, wherein no work can be done. The capacity of a system for doing work is zero at equilibrium because the existence of maximum molecular randomness prevents any organized directed motion from taking place; and so, no further change in the state of the system may occur spontaneously unless *additional* energy is supplied from outside the system through an exchange of mass and/or energy with the environment (flow work). To put it more succinctly, a system in equilibrium has a *minimum* of free energy, a *maximum* of entropy, and is in its most probable spontaneous state corresponding to total disorder.

All of the above considerations may be summarized by the *Second Law of Thermodynamics,* which, for closed systems at uniform temperature, T, may be stated mathematically as follows:

$$ds - \frac{dQ}{T} \geq 0 \qquad\qquad \text{[4-5]}$$

where Q represents heat added to the system, per unit mass, and the equal sign corresponds to a system which is in equilibrium, such that the exchange between dQ and ds is reversible.

The second law of thermodynamics, being an inequality, does more to impose *restrictions* on suitable forms for constitutive equations, than it does to make thermodynamic problems more deterministic. That is, one cannot choose functional forms which lead to results that violate equation [4-5]. However, for the case of equilibrium, one may substitute equation [4-5] into equation [4-4] to get:

$$\delta u = \delta Q - p\,\delta v \qquad\qquad \text{[4-6]}$$

The first law of thermodynamics, written in the form of equation [4-6], is a conservation of energy law stating that the net result of all the heat added *to* a system, minus the work done *by* the system, will be a change in the total energy internal to that system. In its most general sense, the first law of thermodynamics is an energy balance between system and surroundings, and it applies to any open or closed system as long as one accounts for *both* energy changes that result from *mass* flux into or out of the system (*carrying* with it, internal, kinetic, potential, flow, and other forms of energy), and those changes that result from the *pure* exchange of energy between system and surroundings.

It was noted in Chapter 3 that absolute zero defines the state of affairs that exists when there is absolutely no molecular motion at all in the Brownian sense. Moreover, from the definition of heat, we also observe that there can be no such form of energy transfer in a system existing at absolute zero. These considerations suggest the feasibility that $dQ \rightarrow 0$ and $ds \rightarrow 0$ (there is no *disordered* molecular motion if there is *no* molecular motion) as $T \rightarrow 0$, and, indeed, this has been confirmed in experiments performed on a large number of isothermal processes involving pure phases. The change in entropy going to zero for any system or substance in a state of absolute zero temperature (-273.16 degrees Celsius, or -459.69 degrees Fahrenheit) is often referred to as the *Third Law of Thermodynamics,* and is expressed mathematically as:

$$\lim_{T \rightarrow 0} \Delta s = 0 \qquad\qquad \text{[4-7]}$$

The energy laws of thermodynamics (equations [4-2], [4-4], [4-5] and [4-7]), together with Newton's laws of motion (e.g., the momentum equation [3-3]), and the law of conservation of mass (e.g., the continuity equation [3-27]), constitute the "topology" (Mikulecky, 1983), or *Field Equations* that govern any physical process. Added to the constitutive relations that define the function of the individual thermodynamic network elements (i.e., inertial, conservative, or dissipative), and the closure properties of the system (i.e., boundary conditions and/or constraints), we have all of the principles necessary for a complete thermodynamic analysis of any system -- in particular, *physiologic* systems.

The Human Physiologic System is Not a Heat Engine

Physiologic reactions associated with the processes of metabolism usually proceed with either the liberation (exergonic) or absorption (endergonic) of energy, the latter being often referred to as the *heat of reaction.* This is because, at constant pressure and specified temperature, the change in energy of a system undergoing some thermodynamic process is equal to a change in the total heat content, i.e., enthalpy, $h_E = u + pv = c_p T = c_v T + pv$, of the system involved. Observe that for non-flow or isovolumetric systems at a given temperature, the heat of reaction corresponds to a change in internal energy, $u = c_v T = u(s, v)$, of the system.

In an engineering sense, the heat of reaction is called the heat of combustion when oxygen is used to burn fuel with the intent of producing useful work in a heat engine. But heat, per se, liberated during a combustion process can be harnessed to produce useful work only if there is a temperature difference, $\Delta T = T_2 - T_1$, between a heat source (subscript 2) and a heat sink (subscript 1) located within the system producing the work. This can be seen from the simple Carnot heat-cycle relationship:

$$W = Q^* \frac{T_2 - T_1}{T_2}, \qquad \text{[4-8]}$$

where W is the maximum available useful work that can be extracted from the total quantity of heat, Q^*, released when T is reduced from T_2 to T_1. If $T_2 = T_1$ (e.g., for an isothermal process), then $W = 0$ as far as being able to "harness" Q^* is concerned.

The engineering concept of a heat engine which converts combustion heat into useful work presents an interesting problem where the human "engine" is involved. Being essentially an *isothermal* ($37°C =$ constant), *isobaric* (atmospheric pressure $=$ constant) system, the human organism cannot take thermal (heat) energy and convert it directly into work in the sense of the Carnot cycle relationship [4-8]. Indeed, this equation shows that even the *efficiency*, $\frac{W}{Q^*}$, of this very conversion process is directly proportional to ΔT, and goes to zero in the absence of a temperature gradient. Thus, any *heat* energy that is generated by exothermic reactions taking place in the human organism cannot be harnessed, but, instead, must be dissipated to the environment by mechanisms described later in this chapter.

In order to get around this problem, the human body attempts *not* to function as a heat engine, but, rather, to *minimize* the actual nonusable heat released during metabolic reactions -- or, more precisely, to control the *rate* at which exergonic reactions release heat. This is accomplished as follows: rather than proceeding in one big step from raw materials to desired end products -- with a corresponding generation of a great deal of undesired and nonusable heat energy -- metabolic reactions proceed, instead, through a whole sequence of *little* steps. Each little step has associated with it that much less energy. To put it another way, the organism attempts to generate *differential* quantities of heat in each of a *long sequence* of reactions, rather than *finite, measureable* quantities of heat in single raw-materials-to-end-product types of reactions.

The idea here is not to generate less *total* heat. Indeed, since h_E is a *state function*, Δh_E is unaffected by the particular pathway taken between two end states; and, by the first law of thermodynamics, the *overall* heat production of any substrates-to-products reaction pathway will thus be the same whether the reaction takes place in one step or in many steps. But, if we do not choose the *shortest* path between the two end states, then it becomes possible to generate the same amount of heat more *slowly* by "bleeding" it off, i.e., by controlling the *rate* of energy release.

To illustrate this point by a mechanical analog, consider a wheel being driven at high speed by an electric motor. Suppose the motor is now turned off, and we examine the "reaction" or series of reactions that ultimately bring the wheel to rest. The wheel starts out with some rotational kinetic energy (assume that it does not translate, and is mounted in a fixed, smooth bearing of some kind) which is "potential" (bond energy) in the sense that we may transform it into other forms. One way we may transform it is to apply a brake shoe to the rim of the wheel, keep it engaged as hard as we can in a continuous fashion, and maintain the pressure until the wheel stops turning. Our reaction will then have been completed in one big step from high-speed rotary motion ("raw materials") to no motion at all ("desired end-product") with a tremendous generation of heat, smoke, wear and tear on the wheel, the brake, and the brake operator! There would be a corresponding rise in temperature of the system due to the significant amount of heat generated by friction in a relatively short period of time, and the heat would be nonusable in the sense that no subsequent work could be extracted from the dissipative effects of friction.

There is another way to stop the wheel without such dramatic consequences. We could achieve the *same end result* from the *same initial conditions* by "pumping" the brake shoe intermittently against the rim of the wheel and letting it thus decelerate more slowly. Each such pumping action would *gradually* bring us another step closer towards the accomplishment of the same end result -- and, although it would take longer, and require more revolutions of the wheel in this case than it did in the previous case -- the same amount of energy would ultimately be dissipated (i.e., the initial kinetic energy of the wheel) and the same end product (i.e., the stopping of the wheel) would ultimately be achieved. The wheel, however, would not get as hot, the brakes would not overheat, the thermal energy associated with friction would be generated more slowly, it would have more time to be dissipated to the environment, and, indeed, the whole process could be effectively accomplished in a virtual isothermal fashion. This is the idea behind biochemical reactions. They proceed in smaller steps, through more complicated routes, in a more cumbersome fashion to the same end products from the same raw materials. But in so doing, they manage not to cause problems associated with dramatic temperature overshoots in the physiologic system.

Let us take this analogy one step further. Suppose, as the wheel rotates, that it also drives a torque spring that *opposes* the motion. Under these circumstances, when the motor is turned off the torque spring will act to stop the wheel, so that the brake will have to do correspondingly less work to bring the wheel ultimately to rest. Moreover, whereas the brake dissipates energy, the torque spring stores it. So, by putting the torque spring into the system we: (i) reduce the importance of the brake in terms of the same end-product (i.e., bringing the wheel to rest); (ii) reduce the energy dissipated by friction through the brakes; (iii) decrease the heat energy generated by the braking action on the wheels; (iv) *store* some of the kinetic energy of the rotating wheel in a retrievable form as potential energy in the torque spring; and, (v) accomplish the same end result, from the same initial conditions, but in a different way -- one that manages to "save" along the way some of the original kinetic energy of the wheel. In this mechanical example, we used an energy-storing mechanism consisting of the torque spring to stop the wheel, rather than an energy-dissipating mechanism such as brake friction. Again, biochemical reactions employ the same guiding principle.

The human body seeks to *retain* (rather than to dissipate) the energy released during exergonic reactions by using this energy to drive other, endergonic reactions. It thereby keeps the energy stored (as much as is possible) in a retrievable form as *bond* (potential) energy. This is accomplished by *coupling* degradative exergonic (free-energy-releasing) reactions with oxidative synthetic endergonic (free-energy-absorbing) reactions in such a way that much of the energy released by the exergonic reaction (which would otherwise be dissipated as nonusable heat) is reabsorbed and stored in the chemical bonds of the endergonic reaction. Some of the remainder *is* converted into heat, but a good portion of it is extracted to perform the vital physiologic processes that require useful work; and the work is derived from available free (bond) energy, not from heat energy in the sense defined by the Carnot Cycle equation [4-8].

Like our torque spring (endergonic), synthetic energy-storing reactions reduce the total amount of nonusable heat energy (exergonic, friction) that can be potentially lost as a by-product of energy-releasing degradative reactions. There is thus less tendency for body temperature to rise suddenly, and for large-scale temperature fluctuations to accompany metabolic processes such as muscular contraction, nerve conduction, active transport, reproduction, and the synthesis of proteins, nucleic acids, polysaccharides, lipids and high-energy phosphates. Again, the body achieves the same end products, from the same raw materials, but gets there through coupled energy-*storing* mechanisms that are intimately associated with energy-*dissipating* mechanisms. And this philosophy has still a further advantage.

If it takes 100 steps, say, to get from raw materials to end products, each step being crucial to the success of the reaction, then the chain may be broken at any of the 100 steps involved. In other words, *control* can be manifest in several different ways at many levels of the sequential chain of events required for the completion of any series of reactions. This allows the human organism to "fine-tune" biochemical reactions, and to evaluate more variables in determining whether the reactions should go to completion, or not. It allows for great flexibility in the control of metabolism.

"Now, all of this is fine," you may say -- "we have this wonderfully controlled, isothermal, very efficient metabolism wherein reactions are 'stretched out', made more cumbersome, prolonged and delayed in order to conserve energy and keep the temperature constant. But will this philosophy allow the reactions to proceed at a rate consistent with the metabolic *needs* of the organism? After all, in our mechanical example it took more time and required more revolutions of the wheel to stop it in several short, discrete steps, as opposed to one continuous process. Can the human organism *afford* this luxury? And if not, how does it get around this problem?"

Well, in the case of stopping our mechanical wheel, there are several things we can do to accelerate the process. For one thing, we can control the spring constant of the torque spring. A higher spring constant will stop the wheel more quickly, and vice versa. For another thing, we can control the coefficient of friction between the brake and the wheel. A higher coefficient of friction will stop the wheel in less time. For still another thing, we can control the characteristics of the wheel bearing to make it easier or more difficult for the wheel to rotate freely. And so on. In other words, there are very constructive things we can do to facilitate the "reaction" -- and this is precisely what biochemical *catalysts* do. They enhance the efficiency and speed of biochemical reactions to help them progress at rates that ensure physiological well-being.

But biochemical catalysts, known as *enzymes,* also have their limitations. In fact, another reason for the existence of many consecutive reactions in metabolism may very well be the somewhat limited chemistry that can be performed by enzymic protein molecules. To accomplish the net glycolytic reaction:

$$C_6(H_2O)_6 + 6O_2 + 38ADP + 38P_i \longrightarrow 6CO_2 + 6H_2O + 38ATP \quad \text{[4-9]}$$

in a single step, for example, would require a protein molecule and catalytic mechanism of such extreme complexity that it would be unlikely to evolve (Jones, M. N., 1979). Thus, the breakdown of the overall reaction into several consecutive reactions involving relatively simple organic transformations not only makes the reaction kinetically feasible, but also allows glucose utilization to supply intermediates for the synthesis of fats and proteins. Indeed, the reaction pathways can "branch" at any intermediate point to an alternate destination, giving the organism great flexibility in establishing metabolic priorities. It also introduces control possibilities, in that the presence of the right enzyme, in the right place, at the right time, in the right amount, properly "activated", and free to participate in the reaction are all necessary conditions for a biochemical reaction to occur. Note that the enzyme acts *only* to promote the reaction. It is, itself, not changed in any way by the reaction and it is not one of the participating substrates. That is the definition of a catalyst.

Basic Principles of Chemical Kinetics

Bearing the above in mind, and from the basic thermodynamic principles introduced earlier, let us take a closer look at the role played by enzyme catalysts in the physiologic processes associated with metabolism. In order for molecules to react with one another, at least the following criteria must be met:

1) The molecules must have a chemical *affinity* for one another;

2) They must collide, or otherwise come into contact or close *proximity* to one another;

3) They must collide often enough, i.e., the *frequency of collisions* must be high enough for a reasonable number of them to interact and thus generate a desired response;

4) They must have sufficient energy (a function of temperature) to overcome the *energy barrier* for reaction; and,

5) If the reaction is endergonic (positive change in free energy), an *energy source* must be available to drive it. This is not necessary if the reaction is exergonic (negative change in free energy) or "spontaneous".

The Concept of an Energy Barrier for Chemical Reactions

In an everyday sense, we have a qualitative appreciation for what is meant by an "energy barrier" restricting certain chemical or physical processes. Air/fuel mixtures in gasoline engines would not burn unless ignited by the spark plug in the cylinders. The "spark" provides the energy of activation, E_a (or, "primes the pump") for combustion to take place. In high school and college chemistry laboratories, students learn that some chemical reactants need to be heated before they will interact, such heat being often provided by a Bunsen burner or some other energy source. And, in a mechanical analog of this concept, if one envisions a boulder lying in a crevice or groove on the side of a hill, then our need to first "lift" (E_a) the boulder over the crevice before it will roll down the hill by itself represents exactly the activation energy necessary for the reaction (rolling down the hill) to occur.

While resting in the crevice, the boulder is said to be in a *metastable* state, or, in a state of *relative equilibrium*, because it will stay there in the absence of E_a. At the bottom of the hill, the boulder is said to be in a state of *stable equilibrium* because it can go no lower without a significant change in its environmental situation. The change from a state of metastable equilibrium to a state of stable equilibrium represents the total energy released in the "reaction". For open thermodynamic systems, this change is most conveniently expressed in terms of the enthalpy function, $h_E(s, p)$ defined earlier, i.e., $h_E = u + pv$. Thus,

$$\delta h_E = \delta u + p\delta v + v\delta p \qquad \text{[4-10]}$$

or, substituting into equation [4-10] the first law of thermodynamics expressed as equation [4-6],

$$\delta h_E = \delta Q + v\delta p \qquad \text{[4-11]}$$

For an isobaric (constant pressure) process, $\delta p = 0$. Then $\delta h_E = \delta Q$. If we now define a *specific heat* at constant pressure, c_p as the ratio of the heat added per unit mass of a material to the corresponding change in temperature of the material, at constant pressure, i.e., $c_p \equiv \left(\dfrac{\delta Q}{\delta T} \right)_p$, then, by definition, $\delta Q = c_p \delta T$. Thus, for an isobaric process, $\delta h_E = c_p \delta T$, and so we conclude that: $h_E = c_p T$, and, from equation [4-11], that:

$$c_p = \left(\frac{\partial h_E}{\partial T} \right)_p \qquad \text{[4-12]}$$

According to the Law of Dulong and Petit, as the temperature is increased from absolute zero, vigorous atomic vibrations cause the specific heat to increase rather rapidly from zero. It finally levels off to a nearly constant value which, for the elements is about 6 thermodynamic calories, per mole, per degree Kelvin.

For an *isothermal,* isobaric system, the change from a metastable to a stable equilibrium state is accompanied by a change in *Gibbs* free energy, $g(T,p)$, which, you will recall, is equal to enthalpy, h_E, minus the nonusable energy (entropy) portion, sT. That is to say, the state function, or thermodynamic potential that is most appropriate to describe the energetics of isothermal, isobaric systems is the Gibbs free energy, $g = u + pv - sT$; and this is particularly significant for the case of the human body since, in a thermodynamic sense, it is precisely an isothermal

(37°C), isobarometric (atmospheric pressure) open system. Moreover, since the Gibbs free energy is a combination of enthalpy (which derives from the first law of thermodynamics) and entropy (which derives from the second law), it expresses *both* laws in terms of a single function -- which is valid for processes which can be assumed to take place in a "piecewise" quasi-equilibrium manner. We shall return to this a bit later on.

Since enthalpy and entropy are both state functions, Gibbs free energy is also a state function. Thus, we may write:

$$\delta g(T, p) = \left(\frac{\partial g}{\partial T} \right)_p \delta T + \left(\frac{\partial g}{\partial p} \right)_T \delta p \qquad \text{[4-13]}$$

and we see immediately that, for an isothermal (T constant), isobaric (p constant), open system, $\delta g = 0$. We shall define this situation to be *equilibrium* (minimum free energy, maximum entropy) for a thermodynamic system like the human body. To move away from a state of equilibrium we must therefore have some additional gradient in one of our other potential energy functions, or, we must have some mass influx or efflux carrying energy with it. These situations shall be addressed later when we talk about changes in concentrations of certain chemical substrates and end products. In physiologic systems, part of any change in Gibbs free energy is used to do work, part is stored as chemical bond energy following the principles discussed in the previous section on basic thermodynamics, and part is converted into nonusable heat.

The existence of energy barriers for chemical reactions again presents a problem where isothermal systems are concerned. Basically, the energy barrier, or activation energy, represents that *minimum* amount of kinetic energy that molecules must have in order to collide often enough, and with enough energy to interact. That is, all species of molecules have associated with them negatively charged electron "clouds". Thus, under relatively quiescent conditions, any given molecule randomly approaching the "sphere of influence" of another molecule will experience a natural tendency for repulsion, and the molecules will probably never get close enough to "react". In other words, in "gentle" collisions, electrostatic repulsion will cause the colliding molecules to merely "glance" off of one another, and the two will separate while maintaining their original identity.

Given enough *additional* kinetic energy, however, not only does this increase the *probability* that enough molecules will collide enough times for something to happen, but it also allows the colliding molecules to *use* some of that kinetic energy to do work to overcome the repulsive forces. Now, when the molecules collide *hard* enough, they may penetrate each other's sphere of influence far enough to overcome electron repulsion and allow electron *rearrangement* to generate one or more new species. This is precisely what the concept of an energy of activation is all about. But, in practice, the addition of molecular kinetic energy is usually accomplished by heat ("spark plug", or striking a match against a friction surface, or using a Bunsen burner, and so on) which, unfortunately, implies an increase in temperature, as well, assuming a phase-change is not occurring.

One way that we might get around this problem and "reduce" the activation energy is to physically *hold* the molecules together long enough for them to react, and, as much as possible, take out of the picture the element of chance. In other words, instead of depending on having molecules come into contact with one another by chance -- and increasing both *that* probability and the chance that they will have enough *energy* to interact by increasing the kinetic energy (i.e., temperature) of the molecules -- why not devise a means whereby some "match-maker" actually *brings* these molecules into proximity with one another in a systematic "assembly-line" fashion? Why not have the meeting (and mating) of the molecules carefully pre-arranged? Indeed, mother nature has provided isothermal physiologic systems with just such match-makers, which we call biochemical catalysts, or, *enzymes*. Note, especially, that enzymes act *directly* on E_a, which is otherwise not a function of temperature, per se. That is, E_a is just a measure of the kinetic energy *required* by molecules in order for them to react. At higher temperatures, *more* molecules will be in this "active" form; at lower temperatures, *fewer* molecules will have the required activation energy to react. But catalysts actually take charge to actively "by-pass" E_a as described below.

Enzymes are nothing more than protein molecules that have specific, geometrically-shaped "landing" or binding sites ("docks", if you will) that are designed to trap molecules that the catalyst is trying to get to react. In this sense, they are exactly analogous to the carrier molecules which we spoke about in Chapters 2 and 3. In fact, enzyme-catalyzed reactions are really a form of active transport. The enzyme travels around gathering its substrates together, bringing them into proximity with one another in a systematic fashion -- but not for the purpose of *transporting* them. Instead, it directs their *interaction,* and then releases the end-products into the environment. In this fashion, it *decreases* E_a, the activation energy or energy barrier for that particular reaction to take place. While reducing the amount of "heat" necessary to get the reaction to occur, it also allows the reaction to proceed with more precision, at a quicker rate, more efficiently, with a greater likelihood of occurrence, more effectively, and in a more controlled fashion. The latter is due, in part, to the fact that, since enzyme kinetics is essentially a form of active transport, many of the control mechanisms discussed in Chapters 2 and 3 with respect to carrier molecules apply equally well to *this* form of biochemical catalysis. Thus, biochemical kinetics is really the study of *enzyme* kinetics, which we now examine briefly, expanding upon the discussion in Chapter 3 of Michaelis-Menten Active Transport Kinetics.

Enzyme Kinetics

To begin with, we note that descriptions of the isothermal, isobaric metabolic processes associated with physiologic function, concern themselves, as stated above, with changing concentrations of certain substrates and end products. Thus, in the sense that equation [3-30], derived and discussed in Chapter 3, defines the work done to change the concentration of some species S from $[S]_2$ to $[S]_1$, we see that the relevant Gibbs free energy function that applies to physiologic biochemical kinetics problems is:

$$g = RT \ell n [S], \text{ per mole of } S \qquad [4\text{-}14]$$

Equation [4-14] predicts that $g = 0$ if $[S] = 1$ mole per liter, which, of course is not the case except at a temperature of absolute zero. We must therefore define the free energy, g^o, of a solution containing one mole of S per liter of solution at physiologic conditions of temperature and pressure (37°C and 1 atmosphere, respectively), and modify equation [4-14] to read:

$$g = g^o + RT \ell n [S] \qquad [4\text{-}15]$$

Generalizing equation [4-15] still further for a mixture or solution composed of n species $[S_i]$ $(i = 1,2,3,4, \dots , n)$, and letting N_i be the total number of moles of $[S_i]$ present, we may write:

$$g^* = \sum_{i=1}^{i=n} N_i g_i = \sum_{i=1}^{i=n} N_i(g_i^o + RT \ell n [S_i]) \qquad [4\text{-}16]$$

where now g^* is the *total* concentration potential energy contained in the solution, and, again, we are assuming that the *only* change in Gibbs free energy for a biochemical reaction is due to changes in the concentration of substrates and end-products. In other words, the guiding principle in chemical kinetics is that the energy associated with any given reaction may be measured in terms of changes in the concentration of both reactants and products.

Bearing this in mind, consider a simple case involving two substrates, S_1 and S_2, which react to form two products, S_3 and S_4, in a reaction catalyzed by enzyme E:

$$S_1 + S_2 + E \quad \underset{\varepsilon_2{}^*}{\overset{\varepsilon_1{}^*}{\rightleftarrows}} \quad E - S_1 - S_2 \quad \underset{\varepsilon_4{}^*}{\overset{\varepsilon_3{}^*}{\rightleftarrows}} \quad S_3 + S_4 + E \quad \text{[4-17]}$$

Note that we have allowed reversibility of all of the reactions involved, which is somewhat more general than the case considered in Chapter 3 (c.f., equations [3-63] to [3-66]) in our discussion of Michaelis-Menten kinetics of active transport. For the reaction defined in equation [4-17], the rate of formation of the enzyme-substrate complex from the *reactants* is given by: $\varepsilon_1{}^*[S_1][S_2][E]$, and the corresponding rate of dissociation *back* to reactants is given by: $\varepsilon_2{}^*[E - S_1 - S_2]$. At equilibrium the two are equal, i.e., the rate of disappearance of substrate is just balanced by the rate of replenishment of reactants; such that:

$$[E - S_1 - S_2] = \frac{\varepsilon_1{}^*}{\varepsilon_2{}^*} [S_1][S_2][E] \qquad \text{[4-18]}$$

Similarly, the rate of formation of *products* from the enzyme-substrate complex is given by: $\varepsilon_3{}^*[E - S_1 - S_2]$, while the corresponding rate of re-association *back* to the enzyme-substrate complex *from products* is given by: $\varepsilon_4{}^*[S_3][S_4][E]$. Again, at equilibrium, these two are equal, i.e., products are formed at the same rate as the enzyme-substrate complex is resynthesized. Thus,

$$[E - S_1 - S_2) = \frac{\varepsilon_4{}^*}{\varepsilon_3{}^*} [S_3][S_4][E] \qquad \text{[4-19]}$$

Substituting equation [4-18] into equation [4-19], and rearranging, one concludes that:

$$\frac{[S_3][S_4]}{[S_1][S_2]} = \frac{\varepsilon_1{}^*\varepsilon_3{}^*}{\varepsilon_2{}^*\varepsilon_4{}^*} = \frac{k_p}{k_x} = K_e \qquad \text{[4-20]}$$

where k_p is the forward rate constant (i.e., the rate of disappearance of substrate, or formation of products); k_x is the reverse rate constant (i.e., the rate of disappearance of products, or replenishment of substrate); and K_e is the dimensionless *equilibrium constant* for the entire reaction. The enzyme-catalyzed reaction can thus be written in more simplified form as:

$$S_1 + S_2 \;\underset{k_x}{\overset{k_p, E}{\rightleftharpoons}}\; S_3 + S_4 \qquad\qquad \textbf{[4-21]}$$

Note, again, that the enzyme effectively serves the purpose of increasing $k_p = \varepsilon_1{}^* \varepsilon_3{}^*$, and thus driving the reaction to the right faster, more efficiently, and more effectively.

Now, the Gibbs free energy associated with the substrates is, from equation [4-16]:

$$N_{S_1}(g_{S_1}{}^o + RT\,\ell n\,[S_1]) + N_{S_2}(g_{S_2}{}^o + RT\,\ell n\,[S_2]),$$

while the Gibbs free energy associated with the products is:

$$N_{S_3}(g_{S_3}{}^o + RT\,\ell n\,[S_3]) + N_{S_4}(g_{S_4}{}^o + RT\,\ell n\,[S_4]).$$

For the simplest type of reaction, in which one mole ($N_{S_1} = 1$) of S_1 combines with one mole ($N_{S_2} = 1$) of S_2, in the presence of enzyme E, (which is not *consumed* in the overall process, but simply accelerates it), to yield one mole ($N_{S_3} = 1$) of S_3 and one mole ($N_{S_4} = 1$) of S_4, the change, Δg^*, in Gibbs free energy associated with the reaction given in equation [4-21] may be defined as the difference between the total free energy of the products, and the total free energy of the reactants, which, for this case, is:

$$\Delta g^* = g_{S_3}{}^o + g_{S_4}{}^o - g_{S_1}{}^o - g_{S_2}{}^o + RT\,(\ell n\,[S_3] + \ell n\,[S_4]$$

$$\qquad\qquad\qquad\qquad\qquad\qquad\qquad\qquad\qquad \textbf{[4-22]}$$

$$- \ell n\,[S_1] - \ell n\,[S_2])$$

The change in free energy given by:

$$g_{S_3}{}^o + g_{S_4}{}^o - g_{S_1}{}^o - g_{S_2}{}^o = \Delta g^{*o}$$

is called the *standard free energy* of the reacting species, and corresponds to the change in Gibbs free energy associated with reaction [4-21] taking place when $[S_1] = [S_2] = [S_3] = [S_4] =$ one mole per liter. That is, the standard free energy of a chemical reaction is the free energy made available for doing useful work by the reaction of a mole per liter of each reactant to form a mole per liter of each product under standard conditions of temperature and pressure. It also corresponds to the situation when $\ell n [S_4][S_3] = \ell n [S_2][S_1]$, or, when $\dfrac{[S_4][S_3]}{[S_2][S_1]} = K_e = 1$, as shown by equations [4-22] and [4-20]. Thus, the standard free energy defines the situation when the forward rate constant of any given reaction is exactly equal to the reverse rate constant, or, at equilibrium, when the rate of association exactly equals the rate of dissociation in equation [4-21]. Moreover, since:

$$\ell n [S_4] + \ell n [S_3] - \ell n [S_2] - \ell n [S_1] = \ell n \frac{[S_4][S_3]}{[S_2][S_1]} = \ell n K_e,$$

the free energy change associated with the enzyme-catalyzed reaction [4-21] can be written as:

$$\Delta g^* = \Delta g^{*o} + RT \ell n K_e \qquad\qquad \text{[4-23]}$$

and it is valid per mole of reacting species (note that the units of RT are in energy per mole, as are the units of Δg^{*o}).

In a state of equilibrium, the free energy change associated with any given reaction is zero, i.e., the free energy of the products is equal to the free energy of the reactants, and no energy is made available to do useful work. In this case, $\Delta g^* = 0$, so we may write, for a state of equilibrium: $\Delta g^{*o} + RT \ell n K_e = 0$, or,

$$\Delta g^{*o} = - RT \ell n K_e \qquad\qquad \text{[4-24]}$$

Note that if the concentration of products is bigger than the concentration of substrates, then their ratio as defined by K_e is bigger than one, the natural log of their ratio is bigger than zero, the standard free energy of the reaction is *negative* (as defined above) and the process may proceed *spontaneously* from substrates to products. In other words, equation [4-21] is driven to the *right* for Δg^{*o} negative. On the other hand, if the standard free energy of a reaction is *positive,* the implication from above is that the concentration of products tends to be less than the concentration of reactants, such that their ratio as defined by K_e is less than unity, the natural logarithm of their ratio is less than zero (i.e., negative), and the reaction can be made to proceed predominantly to the right only if, in some way, external free energy can be made available to drive the process in that direction.

Now, solving for K_e from equation [4-24], one obtains:

$$K_e = e^{-\frac{\Delta g^{*o}}{RT}} \qquad\qquad \text{[4-25]}$$

This is a specific form of a more generalized expression developed by Van't Hoff and Arrhenius to relate the velocity constant or reaction rate of a spontaneous process to the absolute temperature of the process. That is, in Van't Hoff and Arrhenius' formulation, the standard free energy, Δg^{*o}, corresponds to the energy of activation, E_a, of the reaction, and they write equation [4-24] in terms of the common logarithm as:

$$2.3 \log K_e = A_o - \frac{E_a}{RT} \qquad\qquad \text{[4-26]}$$

where A_o is a dimensionless constant which is a qualitative measure of the frequency of molecular collisions, and of the requirement that molecules have a specific orientation with respect to one another as they collide, in order for a reaction to occur.

We can use the Van't Hoff and Arrhenius formulation [4-26] to illustrate the potency that enzyme-catalysis of biochemical reactions can have on increasing the forward rate constant of the reactions by decreasing the energy of activation, E_a.

The Van't Hoff - Arrhenius Equation

Let us write equation [4-26] in the form:

$$\ell n \left(\frac{k_p}{k_x} \right) = A_o - \frac{E_a}{RT} \qquad [4\text{-}27]$$

and, for the purpose of illustrating a point, let us assume for the moment that k_x is constant. Then, if we define a new constant, A^*_o, equal to $A_o + \ell n\, k_x$, equation [4-27] becomes:

$$\ell n\, k_p = A^*_o - \frac{E_a}{RT} \qquad [4\text{-}28]$$

where k_p is the previously defined forward rate constant for the formation of products from substrates.

Now, for A^*_o, R and T (isothermal) constant, let $(k_p)_1$ represent the rate of reaction for an *uncatalyzed* process having associated with it an energy of activation $(E_a)_1$, and let $(k_p)_2$ represent the corresponding rate of reaction for the *catalyzed* process, having associated with it a lower energy of activation $(E_a)_2$. Then from equation [4-28], one can write:

$$\ell n\, (k_p)_1 - \ell n\, (k_p)_2 = A^*_o - \frac{(E_a)_1}{RT} - A^*_o + \frac{(E_a)_2}{RT} \text{ , or}$$

simplifying and rearranging this equation,

$$\ell n\, \frac{(k_p)_1}{(k_p)_2} = \frac{1}{RT} \left[(E_a)_2 - (E_a)_1 \right] \qquad [4\text{-}29]$$

Suppose $(E_a)_2$ is as little as 20% less than $(E_a)_1$ (in real life, the difference is much more dramatic), such that $(E_a)_2 = 0.8(E_a)_1$. Then, at body temperature ($T = 37°C = 310°K$), with $R = 1.98$ calories per mole, per degrees Kelvin, equation [4-29] yields:

$$\ell n \ \frac{(k_p)_1}{(k_p)_2} = - (E_a)_1 (0.000325839) \qquad [4\text{-}30]$$

and, for $(E_a)_1$ on the order of, say, 25,000 calories per mole, the above equation gives:

$$\frac{(k_p)_1}{(k_p)_2} = 0.00029, \ \text{or,}$$

$$(k_p)_{\text{catalyzed}} = 3.5 \times 10^3 \ (k_p)_{\text{uncatalyzed}}$$

Thus, we see the startling result that just a mere 20% decrease in activation energy can increase the rate of reaction on the order of a thousand-fold! In actuality, enzymes in the human body increase reaction rates by factors as high as *one trillion* times the uncatalyzed rate, and they are able to do this primarily by their first-order effect on E_a -- which, again, is otherwise constant for a given set of reacting species.

Chemical Kinetics Involving Electrically Charged Species

The Gibbs free energy function given by equation [4-14] in terms of concentration potential energy assumes that the reacting species are electrically neutral. Where this is not the case, e.g., in oxidation-reduction reactions, one must use instead the electrochemical potential energy function, μ_c, defined by equation [3-52] in Chapter 3, i.e.,

$$g = \mu_c = \overline{F} ZB + RT \ell n \ [S], \text{per mole of } S \qquad [4\text{-}31]$$

where, again, $\overline{F}$ is the Faraday constant (96,500 coulombs per mole of monovalent ions), Z is the valence of biochemical species S (or, in the case of oxidation-reduction reactions, the *number* of electrons transferred per mole of reacting species), B is the electrical potential (c.f., equation

[3-50]) associated with the lack of charge neutrality in the reacting species, R is the universal gas constant (1.98 calories per mole per degrees Kelvin), T is absolute temperature (310°K), and $[S]$ is the molar concentration of S relative to a solution containing one mole of S per liter. Following reasoning similar to that used to derive equation [4-23], one can now expand the relationship [4-31] to get:

$$\Delta g^* = \Delta g^{*o} + RT \ell n \, K_e + \bar{F} Z(B_p - B_x), \qquad [4\text{-}32]$$

where the subscript p refers to products and the subscript x refers to substrates. In an oxidation-reduction reaction, the products are the *oxidized* form of the reactants, and the substrates are the *reduced* form of the reactants, as, for example:

$$H_2 \underset{\text{reduction}}{\overset{\text{oxidation}}{\rightleftharpoons}} 2H^+ + 2e^- \quad \text{(oxidation reaction} = loss \text{ of electrons)}$$

$$O^{-2} \underset{\text{reduction}}{\overset{\text{oxidation}}{\rightleftharpoons}} \frac{1}{2} O_2 + 2e^- \quad \text{(reduction reaction} = gain \text{ of electrons)}$$

$$[4\text{-}33]$$

In equation [4-33], if we combine the *oxidation* (*increase* in charge from 0 to +1) of hydrogen with the *reduction* (*decrease* in charge from 0 to -2) of oxygen, then:

$$H_2 + \frac{1}{2} O_2 = 2H^+ + 2e^- + \frac{1}{2} O_2 \longrightarrow O^{-2} + 2H^+ = H_2O \quad [4\text{-}34]$$

which, of course, is an equation that defines the formation of water from the substrates hydrogen and oxygen. In the glycolytic reaction [4-9] for the metabolism of glucose, carbon atoms increase in oxidation number from 0 to +4 while the sugar is being oxidized to yield carbon dioxide and water.

In an equilibrium situation, Δg^* in equation [4-32] is equal to zero, while if the products and substrates are present in equal concentrations of one mole per liter at standard conditions of temperature and pressure, $K_e = 1$ (c.f., equation [4-20]), so $\ell n\, K_e = 0$. In this case, we define the *Standard Reduction Potential* from equation [4-32] as:

$$B_p - B_x = -\frac{1}{\overline{F}Z}\, \Delta g^{*o} = \Delta B_o{}' \qquad\qquad \text{[4-35]}$$

where the prime indicates that it is also determined at a neutral pH of 7.00. For the $O_2 - H_2O$ reaction given by equation [4-34], the standard reduction potential is $+0.816$ volts, which means that the oxidation of hydrogen, followed by the reduction of oxygen proceeds spontaneously (i.e., Δg^{*o} is negative in equation [4-35], so $\Delta B_o{}'$, as indicated, has a positive voltage sign) and generates electrons of sufficient quantity (i.e., generates enough "e^- 's"), and with sufficient energy relative to neutral ground, to cause as much current to flow as if a 0.816 volt electromotive force were driving the electrons through a closed electrical circuit.

Substituting equation [4-35] into equation [4-32], and letting $B_p - B_x = \Delta B^*$, one obtains, at equilibrium:

$$\Delta B^* = \Delta B_o{}' - \frac{RT}{\overline{F}Z}\, \ell n\, K_e \qquad\qquad \text{[4-36]}$$

Equation [4-36] is used to define the *redox* ("reduction-oxidation") *potential difference* (or, simply, the redox potential) of a given solution. This is the "electron pressure" that this solution exerts on an inert electrode, just as the pH of a solution represents its "proton pressure". Thus, a system composed of an acid and its salt resists changes in pH (or "buffers" a given solution), while a system composed of oxidant and reductant resists changes in redox potential, ΔB^* (or "poises" the given solution). From equation [4-36], one can define a generalized redox potential as:

$$B^* = B_o{}' + \frac{RT}{\imath \overline{F}}\, \ell n\, \frac{[\text{oxidized product}]}{[\text{reduced substrate}]} \text{ , or,}$$

$$B* - B_o' = \Delta B_r = \frac{RT}{\iota \bar{F}} \, \ell n \, \frac{[\text{oxidized product}]}{[\text{reduced substrate}]} , \qquad [4\text{-}37]$$

where ι is the number of electrons transferred per mole. Equation [4-37] is in the form of a Nernst Equation (c.f., equation [3-54]) for electrochemical equilibrium in oxidation-reduction reactions. Values of B_o' as a function of temperature, and B_o as a function of pH and temperature may be found tabulated in many handbooks, biochemistry texts and other literature references.

Temperature Effects in Enzyme Kinetics

The effect of temperature on enzymic reactions is twofold: with increasing temperature, one first sees an *increase* in reaction rate; however, above some maximal rate, one sees a dramatic *decline* in reactivity with increasing temperature. The first effect can be explained by going back to equation [4-26] and writing, for two temperatures, T_1 and T_2, corresponding, respectively, to equilibrium constants $(K_e)_1$ and $(K_e)_2$, the following relationship:

$$2.3 \, \frac{R}{E_a} \log \frac{(K_e)_1}{(K_e)_2} = \left(\frac{1}{T_2} - \frac{1}{T_1} \right) \qquad [4\text{-}38]$$

where, as explained earlier, the energy of activation is *not*, itself, a function of temperature (only the *number* of molecules possessing E_a depends on T), and neither is the dimensionless constant A_o. Thus, whereas enzymic activity accelerates reactions by its effect on E_a as illustrated in equation [4-29], temperature accelerates reactions by its effect on K_e as illustrated in equation [4-38]. That is, if T_2 is bigger than T_1, then $\left(\frac{1}{T_2} \right)$ is less than $\left(\frac{1}{T_1} \right)$, the expression on the right in equation [4-38] is negative, which means that we are taking the logarithm of a ratio less than unity, which, in turn, means that $(K_e)_2$ at the higher temperature is correspondingly bigger than $(K_e)_1$ at the lower temperature. This is because temperature can increase the kinetic energy of molecules, thus allowing more of the molecules of the substrate to possess the required energy to react -- and, thereby increasing the rate of formation of the end-products.

However, as is commonly understood, physical substances generally expand with increasing temperature and contract with decreasing temperature. In the case of enzymes, at higher temperatures the protein molecules undergo a complex rearrangement of structure due both, to a geometric "warping", and to the disassociation of many weak secondary polar bonds in the structure. Since the geometry of the molecule is crucial to its being able to function as a biochemical catalyst, excessive increases or decreases in temperature can "denature" it, in the sense that its molecular configuration takes on the wrong geometry for the substrates upon which it is expected to act. In other words, enzymes can be neutralized as biochemical catalysts by "thermally deactivating" the necessary "stereo-specific geometric congruence" that must exist between reactants and reaction "facilitators".

Our 98.6°F (or 37°C) body temperature is thus a "compromise" between the *enhancing* effect of temperature on chemical reaction rates-- about two to three times for every 20°F increase--and the *adverse* effects of temperature on the activity of enzyme systems--outside a $\pm 10\%$ change in temperature from 37°C, the destructive effects on enzymes and other proteins may exceed the ability of the body to repair or replace them by a factor of 40 or more for every degree change in temperature. Thus, the need to maintain geometric compatibility between enzymes and substrates is one of the reasons that we are homeothermic (warm-blooded) rather than cold-blooded animals. That is to say, in order to preserve the functionality of our various enzyme systems, our metabolic processes must proceed along relatively isothermal pathways; and the reason the relevant temperature involved is 37°C is that this is the temperature at which our enzymes "work" best, i.e., fastest and most efficiently. This brings us to the subject of how the organism handles effectively the heat generated by biochemical reactions, and, how it protects itself from dangerously low or high environmental temperatures. But first, let us find out exactly how much metabolic heat the body has to deal with on an everyday basis.

Basic Metabolic Considerations

The Concept of Basal Metabolism

The absolute *lowest* level of metabolism required simply to maintain the life of an individual is called the *basal metabolic rate*. For an "average" person, this amounts to some 1000 calories (i.e., one kilocalorie, which is a dietary calorie, written with an upper-case "C", or abbreviated Cal) per hour, per kilogram of body weight. This translates into anywhere from 1500 to 2000 Calories (kilocalories) per day, with an average of 1800 being quite typical. We shall use the latter for illustrative purposes.

About 80% of an individual's basal energy requirement (some 60,000 thermodynamic calories per hour) is needed to keep alive the "core" or internal organs, which include, for example, the heart, lungs, liver, kidneys, and brain. The remaining 20% (15,000 calories per hour) is needed by the peripheral tissues (arms, legs, and so on). Moreover, about 20% of the basal energy is utilized by the liver, 15% by the brain (up to as much as 20%), 12% by the heart, and the remainder elsewhere. Thus, of the 100% total, the brain utilizes up to 20%, the extremities up to 20%, the liver up to 20%, the heart up to 12%, and the remaining organs of the trunk up to 28%. Basal metabolic rate decreases with age (most dramatically between ages 0 and 20 for both males and females) and increases almost linearly with an individual's weight.

Diet: The Energy Source (Fuel) for our Engine

The energy for basal metabolism (indeed, for *all* physiologic function) is derived from the three basic foods we eat, which are classified as protein, carbohydrate and fat. As mentioned in Chapter 3, physiologic oxidation of one gram of each of these provides the following energy yields: *Protein*: 4100 thermodynamic calories per gram; *Carbohydrate*: 4100 cal/gram; and *Fat*: 9300 cal/gram. Thus, on a *balanced* diet of 341.5 grams of carbohydrate, 96.75 grams of fat, and 73.25 grams of protein (total, 511.5 grams or about 1.125 pounds of dry food daily), an "average individual" consumes, in a typical daily diet, some 1400 dietary Calories of carbohydrate (range, 35-60% of diet), 900 Calories of fat (range, 25-40% of diet), and 300 Calories of protein (range, 15-25% of diet), for a total of 2600 dietary calories. Of these, it is estimated that 5% is lost in the process of being broken down into usable cell substrates (i.e., ingestion, digestion, mass transport, and so on), leaving 2470 Calories available to the cells following digestion (Ruch and Patton, 1965).

An additional 50% or so (i.e., around 1300 Calories) of the total consumed enters a free-energy pool (e.g., ATP) to be stored and subsequently retrieved to drive cellular processes (see Chapter 5). Of the remaining 45% (1170 Calories), half or more, i.e., at least 585 to 650 Calories is utilized immediately to meet the internal requirements of the organism (e.g., structural, functional and chemical integrity of the body), leaving around 20% or less (some 520 Calories) of potential energy available to do external work (e.g., skeletal muscle contraction). Again, some 1800 of the original 2600 Calories are generally required for "housekeeping" (basal metabolism), 200 are necessary for "miscellaneous" dynamic activities (e.g., processing sensory information) and "other" bodily functions that are not related to the maintenance of life, itself (e.g., procreation), leaving 600 Calories or so available for performing external physical activities.

The above calculations, of course, do not take into account dietary intake of water (700 mℓ, or around 700 grams, or some 1.5 pounds), which is inherently mixed in with the food, but has no caloric value (c.f., Chapter 3 discussion surrounding Figure 3-4). Nor do they account for undigested (or undigestible) sewage that passes right on through the gastro-intestinal system without ever making it into the organism. This may amount to another quarter-to-half-a-pound of food, such that the average total food intake per day is around 3 pounds (1.36 kg) or on the order of 1.5 to 2.0 percent of body weight. To put this in perspective, consider yourself insulted if you are accused of "eating like a bird" (a comment normally intended to compliment "light" eaters). Birds consume daily as much as six times their body weight!

Atmosphere: The Oxygen Source (Air) For Our Engine

Let us now relate the above considerations to the oxygen requirements of the body. To begin with, one gram-mole of glucose ($C_6H_{12}O_6$) weighs (6)(12) + (12)(1) + (6)(16) = 180 grams. It reacts with six gram-moles of oxygen weighing (6)(32) = 192 grams, to yield 6 moles of carbon dioxide (264 grams) and 6 moles of water (108 grams) in a reaction such as that given by equation [4-9]. Thus, 6 moles of O_2 will oxidize 180 grams of glucose to yield (using 4100 cal/gm for carbohydrates) 738,000 thermodynamic calories under physiologic conditions. This figure is comparable to experimentally measured values for *pure* glucose of 675,000 reported at 0°C, 688,160 reported at 25°C, and 686,000 reported in some textbooks in biochemistry. According to equation [4-9], the 6 moles of CO_2 produced in this reaction is such that the volume of carbon dioxide produced will exactly equal the volume of oxygen utilized -- since, according to Avogadro's principle, under identical conditions, all gases have equal molar volumes. This ratio of volumes (carbon dioxide to oxygen) is called the *respiratory quotient* ("RQ", in the trade), which, on a mixed diet of carbohydrate, fat and protein, is generally equal to around 0.82.

Going still one step further, from the perfect gas law, one can estimate the volume that 192 grams of oxygen occupies. That is, with physiologic pressure, $p = 1$ atmosphere = 1.01325×10^6 dynes/cm^2 at the cellular level, $N = 6$ moles, $R = 8.31 \times 10^7$ dyne-cm/deg-mole, and $T = 310.15°K$, $V = \dfrac{NRT}{p}$ comes out to be equal to 152.6 liters. Thus, at atmospheric pressure, 152.6 liters of oxygen will react with 180 grams of glucose at body temperature, giving a ratio of 0.85 liters of oxygen per gram of glucose. And, since each gram of glucose yields 4100 calories while using up this 850 milliliters of oxygen, one *liter* of oxygen would yield $\frac{1000}{850}(4100) = 4823.5$ calories at 37°C. This value is called the *caloric equivalent* of one liter of oxygen. Again, for a *mixed* diet containing all three major types of nutrients, the caloric equivalent is rounded off to some 5000 calories per liter of oxygen, or, 5 thermodynamic cal. per mℓ of O_2.

Suppose, then, that an individual has a basal metabolic rate of 1800 dietary Calories per day $= \frac{1,800,000}{24} = 75,000$ thermodynamic calories per hour = 1250 cal/min. This individual's cells would then need to have delivered to them, $(1250)\frac{cal}{min} \times \frac{1}{5}\frac{m\ell\ O_2}{cal} = 250$ milliliters of oxygen per minute. Now, the normal resting cardiac output is 5 liters of blood per minute, which is to say, blood flows by the cells and tissues of the body at a mean volumetric flow rate of 5000 mℓ/min. Thus, on the average, cells must extract 0.05 mℓ of oxygen for every 100 mℓ of blood that goes by, in order to get 250 cc of O_2 per minute, i.e., $(250)\dfrac{\frac{m\ell\ O_2}{min}}{\frac{minute}{5000\ m\ell\ Blood}} = 0.050$ m$\ell\ O_2$ per mℓ of blood, or, 5 mℓ of oxygen per 100 milliliters of blood. This confirms the arterio-venous $AV - \Delta O_2$ difference of 5 volumes-percent mentioned in Chapter 3 (pg. 241) in the discussion of substrate control of transport. Thus, on the average, arterial blood contains 19.6 cc O_2 per 100 mℓ of blood (c.f., pg. 159), and, at rest, venous blood contains some 14.6 cc O_2 per 100 cc of blood.

Metabolic Heat Generation as a Product of Biochemical Reactions

When an individual does work, the metabolic rate can increase to 500,000 calories per hour (a six-to-seven-fold increase compared to the basal metabolic rate), and, at maximum activity, this figure can be as high as 20 times the basal metabolic rate! While the rate of energy utilization in the organs rises slightly (and the need for additional oxygen is met partially by the oxygen-reserve mechanism discussed in Chapter 3, pgs. 241-242), most of this increase in metabolic activity is attributable to complex chemical reactions taking place in the peripheral musculature of the body, and more than 75% of it shows up ultimately as heat energy. In other words, of the average daily caloric intake (energy *potentially* available to the physiologic system) of about 2,600,000 thermodynamic calories (1.09×10^7 Joules), more than 1,950,000 gram-calories (0.8×10^7 Joules) winds up as heat energy following a complicated sequence of life-sustaining and other chemical reactions associated with the activities of daily living. This amounts to a rate of generation of heat on the order of 81,250 calories per hour.

If we consider the above rate of heat generation as contributing to the enthalpy pool of the organism (c.f., equation [4-12]), then, per unit mass, $\delta Q = c_p \delta T$, and so, for a "typical" 75-kg individual,

$$\frac{\delta Q}{\delta t} = M c_p \frac{\delta T}{\delta t} \qquad [4\text{-}39]$$

Let $(\delta Q/\delta T) = 81{,}250$ cal/hr, $M = 75$ kg, and $c_p = 0.92$ Cal/Kg $-$ °C (see later in this chapter). Then, from equation [4-39] we calculate $\dfrac{\delta T}{\delta t}$ to be on the order of 1.18°C/hr. Thus, in the absence of proper heat-dissipating mechanisms, even under the simplest basal, resting conditions, body core temperature could rise as fast as 1 to $1 - \dfrac{1}{4}$ °C (1.8 to 2.25°F) per hour -- which, in the short span of just three hours, would raise the temperature to critical, near-death levels. Since the human organism must function as an isothermal engine, the heat generated by such activities of daily living, basal metabolism and muscular activity *must* be disposed of *quickly* and *efficiently* to prevent overheating and breakdown of the system. Thus, let us return to our generic transport equation [2-4] and apply it this time to the physiologic transport of heat energy. That is to say, with:

$$\Phi = - \frac{A_t \rho \alpha (\nabla \phi)}{(2\phi)^n} \qquad [4\text{-}40]$$

we seek, now, to find appropriate quantities A_i, ρ, α, and ϕ, such that we may calculate, or at least get a reasonable approximation of Φ for *heat* flux, when $n = 0$ for energy. As an example, consider the simple transport of heat by conduction (a *thermal*, as opposed to *mass* diffusion problem such as was considered in Chapter 3). For this case, the appropriate value of ϕ is $c_p T$, the enthalpy or total heat content of an open thermodynamic system. The appropriate kinematic transport coefficient is $\alpha = \dfrac{k_t}{\rho c_p}$, the thermal diffusivity of the system (where k_t is its thermal conductivity). And, since $n = 0$ for energy, setting A_i equal to unity in equation [4-40] we arrive at the well-known Fourier Law for the heat-flux that results from the existence of a temperature gradient in the system, i.e.,

$$\Phi = -\frac{\rho k_t}{\rho c_p} \nabla(c_p T) = -k_t \nabla T, \qquad [4\text{-}41]$$

where c_p is assumed to be constant with respect to the gradient operator.

The Fourier Law is a typical example of one type of constitutive equation (see earlier discussion) -- i.e., a thermal heat transport equation for conduction -- that is necessary to develop a complete thermodynamic formulation for the network branches of a system. In general, one needs at least three of these: one for stress (e.g., Hooke's Law for a linearly elastic solid or Newton's Law of Viscosity for a linearly elastic fluid), one for heat flux (e.g., Fourier's Law), and one for internal energy (e.g., a state equation such as the Perfect Gas Law or an energetic equation of state relating internal energy to kinematic variables).

As mentioned in Chapters 2 and 3, where the appropriate transport coefficient or energy function required for equation [4-40] cannot be defined uniquely for a specific situation from the properties of the system involved and/or related parameters of physics, an order of magnitude analysis can yield a generalized transport coefficient and intensive energy function based on the characteristic length and time scales for the system. A simple example of this for flow through a capillary was given at the end of Chapter 3, where we were able to formulate a meaningful nondimensional ratio between a measure of the transport of momentum by *conduction* in capillaries, and a measure of the transport of momentum by *convection* in capillaries.

The establishment of such ratios among transport coefficients, charge, transport driving forces (proportional to gradients in charge), and geometrical scales that characterize a physical system fall within the realm of the concept of Dimensional Analysis.

The Significance of Dimensional Analysis

In dimensional analysis, one takes quantities such as ϕ, ψ, $\nabla\phi$, Φ, and L, and arranges them into ratios (or products of ratios) which, since the numerator and denominator both have the same dimensions, are pure numbers with no associated units. The significance of these numbers is that the generalized concept of a physical configuration -- i.e., the state of affairs of a system as defined in Chapters 2, 3, and, so far, 4 -- is uniquely characterized by a corresponding matrix of dimensionless parameters, each element of which consists of one of the above-mentioned ratios (or product of ratios) of dimensionally homogeneous parts. The process of finding, quantifying, and interpreting this set is called configurational analysis (Becker, 1976).

One can construct configurational matrices that define physiologic transport as follows. Recall that in Chapter 2 we defined a whole series of ϕ's and α's corresponding to transport driven by field forces, convective body forces, inertial forces, surface forces, and active transport mechanisms; leading to conduction-type, convection-type, and radiation-type fluxes. Using these definitions, one can set up a table wherein the quantities ϕ and α form *both* the columns *and* the rows of the table. Then, the elements of this matrix establish the corresponding nondimensional parameter defined as the ratio of the ϕ or α indicated in any given column, divided by the ϕ or α specified in the intersecting rows. Many examples illustrating the construction of transport matrices may be found elsewhere (Weast, 1969, Becker, 1976, Schepartz, 1980, and Schneck and Starowicz, 1988b).

Some of the ratios derived as above from configurational matrices are well-known nondimensional parameters in fields such as heat transfer and fluid mechanics. They include, for example, for the transport of *mass*, (see equation [3-11]), the Lewis number,

$$Le = \frac{\alpha_{\text{mass diffusivity}}}{\alpha_{\text{thermal diffusivity}}} = \frac{\alpha_S \rho c_p}{k_t}$$

the mass Peclet number,

$$Pe = \frac{\alpha_{\text{mass convection}}}{\alpha_{\text{mass diffusion}}} = \frac{\ell v}{\alpha_S},$$

and the Schmidt number,

$$Sc = \frac{\alpha_{\text{momentum diffusivity}}}{\alpha_{\text{mass diffusivity}}} = \frac{\mu}{\rho \alpha_S};$$

for the transport of *momentum*, the Reynolds number, $Re = \frac{\rho v \ell}{\mu}$, Euler

number, $Eu = \frac{\phi_{\text{pressure}}}{\phi_{\text{kinetic}}} = \frac{p}{\frac{1}{2} \rho v^2}$, and Strouhal number,

$$St = \frac{\text{Vibrational velocity}}{\text{Translational velocity}} = \frac{f\ell}{v};$$

and, for the transport of *energy*, the Eckert number,

$$Ec = \frac{\phi_{\text{kinetic}}}{\phi_{\text{enthalpy}}} = \frac{1}{2} \frac{v^2}{c_p \Delta T},$$

Stanton number, $Sn = \frac{\phi_{\text{convection}}}{\phi_{\text{kinetic}}} = \frac{h}{\rho v c_p}$, and Graetz number,

$$Gr = \frac{\alpha_{\text{mass convection}}}{\alpha_{\text{thermal diffusivity}}} = \frac{\rho c_p v \ell}{k_t}.$$

Others of the dimensionless ratios have no specific names, and are based on empirical relationships which are specific to physiological processes. These include primarily those transport coefficients associated with sensible and insensible respiratory and perspiratory heat losses (see later, and Cooney, 1976); and those associated with passive, facilitated, and active transport processes across biological membranes (see Chapter 3).

To construct dimensionless matrices when information such as that given in Chapters 2 and 3 is not available, one follows a well-defined rigor involving configurational analysis (Becker, 1976, and Schneck and Starowicz, 1988b). To illustrate the process, let us get back to a point made in Chapter 1 (pgs. 28-29), namely, that central to the mechanisms involved in physiologic thermoregulation is the role played by the cardiovascular system. Because of its complexity, this system is difficult to analyze with complete mathematical rigor and generality. In many ways, it can be studied more effectively using techniques of nondimensional analysis. Thus, for example, suppose one wished to examine the heat-transport characteristics of whole, human blood, in an effort to assess the cardiovascular response to thermal stress. One could proceed to gain some insights into the physics of this process as follows (see Schneck and Starowicz, 1988b).

Configurational Analysis Using the Buckingham Pi-Theorem to Study the Heat-Transport Characteristics of Whole Human Blood

The human cardiovascular system may be examined in terms of its being composed of four basic subsystems (see Chapter 1), i.e., the fluid, blood (or, the hematologic subsystem), the pump, heart (or, the cardiac subsystem), the flow pathways, including arteries, arterioles, capillaries, venules, and veins (comprising the vascular subsystem); and, the control mechanisms or subsystems concerned with control through the central and autonomic nervous systems, the endocrine system, and various intrinsic factors. Then, taking body core temperature, T_c, to be the *dependent variable*, one can proceed to identify over 600 *independent variables* that are capable of affecting T_c through each (or combinations of all) of the four basic subsystems of the cardiovascular system. These independent variables can be classified into groups -- such as kinematic and kinetic quantities, thermodynamic, physical and chemical properties of the system, a wide variety of biochemical constituents, electromechanical and electrothermodynamic variables, many associated environmental factors, and so on. Although too numerous to be listed and described here, the variables are all identified and discussed in their entirety in Schneck, D. J. (1983e). Here, we will address only those variables upon which the heat-transport characteristics of blood can be assumed to depend.

Once the variables affecting T_c are identified in their entirety (and the list *must* be complete and all-inclusive for the subsequent nondimensional analysis to be accurate and meaningful), all of them, excluding those that are already inherently dimensionless (by definition) are expressed in terms of the five fundamental dimensions which suffice to measure all of the perceptual quantities of physics. As discussed in earlier portions of this text, these include three fundamental dimensions of mechanics: length, L, and time, t, forming the fundamental aspects of space-time sensual perception, and mass, M (or force, F) characterizing the fundamental measure of mechanical inertia; one fundamental dimension of thermodynamic perception: temperature, T; and, one fundamental dimension of electromagnetic perception: electric charge, q. Having done this, one can then employ the Buckingham Pi Theorem (Buckingham, E., 1914, 1915) to group the $n = 600$ variables, employing $j = 5$ dimensions, into $(n - j) = 595$ independent dimensionless Pi-, or, ¶-Parameters that constitute the elements of at least four configurational matrices. These may correspond, for example, to 15 characteristic ϕ's and/or ψ's, 15 α's, 9 Φ's, and 8 L's, the squares of which total 595 parameters that relate to the general thermoregulatory function of the cardiovascular system (Starowicz and Schneck, 1986). Determination of the ¶-Parameters proceeds as follows.

Determination of the Buckingham Pi-Parameters

Typically, the method for determining the ¶-Parameters is to select j of the n quantities D_i ($i = 1$ to n, assuming, for now, that all n quantities have definable dimensions), such that the j quantities have *different* dimensions, but contain among them *all j* dimensions as listed above. These are then used as "repeating variables" which can be grouped separately with any one of the other D_i—quantities in the determination of each ¶-Parameter. The term "repeating variables" is actually a misnomer, in that the "variable" may, in fact, be *any* quantity (variable, or not) of which the *dependent* variable may be a function -- including material properties, universal constants (such as the speed of light, the Faraday constant, the gravitational constant, the universal gas constant, Planck's constant), and so on. The only *specific* requirement is that no one of the j selected quantities used as repeating "variables" may be derivable from the other repeating variables. That is, no combination of $(j-1)$ of the repeating variables can be used to derive the j'th repeating variable. Other than this one absolute restriction, the choice of repeating variables is at the discretion of the investigator, and is usually based on the physical nature of the information being sought from any given problem.

    In the present case, for example, since the problem is to address the heat-transport characteristics of whole human blood, and since $j = 5$, it makes sense to choose as repeating variables five fluid properties that are related to heat transport in some relevant way, that are not derivable from one another, and that contain among them the dimensions $M, L, t, T,$ and q. The ones so picked for this study are, logically, the total thermal capacity, C_T, of blood, its thermal conductivity, k_t (to account for heat transfer by conduction), its convection coefficient, h (to account for heat transfer by convection), its specific heat at constant pressure, c_p, and, to account for electrochemical activity, the Faraday Constant, $\overline{F}$ (Note that the radiation emissivity, κ, which accounts for heat transfer by radiation, is *already* dimensionless by definition -- see later in this chapter). Again, these are, to a certain extent, arbitrary, and others could certainly have been selected with equal validity. This author, however, elected to concentrate on C_T, k_t, h, c_p, and $\overline{F}$, as being representative of the phenomena associated with the heat-transport characteristics of blood as they relate to cardiovascular thermoregulation.

The total thermal capacity of blood is defined to be the product of three quantities: the fluid mass density, ρ, its specific heat at constant volume, c_v, and total blood volume, V^*_B. It thus has dimensions of energy ($F \times L$, or ML^2/t^2) per degree temperature, or, $\phi M/T$, and it represents the total quantity of heat required to change by one degree the temperature of the blood contained in the entire cardiovascular system. The mass density of blood generally ranges from 1.05 gms/cm^3 to 1.07 gms/cm^3 at 37°C, with a reasonable mean value being 1.06 gms/cm^3 at body temperature.

The average adult male possesses between five and six liters of whole blood, while the average adult female generally has about a liter less, and the total volume of blood in any given individual is kept conspicuously constant at about 69 cc per kg of body weight for males and 64 cc per kg of body weight for females. Using a mass density of 1.06 gms/cm^3, this converts into 73.14 gms/kg for males and 67.84 gms/kg for females, with an average of 70.5 gms/kg across the board. This is somewhat lower than the value of 78 gms/kg given by Cooney (1976), but more in line with clinically measured values, although there is a $\pm$ 10% range in this mean and it depends very much on the corresponding amount of fat tissue contained in any given individual, since fat tissue has very little vascular volume (Guyton, 1981). A better rule-of-thumb is to calculate blood mass at approximately 8.0% of total body mass, so, for our "average" 75-kg individual that has a basal metabolic rate of 1800 Calories per day, we would have 0.08(75) = 6 kg of blood, which, at 1.06 kg/liter, gives us $V^*_B = 6/1.06 = 5.66$ liters. It would appear from this calculation that our "average" individual is probably a male (or, in the male range of clinical measurements), but, offering appropriate apologies to the distaff side of the population, we will use these values for illustrative purposes.

The specific heat of a fluid at constant volume is always less than (for gases) or approximately equal to (for liquids) that at constant pressure, and both are functions of temperature. For blood, c_p at 37°C is given in the range 0.86 cal/gm-°C (Cooney, 1976) to 0.98 cal/gm-°C (Victor and Shah, 1975), with a mean value of 0.92 cal/gm-°C (Altman and Dittmer, 1971). Assuming that the specific heat ratio, c_p/c_v, for blood is essentially the same as that for water, and that the latter is of order unity, c_v for blood may be set approximately equal to 0.92 cal/gm-°C = 3.85×10^3 Joules/kg-°K.

Putting all of the above information together, then $C_T = \rho c_v V^*{}_B = \left(1.06 \dfrac{\text{gms}}{\text{cm}^3}\right) \times \left(0.92 \dfrac{\text{cal}}{\text{gm-°C}}\right) \times (5{,}660 \text{ cm}^3) = 5{,}520 \text{ cal/°C}$, or, approximately 5.5 dietary Calories per degree centigrade for a 75-kg individual. This means that each 5.5 Kcal of heat entering the cardiovascular system, acting as a huge heat sink, has the potential to raise the temperature of the sink by one degree centigrade, assuming this heat is not somehow released to the environment (which, of course it *is*, as described later).

The thermal conductivity, k_t, of blood increases slightly with decreasing temperature (especially at temperatures below 0°C, Mitvalsky, 1965, Wessling and Blackshear, 1973), but it has a value reported to be in the relatively narrow range from 0.30 Btu/hr-°F-ft (Charm, et al., 1968, Spells, 1960, Victor and Shah, 1975, Wessling and Blackshear, 1973) to 0.351 Btu/hr-°F-ft (Charm, et al., 1968), which corresponds in the metric system to 1.24×10^{-3} to 1.45×10^{-3} cal/sec-°C-cm, or, 0.52 to 0.61 Joules/sec-°K-m, or, 0.4464 to 0.522 Kcal/hr-°C-m at 37°C. At least one author (Mitvalsky, 1965) reports a value as low as 0.33 Kcal/hr-°C-m, but in the work described here, the mean of the ranges quoted above will be used, such that $k_t = 0.326$ Btu/hr-°F-ft = 1.35×10^{-3} cal/sec-°C-cm = 0.57 Joules/sec-°K-m = 0.484 Kcal/hr-°C-m. A general rule of thumb is that k_t for blood varies with temperature in a manner similar to that for water, and has corresponding values which are about 86% those of water (Wessling and Blackshear, 1973).

The convection coefficient, h, for blood, as is the case for all fluids, depends on the geometry of the contiguous surfaces involved in the heat-transfer, the velocity of flow, the physical properties of the fluid, and often even on the temperature difference between the heated surface and the fluid far away from the surface. For these reasons, one must distinguish between a *local* and an *average* convective heat transfer coefficient, the latter being of greater practical interest for most engineering applications. Still, even "average" convective coefficients are quoted over very wide ranges -- that for forced convection of water, for example, being given as lying between 50 and 2000 Btu/hr-ft²-°F.

Charm, et al. (1968) give the convection coefficient for blood in the range 26.2 to 2580 Btu/hr-ft²-°F, which is somewhat comparable to water, but the range encompasses two orders of magnitude in h. Thus, at best, all one can say about h is that it can vary by as much as a factor of 100. Victor and Shah (1975) do give a more precise value of 4.86×10^{-3} cal/sec-cm²-°C (35.8 Btu/hr-°F-ft²) for the case of fully-developed, equilibrated blood flow in a tube subjected to a uniform heat flux and uniform wall temperature. Since this simulates the situation we are trying to analyze here, and, for want of any reason to the contrary -- and, because it yields plausible results, this value for h will be used as a reference quantity in the investigation reported here. It converts to 175 Cal/hr-m²-°K, or, 203 Joules/sec- °C-m².

Finally, as previously defined, the Faraday Constant, $\overline{F}$, has a fixed value of 96,500 coulombs per mole of monovalent ions. Table 4-I summarizes the dimensions of the five repeating variables, and lists the representative values of these which are being used in this study.

Determination of the Configurational Matrix Elements

Having determined the repeating variables as described above, one can now begin systematically to develop all of the ¶-Parameters that relate to

TABLE 4-I

Listing of Repeating Variables, Together With Their Corresponding
Dimensions and Representative Values Used In This Study

REPEATING VARIABLE	DIMENSIONS	VALUE USED IN THIS STUDY
Total Thermal Capacity of Blood, C_T	$\dfrac{ML^2}{Tt^2}$	5.520 Kcalories/$^\circ$C *
Thermal Conductivity of Blood, k_t	$\dfrac{ML}{Tt^3}$	0.484 Kcal/$^\circ$C-hr-m
Convection Coefficient of Blood, h	$\dfrac{M}{Tt^3}$	175.0 Kcal/$^\circ$C-hr-m^2 **
Specific Heat at Constant Pressure For Blood, c_p	$\dfrac{L^2}{Tt^2}$	0.920 Kcal/Kg-$^\circ$C
Faraday Constant, $\bar{F}$	$\dfrac{q}{M}$	96,500 Coulombs Per Mole of Monovalent Ions

---

M = Mass   * Average For a 75-Kilogram Individual

L = Length   ** Average For Fully-Developed, Equilibrated

t = Time   Blood Flow In A Tube Subjected To A

q = Electric Charge   Uniform Heat Flux and Uniform Wall

T = Temperature   Temperature

---

the problem under consideration, and that represent the elements of the configurational matrix that defines the state of affairs of the cardiovascular system (Starowicz and Schneck, 1986). For example, the first such parameter (or matrix element) may be derived from the relation:

$$\P_1 = C_T^{y_1} k_t^{y_2} h^{y_3} c_p^{y_4} \overline{F}^{y_5} D_1 \qquad [4\text{-}42]$$

where D_1 is any one of the remaining (unused) dimensional variables that have yet to be examined, and the undetermined exponents, $y_j(j = 1,2,3,4,5)$ are to be defined from the requirement that $\P_1$ be dimensionless. Thus, the appropriate dimensions of C_T, k_t, h, c_p, and $\overline{F}$, as given in Table 4-I, are now substituted into the assumed form for $\P_1$, i.e., equation [4-42], together with the corresponding dimensions of D_1 (whatever they happen to be for the particular D_1 chosen), and the respective exponents associated with L, t, M (or F), T and q are collected and set equal to zero, according to the laws for addition of exponents when they appear as products in an equation. This procedure systematically generates five equations (one for each dimension) in the five unknowns, y_j, so that one may solve them simultaneously for the values of the unknown quantities, and hence for $\P_1$.

As an example, let D_1 be the dimensional variable ω, which is the basic circular frequency (2π times the linear frequency, f) of pulsatile blood flow. This variable has dimensions of reciprocal time (radians per second for ω or cycles per second for f), so, from the definition of $\P_1$ given in equation [4-42], and the information provided in Table 4-I, one can write the following equation:

$$\P_1 = \left[\frac{ML^2}{Tt^2} \right]^{(y_1)} \left[\frac{ML}{Tt^3} \right]^{(y_2)} \left[\frac{M}{Tt^3} \right]^{(y_3)} \left[\frac{L^2}{Tt^2} \right]^{(y_4)} \left[\frac{q}{M} \right]^{(y_5)} \left[\frac{1}{t} \right]$$

$$[4\text{-}43]$$

In order for equation [4-43] to be dimensionless, the following equalities are required:

For M: $y_1 + y_2 + y_3 + 0y_4 - y_5 = 0$ [4-44]

L: $2y_1 + y_2 + 0y_3 + 2y_4 + 0y_5 = 0$ [4-45]

t: $-2y_1 - 3y_2 - 3y_3 - 2y_4 + 0y_5 - 1 = 0$ [4-46]

T: $-y_1 - y_2 - y_3 - y_4 + 0y_5 = 0$ [4-47]

and, for q: $0y_1 + 0y_2 + 0y_3 + 0y_4 + y_5 = 0$ [4-48]

Solving the above set of equations simultaneously yields:

$$y_4 = y_5 = 0; \qquad y_1 = y_3 = 1; \quad \text{and,} \quad y_2 = -2,$$

so that, equation [4-42] becomes:

$$\P_1 = \frac{C_T h}{k_t^2}\, \omega = \frac{\omega}{\left[\dfrac{k_t^2}{C_T h}\right]} \qquad \text{[4-49]}$$

We recognize the $\P$-Parameter in equation [4-49] as the ratio of two time scales: one, $\frac{1}{\omega}$, being associated with the time-period of the cardiac cycle (scaled by the factor $\frac{1}{2\pi}$), and the other, $\frac{C_T h}{k_t^2}$, being the time scale associated with the heat transport characteristics of blood. Putting in the appropriate values from table 4-I, one gets for the latter, $\frac{C_T h}{k_t^2} = \frac{(5.52)(175)}{(0.484)^2} = 4{,}123.7$ hours, or, 1.485×10^7 seconds. Assuming a normal heart rate of about 80 cycles per minute, the time scale for heat transport compares with the following time scale for the transport of unsteady inertial phenomena: $\frac{1}{\omega} = \frac{1}{2\pi f} = \frac{1}{2(3.14159)(80)} = 0.002$ minutes, or, 0.12 seconds. One concludes, therefore, that events associated with the convective transport of blood (i.e., flow) through the cardiovascular system occur some 100 million times faster than do those associated with the ability of the fluid to transport heat through its own thermal properties.

In other words, the value $\P_1 = \frac{1.485 \times 10^7}{1.200 \times 10^{-1}} = 1.24 \times 10^8$ indicates that if a given amount of heat were to be infused into the fluid at some upstream location, it would take on the order of 10^8 times as long for that heat to reach some downstream location by *thermal* events (molecular transport or conduction) taking place in the fluid at *rest*, as it does to get there by *mechanical* events (continuum transport or active flow convection) taking place in the fluid in *motion*. This fact has been appreciated for quite some time in a *qualitative* sense, but the nondimensionalization scheme so far developed allows one to get a *quantitative* estimate of the magnitude of this effect. Indeed, it has long been recognized that we have a significant cutaneous circulation because the thermal properties of blood and tissue are far too inadequate to move heat effectively by conduction or passive convection alone. This may explain why various estimates suggest that as much as 95% of the 1,950,000 calories of heat generated daily by the body, at any given time *flows* to the surface (skin and lung alveoli), where 60% of it (1,112 Cal per day) is lost by *radiation* (up to 50,000 calories per hour), 25% by evaporation (463 Cal per day, or up to 20,000 calories per hour lost through sweating and from the lungs), and only 12% (222 Cal per day, up to 10,000 cal/hr, depending upon external air velocities) and 3% (55.5 Cal per day, up to 2,500 cal/hr), respectively, by thermal convection and conduction to the environment. These percentages may, of course, vary with ambient conditions and mode of dress. During periods of severe exercise, respiratory losses (as by panting and sensible heat loss due to heavy breathing) can add as much as 8000 calories per hour to this total, but this is normally considered to be a negligible avenue of heat loss compared to the others mentioned. The remaining 5% or so of heat that is not conveyed to the body surface is lost by other mechanisms, such as the expulsion of warm urine and feces.

The various modes of heat loss shall be addressed further later in this chapter. The bottom line, however, is that the cardiovascular system does a better job of physically *carrying* heat (an event associated with the inertial time scale, $1/\omega$), than it does of *dissipating* it by itself (an event associated with the conductive time scale $C_T h/k_t^2$). The quantity $\left[\dfrac{h}{k_t^2} \right]$ is called the *thermal* resistance of the fluid, and is designated as R_T^*. Since blood has a rather low thermal conductivity, k_t, and comparatively high convection coefficient, h, it is a better insulator (i.e., has a high thermal resistance) than it is a conductor (which would have a low thermal resistance) of heat. Heat diffuses intramurally *through* blood extremely slowly, compared with the rate at which the fluid *carrying* heat is moving inertially through the cardiovascular system. On the other hand, blood has a rather high *thermal* capacity, C_T, which suggests that the fluid has a good ability to *absorb* large quantities of heat for subsequent transport.

Observe that the characteristic *thermal* transport time scale obtained as the product of *thermal* capacity, C_T, and *thermal* resistance, $R_T{}^*$ (compare this with the cell membrane *electrical* resistance-capacitance time scale of order 10^{-3}, calculated as the product of *electric* capacitance, $C_m{}^*$, and *electric* resistance, $R_m{}^*$ in Chapter 3, pg. 214) came out naturally from the derivation of the dimensionless ¶-Parameter, ¶$_1$. In similar fashion, scales may be derived for length, mass (or force), electrical charge, and temperature -- all associated with hemodynamic thermal phenomena. From these, corresponding scales may be developed for kinematic quantities (velocity, L/t, acceleration, L/t^2, "jerk", L/t^3), kinetic quantities (charge, L^2/t^2, transport coefficients, L^2/t, power, L^2/t^3), geometric quantities (area, L^2, volume, L^3), material quantities (density, M/L^3, flux, M/L^2t), and so on. These scales can then be inserted into the generic transport equations [2-1] to [2-4] in order to get an estimate of the order of magnitude of the transport process under consideration. The entire analysis of mass, energy, or momentum transport is thus reduced to performing a *dimensional* study of the problem involved (to get appropriate scales for L, M, and t), and then using any of the generic transport equations [2-1] to [2-4] to solve for the corresponding measure of Φ.

Various Scales Derived From This Study

If the process described above is repeated, keeping the same five original quantities, C_T, k_t, h, c_p, and $\bar{F}$ (hence their designation as "repeating" variables) but going to a *new* sixth variable, say D_2 = total blood volume,

$V_B{}^*$, in the cardiovascular system, then a new Pi-Parameter is generated. Going through the same exercise as before, one now concludes that: $y_1 = y_4 = y_5 = 0$; $y_2 = -3$; and, $y_3 = +3$, so that:

$$\P_2 = \frac{h^3}{k_t^3} V_B{}^* = \frac{V_B{}^*}{\left[\dfrac{k_t}{h}\right]^3} \qquad [4\text{-}50]$$

We recognize *this* Pi-Parameter as the ratio of the cube of two length scales: one, $(V_B{}^*)^{1/3} = \ell = (5{,}660 \text{ cm}^3)^{1/3} = 17.82 \text{ cm} = 1.78 \times 10^2 \text{ mm}$ being associated with a characteristic geometric dimension of the cardiovascular system, and the other, $\dfrac{k_t}{h} = \dfrac{0.484}{175.0} = 0.00277 \text{ m} = 2.77$ mm, being the length scale associated with the heat transport characteristics of blood. Note that the latter is of the order of magnitude of the length of a capillary, or the mean radius of a medium size artery (or vein), or the average wall thickness of any major blood vessel, such as the Aorta or the Vena Cava.

The cube-root of $\P_2$, i.e., $\dfrac{(V_B{}^*)^{1/3}h}{k_t} = \dfrac{h\ell}{k_t}$, defines a well-known nondimensional parameter in thermodynamics and heat transfer. This parameter, known as the *Nusselt Number, Nu*, can be written as follows:

$$Nu = \frac{h\ell}{k_t} = \frac{h\ell}{\rho c_p} \times \frac{\rho c_p}{k_t} = \frac{\alpha_h}{\alpha_k} \qquad [4\text{-}51]$$

We recognize the quantity $\rho c_p / k_t$ as the reciprocal of the thermal diffusivity, $\alpha_{\text{conduction}}$, defined earlier (c.f., discussion of equation [4-41] and Chapter 2). Thus, we conclude that the Nusselt number, as shown by equation [4-51], represents the ratio of two kinematic transport coefficients, i.e., a *convection* thermal diffusivity, $\dfrac{h\ell}{\rho c_p} = \alpha_h$, and a *conduction* thermal diffusivity, $\dfrac{k_t}{\rho c_p} = \alpha_k$. Using the values from Table 4-I, we see that $Nu = \dfrac{178}{2.77} = 64.26$ for the illustrative problem under consideration, suggesting (or confirming still further) that hemodynamic convective diffusivity is at least an order of magnitude or two larger than is hemodynamic conduction diffusivity. In fact, the small length scale, (k_t/h), of order 10^{-3} meters, coupled with the very large time scale, $(C_T h/k_t^2)$, of order 10^7 seconds, yields a heat-propagation characteristic velocity on the order of $\dfrac{10^{-3}}{10^7} = 10^{-10}$ meters per second, or 1 Angstrom unit per second, compared with blood flow rate velocities on the order of 10^{-3} m/sec in even the tiniest of capillaries. Again, the poor conductivity of the fluid is re-emphasized as one examines the molecular (mean-free-path) scales associated with conduction heat transfer, and compares them with the continuum geometric scales associated with momentum transfer (see concluding remarks in Chapter 3).

Moreover, we have here a further example of LaBarbera and Vogel's (1982) "first design principle", discussed briefly in Chapter 2 with respect to the transport of mass. Recall this principle asserts that: "While diffusion is always used for short-distance transport, it is augmented by bulk flow ... for any long-distance transport." In the case of physiologic heat dissipation, the heat generated in one region of the body (i.e., the trillions of cells considered en mass), which is remote from the environment, is transferred mainly by conduction to a second "compartment" (i.e., the blood coursing through the cardiovascular system), to be momentarily stored in the molecules of the receiving region, while being carried by convection to still a third site (i.e., the skin and lungs), where it can be released to the environment mainly by radiation, sensible and insensible perspiration, and sensible and insensible respiration (see later).

Continuing, then, according to the method described thus far, corresponding scales for mass, M, temperature, T, and electric charge, q, can be established to characterize the heat-transport properties of whole human blood. These are summarized in Table 4-II, together with representative values that were calculated from the information contained in Table 4-I.

The scale for mass is just the total mass of blood in the system. In fact, if we substitute for C_T its corresponding definition, $\rho c_v V^*_B$, and let

TABLE 4-II

Derived Scales For The Five Fundamental Dimensions Required To
Measure All Of The Basic Quantities Of Physics, Together With
Representative Values Used In This Investigation

DIMENSION	SCALE	VALUE BASED ON TABLE 4-I
Time, t	$\dfrac{C_T h}{k_t^2}$	1.485×10^7 Seconds
Length, L	$\dfrac{k_t}{h}$	2.77×10^{-3} meters
Mass, M	$\dfrac{C_T}{c_p}$	6.00 Kilograms
Temperature, T	$\dfrac{k_t^6}{C_T^2 c_p h^4}$	9.00×10^{-24} $^\circ$C, or, Nearly 0°C *
Electric Charge, q	$\dfrac{C_T \overline{F}}{c_p} \times \dfrac{Z}{MW}$	0.178×10^5 Coulombs **

---

* Includes conversion factors from Kcal to Joules and Hours to
Seconds to ensure total dimensional homogeneity in the results

** The correction factor (Z/MW) is required to convert Coulombs per
mole of monovalent ions into Coulombs per Kilogram. Calculation
assumes $Z = 1$ and $MW = 32.5$ (the mean of the monovalent ions
listed in Table 3-I).

$m_B = \rho V^*{}_B$, we see that the scale $C_T/c_p = m_B c_v/c_p = m_B/(c_p/c_v)$ actually defines a third Pi-parmeter, which is the specific heat ratio of the fluid:

$$\P_3 = \frac{c_p}{c_v} = \text{Order Unity for blood.} \qquad\qquad [4\text{-}52]$$

The scale for electric charge is of the order of the total ionic charge contained in the cardiovascular system. That is, recall from Chapter 3 (section on Gibbs-Donnan Equilibrium and Capillary Mass Transport), that in an "average" individual, one normally finds about 146.260 mMoles per liter of negatively monovalent chlorine anions in plasma, 164.260 mMoles per liter of positively monovalent sodium cations in plasma, and 1 millimole per liter of plasma proteins having a valence of around −18. The latter is equivalent to 18 millimoles per liter of a negatively charged monovalent anion, so, in effect, plasma contains 146.26 + 164.26 + 18 millimoles per liter = 0.32852 moles per liter of monovalent ions. At 96,500 coulombs per mole of monovalent ions ($\overline{F}$), this converts to 31,702.18 coulombs per liter of plasma. Assuming a hematocrit of 45%, and total blood volume of 5.66 liters, this "average" individual has about 0.55(5.66) = 3.11 liters of plasma, and thus has a total cardiovascular fluid (blood) charge of 0.986×10^5 coulombs, which is at least in the same ball park as the value given in Table 4-II. In other words, the value seems reasonable.

The temperature scale, nearly 0°C, or 273°K is the "standard" thermodynamic reference temperature (as in "Standard Temperature and Pressure," STP) for the measurement of most heat transfer processes.

One may therefore conclude that all of the reference values shown in Table 4-II are *reasonable* estimates of cardiovascular function as it relates to the heat transport characteristics of blood. They certainly show a level of self-consistency that suggests the analysis to be plausible. With this in mind, and from the second column of Table 4-II, one can now go on to derive corresponding reference values for *any* other physical quantity associated with hemodynamic heat transport (as was done, for example, in our earlier discussion of a representative scale for the heat-propagation velocity in angstrom units per second). Consider, for instance, the dimensional dynamic viscosity, μ, which, from Newton's Law of Viscosity, $\sigma = \mu\dot{y}$, has derived dimensions M/Lt. Going to the second column of Table 4-II, one can write the scale for dynamic viscosity as:

$$\mu \stackrel{m}{=} \frac{M}{Lt} \stackrel{m}{=} \frac{C_T}{c_p} \times \frac{h}{k_t} \times \frac{k_t^2}{C_T h} = \frac{k_t}{c_p} \qquad \text{[4-53]}$$

and, from the third column of Table 4-I, one calculates the reference scale for μ to be: $(0.484)(1/0.92) = 0.526$ Kg/hr-m $= 0.146$ centipoise, or milli-Pascal-seconds.

Furthermore, nondimensionalizing μ in equation [4-53] with respect to its corresponding thermodynamic scale, k_t/c_p, one gets the Pi-parameter:

$$\P_4 = \frac{\mu c_p}{k_t}, \qquad \text{[4-54]}$$

which is recognized to be the Prandtl Number, Pr, associated with heat transfer processes. the Prandtl Number, written as:

$$Pr = \frac{\mu c_p}{k_t} = \frac{\mu}{\rho} \times \frac{\rho c_p}{k_t} = \frac{\alpha_\mu}{\alpha_k} \qquad \text{[4-55]}$$

can actually be seen to give essentially the ratio of momentum diffusivity (i.e., the transport of momentum by conduction through shear), $\alpha_\mu = \frac{\mu}{\rho}$, to thermal diffusivity (i.e., the transport of heat by conduction through direct molecular interaction), $\alpha_k = \frac{k_t}{\rho c_p}$. Like the Nusselt Number defined earlier (c.f., equation [4-51]), this parameter, too, is then a ratio of two kinematic transport coefficients. Moreover, the Prandtl Number gives the ratio of the hydrodynamic boundary layer thickness associated with fluid *momentum* transport and the thermal boundary layer thickness associated with the transport of *heat energy* by conduction.

Assuming an average blood viscosity of some 5.09 milli-Pascal-sec at 37°C, and, values for k_t and c_p from Table 4-I, one calculates the Prandtl Number for blood flow to be approximately 35, which is quite large. In other words, momentum diffusivity, as measured by α_μ, is much larger than is thermal diffusivity, as measured by α_k -- illustrating still again that conductive heat transfer in the cardiovascular system is much smaller than is that due to bulk flow. While the Prandtl Number for gases is of the order of unity (e.g., $Pr = 0.7$ for air), it is not unusual for liquids to have very large Prandtl Numbers, and blood appears to be no exception to this.

Another important dimensionless parameter that arises in this study is also a ratio of two kinematic transport coefficients, one, $\alpha_v = v\ell$, being a measure of momentum transport by *convection*, and the other, $\alpha_\mu = \frac{\mu}{\rho}$, being the previously defined measure of momentum transport by *conduction*. This parameter, then, is the mechanical analogue of the thermodynamic Nusselt Number, and is called the *Reynolds Number*. It is defined as follows:

$$\P_5 = Re = \frac{\alpha_v}{\alpha_\mu} = \frac{\rho v\ell}{\mu} \qquad \text{[4-56]}$$

The product of the Reynolds Number and the Prandtl Number is called the thermodynamic Peclet Number (not to be confused with the *mass* Peclet Number, $\frac{\ell v}{\alpha_S}$), which, from equations [4-55] and [4-56], is seen to be equal to:

$$\P_6 = \frac{\rho v\ell}{\mu} \times \frac{\mu c_p}{k_t} = \frac{\rho v\ell c_p}{k_t} = \frac{\alpha_v}{\alpha_k} = Pe \qquad \text{[4-57]}$$

The Peclet Number, then, is a ratio of convected mechanical inertia to thermal diffusivity.

One can define further a dimensionless Stanton Number, as the Nusselt Number divided by the Peclet Number:

$$\P_7 = \frac{h\ell}{k_t} \times \frac{k_t}{\rho v \ell c_p} = \frac{h}{\rho c_p v} = \frac{\alpha_h}{\alpha_v} = St \qquad [4\text{-}58]$$

and we see that *this* parameter gives a ratio of the kinematic transport coefficient associated with *thermal* convection, to that associated with *inertial* convection.

Proceeding in like fashion, then, we can continue to systematically develop our configurational matrix of nondimensional parameters that characterize the heat-transport characteristics of blood. That is, all of the variables identified in Schneck (1983e) can be reduced to the basic five dimensions listed in Table 4-II. In a very straight-forward manner (as illustrated thus far), corresponding reference variables and associated Pi-parameters can be developed for each dimensional variable, a complete tabulation of these appearing in Starowicz and Schneck (1986). Space precludes being able to reproduce this tabulation here in its entirety, but Table 4-III shows some of the results obtained in the derivation of scales for corresponding dimensions of state and other significant transport quantities. Note, in particular, the very small driving force per unit mass (of order 10^{-17} Newtons per Kilogram) available to propel heat flux by conduction; and the fact that the kinematic transport coefficient (of order 10^{-7} mm^2/sec) for the latter is orders of magnitude smaller for *thermal* conduction than it is for the conduction of *momentum* (of order 10^0 mm^2/sec) as calculated at the end of Chapter 3 for capillaries. In other words, we again get self-consistent confirmation that the role of blood in the thermoregulatory process appears to be as a *carrier* of heat by active transport, rather than as a *diffuser* of heat by conduction.

Among the "variables" defined in Schneck (1983e) that are already in $\P$-form, i.e., that are inherently dimensionless by definition, are: the alkalinity or acidity of blood ($\P_8 = $ pH); the volume fraction of red blood cells in blood ($\P_9 = $ Hematocrit); the ambient humidity ratio ($\P_{10} = $ Relative Humidity); heat radiation emissivity ($\P_{11} = \kappa$); the dielectric constant for blood ($\P_{12} = \kappa^*$); the non-Newtonian index of blood ($\P_{13} = \bar{n}$); and,

TABLE 4-III

Derived Scales For Dimensions Of State

And Other Transport Quantities

DERIVED QUANTITY	DIMENSIONS	ASSOCIATED SCALE FOR THE HEAT TRANSPORT CHARACTERISTICS OF BLOOD
Transport Velocity, v	$\dfrac{L}{t}$	1.865×10^{-10} meters/second
Transport Acceleration, a^*	$\dfrac{L}{t^2}$	1.256×10^{-17} meters/second^2
Kinematic Transport Coefficient, α	$\dfrac{L^2}{t}$	5.167×10^{-13} meters^2/second
Intensive Energy Potential, ϕ	$\dfrac{L^2}{t^2}$	3.479×10^{-20} meters^2/second^2
Energy Flux, Φ	$\dfrac{M}{t^3}$	1.832×10^{-21} Kilograms/second^3

All Calculations are based on values given in Table 4-II for M, L, and

t. Dimensions are based on the definition of the quantities identified

in the first column of the Table.

solution activity factors ($\P_{14} = \alpha^*$), to name but a few. Calculation of these and all the rest of the $\P$ -parameters developed in this investigation ultimately depends upon how much information is available to describe the physical, chemical and thermodynamic properties of the cardiovascular system. In this respect, the current study revealed large gaps in the data available to obtain numerical values for many of the dimensionless parameters derived.

Such quantities, for example, as the Joule Coefficient, Magnetic Permeability, Electrical Permeability, Resistivity, and other electromagnetic properties and time-dependent visco-elastic parameters related to blood and the cardiovascular system remain essentially unquantified. Perhaps the need to know them has not surfaced before now. Perhaps they are too difficult to measure in a definitive, meaningful way. Perhaps the appropriate instrumentation for making these measurements is not yet state-of-the-art. Whatever the reason, there is still much to be learned about the physical and constitutive characteristics of the human cardiovascular system. To the extent that nondimensional techniques may contribute to directing future research efforts by identifying the need for certain measurements to be made (a need that might otherwise not have been noticed, or appreciated), they should be pursued and exploited much more so than has been the case in the history of this field.

To the extent that such techniques also: (a) provide a systematic means for generating appropriate scales for certain physical processes that might not have been immediately obvious from a cursory view of the problem; (b) allow an investigator to extract the fundamental physics and essential features of physiologic processes (and to gain some useful insights into those processes) prior to attempting to pursue a more complicated (and perhaps impossible) mathematical formulation; (c) help to identify explicitly those variables that need to be measured in physiologic experimentation in order to model successfully the physiologic process under investigation; (d) allow an investigator, by a proper order-of-magnitude analysis, to effectively distinguish between first-order and higher-order effects in the evaluation of data obtained from well designed physiologic experiments; (e) can lead (from the results of analyses such as are described in (d)) to the establishment of objective criteria that can be used to "screen" individuals for their susceptibility to environmental stress; (f) lead to the natural establishment of those design parameters which can be used either to develop personnel training maneuvers and acclimatization exercises, or personnel protective gear, or both; ... for all of these, and for still more reasons, the inherent value of nondimensional analysis in physiologic research needs to be appreciated in our continuing efforts to improve the health, comfort and understanding of man.

Having made this point, and the point that *most* of the heat generated by physiologic metabolic processes must be ultimately dissipated by radiation and evaporation through the skin and lungs, let us now examine the latter in somewhat more detail as we continue our discussion of thermoregulation.

Heat Dissipation by Radiation

Radiation is a process by which heat flows from a high-temperature body or region to a region or body at a lower temperature when the bodies are physically separated in space -- in fact, even when a vacuum exists between them. In other words, like field forces and electromagnetic radiation, heat transport by radiation can be "at-a-distance" and does not require contiguous body contact. All bodies, the human organism included, emit radiant heat continuously, which travels at the speed of light in the form of finite batches, or quanta of energy, the flux of which can be defined by the Stefan-Boltzmann Law:

$$\Phi_r = k_s T^4 \qquad\qquad [4\text{-}59]$$

where Φ_r is that heat flux (in Btu/unit time/unit area) of energy radiated from a "perfect radiator" (or "black" body) existing at absolute temperature T, and k_s is a constant known as the Stefan-Boltzmann coefficient, having a value of 0.1714×10^{-8} Btu/hr-ft^2-$^\circ$R^4. *Real* bodies do not meet the specifications of an ideal radiator but emit radiation at a lower rate than do black bodies. The ratio of real-body radiation to black-body radiation is called the *emissivity*, κ, of the real body. Thus, for a *real* body:

$$\Phi_r = \kappa k_s T^4 \qquad [4\text{-}60]$$

The *net* emissivity of a system of bodies exchanging heat by radiation is equal to the product of the emissivity of each. In the case of human-body-environment interaction, $\kappa = \kappa_B \kappa_E$, where κ_B is the emissivity of human skin -- generally having a value between 0.94 and 0.98 -- and κ_E is the ambient emissivity, which, depending upon a variety of environmental factors, may be as low as 0.01 or as high as 0.98, with values above 0.72 being most common. Also, where the bodies exchanging heat exist at different temperatures, T_B, and T_E, equation [4-60] takes on the form:

$$\Phi_r = \kappa_B \kappa_E k_s (T_B^4 - T_E^4) \qquad [4\text{-}61]$$

where T_B is absolute skin temperature (°K) and T_e is absolute ambient temperature. The product $\kappa_B \kappa_E k_s$ is often called the *radiation constant*, which is on the order of 4.92×10^{-8} Kcal/hr-m²-°K⁴ for radiation from the skin surface.

Now, for the 3% or so of radiated heat that is carried away from the body by conduction, we may equate equation [4-41], in a finite form with respect to a characteristic length, ℓ, to equation [4-60], where T_E is momentarily-neglected, to get:

$$- k_t \frac{(T_E - T_B)}{\ell} = \kappa k_s T_B^4,$$

or, letting $(T_B - T_E) = \Delta T$, where $T_B > T_E$, and rearranging somewhat:

$$\P_{15} = \frac{\Delta T}{T_B} = \frac{\kappa k_s T_B^3 \ell}{k_t} \qquad [4\text{-}62]$$

Equation [4-62] defines a dimensionless number known as the Stefan Number, Sf. Written as the product:

$$\frac{\kappa k_s T_B^3 \ell}{\rho c_p} \times \frac{\rho c_p}{k_t} = \frac{\alpha_r}{\alpha_k}$$

the dimensionless Stefan Number can be seen to be the ratio of a "radiation diffusivity," or radiation kinematic transport coefficient, $\alpha_r = \frac{\kappa k_s T_B^3 \ell}{\rho c_p}$, to the previously defined conduction thermal diffusivity, α_k.

Likewise, for the 12% or so of radiated heat that is carried away from the body by convection, we may equate convected thermal flux, as defined by the *Newtonian Equation,*

$$\Phi_h = h\Delta T \qquad\qquad\qquad [4\text{-}63]$$

to radiated heat flux, as defined by equation [4-60], to get:

$$h\Delta T = \kappa k_s T_B^4, \text{ or, again rearranging,}$$

$$\P_{16} = \frac{\Delta T}{T_B} = \frac{\kappa k_s T_B^3}{h} = \frac{\kappa k_s T_B^3 \ell}{\rho c_p} \times \frac{\rho c_p}{\ell h} = \frac{\alpha_r}{\alpha_h} \qquad [4\text{-}64]$$

Equation [4-64] defines still another dimensionless parameter known as the Radiation Number, which gives the ratio of a radiation diffusivity, α_r, to the previously defined convection thermal diffusivity, α_h, as shown above. The ratio of *Bulk Heat Transport,* $(\rho c_p \Delta T) \times (v\ell^2)$, having dimensions of energy per unit time (as in Btu/sec, for example) to *total* radiation heat transport, $\kappa A k_s T_B{}^4$, is a related parameter called the Boltzmann Number,

$$\P_{17} = \frac{\Delta T}{T_B} = \frac{\rho c_p v}{\kappa k_s T_B^3} = \frac{\rho c_p}{\kappa k_s T_B^3 \ell} \times v\ell = \frac{\alpha_v}{\alpha_r} \qquad [4\text{-}65]$$

which gives the ratio of the previously defined momentum convection kinematic transport coefficient, $v\ell$, to the radiation diffusivity.

The Boltzmann number also suggests an empirical relationship between the convection coefficient, h, the properties of the material, ρ, c_p, and the ambient flow velocity, v, i.e.,

$$h \propto \rho c_p v \qquad [4\text{-}66]$$

as is suggested by combining Pi-parameters $\P_{16}$ and $\P_{17}$, i.e.,

$$\P_{16}\P_{17} = \P_{18} = \frac{\rho c_p v}{h} = \frac{\rho c_p}{\ell h} \times v\ell = \frac{\alpha_v}{\alpha_h} \qquad [4\text{-}67]$$

An empirical functional relationship for equation [4-66] was developed at Fort Knox for horizontal ambient air flow between 10 and 1,000 feet per minute. It is given by:

$$h = (0.53)\sqrt{v} \qquad [4\text{-}68]$$

where appropriate conversion factors have been applied so that with v actually expressed in ft/min, h has units of Kcal/°C-hr-m^2.

In addition to the 15% of radiated heat that leaves the body by conduction (3%) and convection (12%) as described above, another 60% leaves by *pure* radiation as defined by equation [4-61], accounting for three-quarters of the total amount of heat to be dissipated on a daily basis from the body surface. The remaining quarter leaves by evaporation and its associated phase changes, as described below.

Heat Dissipation by Evaporation

An individual "sweats" (dissipates heat by evaporation) constantly, even in the dead of winter, where as much as several pints a day of sweat may be secreted almost imperceptively ("insensible perspiration"). In a hot environment, or during severe exercise, an average individual can secrete well over one liter of sweat per hour (about 3 pints, but with a maximum of 21 pints in any given 24-hour interval) -- all of it intended to dispose of exothermic biochemical heat production that is of no use to the organism.

Heat transfer by evaporation involves a change of phase, wherein the heat absorbed by a fluid (in the physiologic case, mostly water or sweat) goes towards changing the state of the fluid from a liquid (water) to a gas (water vapor). As the fluid evaporates off of the skin surface, it removes the heat, in a vaporization process that is aided by convection when the air near the skin surface is in motion (due to wind, fanning, and so on). Evaporative heat losses from the physiologic system occur by one (or all) of the following mechanisms:

1. "Insensible" Perspiration (uncontrolled ordinary diffusion of water through the porous skin surface to the environment);

2. "Sensible" Perspiration (sympathetic nervous-system-controlled secretion of sweat from the sweat glands, followed by evaporation from the skin surface); and,

3. *Both* insensible (uncontrolled latent) and sensible (controlled deep breathing and panting) evaporation of water into inspired air, to be released to the environment during expiration (respiration heat losses).

Insensible Perspiration

One of the distinctions between "sensible" and "insensible" heat losses has to do with whether (sensible) or not (insensible) the loss is the result of thermoregulatory mechanisms that are triggered by body receptors which "sense" and respond to significant changes in core temperature (the temperature of the visceral organs; see Chapter 1). Thus, ordinary water diffusion through the intact human skin is "insensible" to the extent that 300-500 mℓ of water per day (in an average sedentary person, under routine activities of daily living) normally diffuse via this route out of the body into the environment, *regardless* of anything else that might be happening to require regulation of body temperature. Such diffusion inherently carries a certain amount of heat with it, and this loss is basically unregulated.

On the other hand, noticeable sweating, per se, as a function of the sweat glands, the secretions of which are under the extrinsic control of the sympathetic nervous system, is strictly a thermoregulatory response to a perceived excess of body core temperature, and, as such, is not otherwise manifest. This makes it a "sensible" mechanism of heat loss. Similarly, a certain amount of heat is lost *naturally* by evaporation in the process of breathing (latent heat loss due to warming of the air taken into the lungs), but, depending on the alveolar air temperature, the respiration rate (e.g., panting and other thermoregulatory controls on breathing), blood flow rate through the lungs, and so on, *more* heat may be lost by convection if the outside air temperature is significantly below body core temperature, or *less*, if the reverse is true. This is a control mechanism that is termed sensible respiratory heat-transfer.

With this in mind, and because the mathematical theory of heat transfer accompanied by phase changes is not nearly as well developed as is that for heat transfer by conduction, convection, and radiation, the equations that have appeared to describe evaporation as a physiologic heat transfer mechanism are mostly phenomenological, descriptive, and empirical. For example, average insensible diffusional heat loss through the skin has been approximated by the relation (Cooney, 1976):

$$\Phi_d = 0.35(p_s - p_a) \text{ Kcal/hr/m}^2 \qquad \textbf{[4-69]}$$

where p_s is the vapor pressure of water (in mm Hg) at skin temperature, and p_a is the ambient partial pressure of water vapor, also expressed in millimeters of mercury (0 for "dry" air). The former can be approximated in the 27°C to 37°C range by the formula:

$$p_s = 1.92T_s - 25.3 \text{ mm Hg} \qquad \textbf{[4-70]}$$

where T_s is skin temperature in degrees centigrade. To convert equation [4-69] into equivalent milliliters of water diffusion, let the latent heat of vaporization of water be c_L kilocalories per kilogram (having dimensions of $\phi \overset{m}{=} L^2/t^2$) at temperature T_s. Then $\Phi_d = c_L \times$ (Kg/hr-m^2 of body surface area) $= c_L \times \dfrac{\text{kilograms}}{\text{liter}} \times \dfrac{\text{liters}}{\text{hr-m}^2}$. Now, the quantity kilograms/liter is just the density, ρ_f, of sweat. Furthermore, if we let liters/hr-m^2 = $\dot{V}$ = the volumetric diffusion rate per unit area of skin surface (note that this has dimensions of a characteristic velocity), then:

$$\Phi_d = 0.35(p_s - p_a) = 0.35(1.92T_s - 25.3 - p_a) = c_L\rho_f\dot{V} \qquad \text{[4-71]}$$

from which,

$$\dot{V} = \frac{350(1.92T_s - 25.3 - p_a)}{c_L\rho_f} \qquad \text{[4-72]}$$

milliliters of fluid per hour, per square meter of skin surface. As an example, consider the situation that would exist for an ambient temperature of 10°C (i.e., a "cool" 50°F day), for which $p_a = 9.1$ mm Hg (Guyton, 1981, pg. 492), $c_L = 540$ Kcal/Kg, and $\rho_f =$ about 1 Kg per liter. Then, assuming a skin temperature of some 30°C, $\dot{V}$ as calculated from equation [4-72] comes out to be around 15 mℓ/hr-m^2, which converts to 450 mℓ/day for an individual having a total body surface area of some 1.25 meters^2. The latter can vary for naked individuals from 0.71 to 1.8 or more depending not only on physical size, but also on body orientation (i.e., sitting, standing, squatting, and so on).

We see, then, that insensible perspiration, in this case, removes 0.450 (liters/day) $\times$ 540 (Cal/kg) $\times$ 1 (kg/liter) = 243 Cal/day, which is some 13.5% of the basal metabolic rate of 1800 Cal/day for this particular individual. As a general rule of thumb, under normal conditions, this mode of heat loss rarely exceeds 15%.

If we multiply equation [4-72] by some characteristic length, ℓ, which may, perhaps be the ratio of total body surface area, A_N, in square meters, to the height of the individual, ℓ^*, in meters, (or, total body volume to total body surface area) we can derive an empirical "insensible perspiration, evaporative water diffusion, kinematic transport coefficient, α_d, " which, applying the proper conversion factors to ensure dimensional homogeneity, may be given by:

$$\alpha_d = \dot{V}\frac{A_N}{\ell^*} = v\ell = \frac{3.5(1.92T_s - 25.3 - p_a)A_N}{c_L\rho_f\ell^*} \qquad [4\text{-}73]$$

This empirical perspiration "diffusivity" for normal water diffusion through the skin, along with the intensive scalar potential, $\phi = c_L$, and the fluxes Φ_d for energy (equations [4-69] to [4-71]) and $\rho_f\dot{V}$ for mass suffice to define the thermodynamic state of the body as it relates to this form of heat dissipation. This diffusion process is normally viewed as being independent of sweat secretion, i.e., the wetted area of skin due to sensible sweat secretion under normal conditions is considered to be but a small proportion of the total skin area, A_N, and thus does not significantly affect the total diffusion area.

Sensible Sweating

Sweat secretion occurs in response to the organism's need to dissipate more heat than can normally be accomplished by other heat-transfer methods. In this process, numerous specialized glands embedded in the surface of the skin, on the command of the sympathetic nervous system, actively secrete a dilute electrolyte solution onto the epidermal layer of this largest organ of the body. Evaporation then occurs from the wetted skin surface, removing anywhere from 500 to 600 calories of heat per gram of water evaporated ($\phi = c_L$) -- the latter varying with T_s. When necessary, the body can secrete up to 5 or more liters of sweat per day, removing a quantity of heat the flux of which can be approximated by the equation (Cooney, 1976):

$$\Phi_s = 3.37(T_s - T_a)^{0.258}(p_s - p_a), \text{ Kcal/hr/m}^2 \text{ of wetted} \qquad [4\text{-}74]$$
$$\text{skin surface area}$$

Equation [4-74] is valid when the difference between skin temperature, T_s (in °C) and ambient temperature, T_a (in °C) lies between 1 and 20 degrees centigrade, and when the ambient air flow is *calm* (i.e., there is no wind and a condition of free convection can be assumed). If this is *not* the case, then one may apply another empirical equation:

$$\Phi_s = 1.84v^{0.4}(p_s - p_a) \qquad [4\text{-}75]$$

where, again, Φ_s is expressed in Kcal/hr/m^2 of wetted skin surface area (assuming the skin surface is *saturated* with perspiration), v is the ambient air flow velocity in cm/sec, p_s is the saturated vapor pressure (in mm Hg) of water at skin temperature T_s(°C), and p_a is the saturated vapor pressure of water (also in mm Hg) in air at an ambient temperature T_a(°C) and relative humidity RH. The latter vapor pressure is equal to $(RH) \times$ the saturated vapor pressure of water vapor at 100% relative air humidity ratio and air temperature T_a. If the skin surface is not saturated with perspiration, then the coefficient 1.84 is less than this value, and dependent upon T_s, T_c (body core temperature), T_a, and time, t (Bligh, 1972, 1973, Atkins, 1962).

If we now let $\dot{V}_p$ represent the volumetric rate (per unit area of wetted skin surface) of fluid secretion by the sweat glands, then we may again write: $\Phi_s = \rho_f \dot{V}_p c_L$, as we did in the previous section. Also, defining a characteristic length, ℓ^*_p, for perspiration as the ratio of wetted skin surface area, A_w, to, say, the height of the individual, ℓ^*, as we also did in the previous section, we may define a "sweat diffusivity," or kinematic transport coefficient associated with sensible perspiration from equations [4-74] or [4-75] as follows:

$$\alpha_p = \dot{V}_p \frac{A_w}{\ell^*} = \frac{3.37(T_s - T_a)^{0.258}(p_s - p_a)A_w}{\rho_f c_L \ell^*} \qquad [4\text{-}76]$$

for perspiration and evaporation involving only free convection; and,

$$\alpha_p = \frac{1.84v^{0.4}(p_s - p_a)A_w}{\rho_f c_L \ell^*}$$ [4-77]

for perspiration and evaporation involving forced convection.

Note that the latter considerations apply only to the specific case that assumes *all* the sweat reaching the skin surface will be vaporized as it arrives. This is not necessarily the case if the surrounding air is not perfectly dry ($RH = 0$, $p_a = 0$) and moving well (v large) -- or, if one is sweating so copiously that much of it runs off the body as liquid, rather than evaporating. In the latter case, not only does the sweating process, per se, accomplish little to dissipate heat, but it also contributes in an adverse way to body dehydration and electrolyte loss. Thus, depending upon how serious this loss is, one may experience muscle cramping, dizziness, delirium, convulsions, shock, coma, and even death in extreme cases. Interestingly, dehydration can occur *both* from *heat* stress that leads to copious sweating and excessive fluid losses, and from *cold* stress that leads to copious urination with excessive fluid losses (see end of section entitled, *Compartmental Kinetics Analysis of Physiologic Transport* in the previous chapter and later in this chapter).

Respiratory Evaporation

Because the ambient air we breath in (inspiration) is at a temperature T_a which is different from body core temperature, T_c, there will be a convective heat loss (if $T_c > T_a$) or heat gain (if $T_c < T_a$) which may be described approximately by the equation (Cooney, 1976):

$$\Phi_L = \rho_a V_a \dot{N} c_p (T_c - T_a) \frac{1}{A_a}$$ [4-78]

where Φ_L is expressed in Kcal/hr/m^2 of contact area between ambient air and the alveolar air sacs, ρ_a is the air density (Kg/liter) at ambient temperature and pressure, V_a is the tidal volume (the amount of air inspired or expired with each breath $\simeq$500 mℓ at rest), $\dot{N}$ is the ventilation rate or respiration rate (the number of breaths per minute $\simeq$12 at rest, or 720 per hour), c_p is the specific heat at constant pressure for air (in Kcal/Kg-°C), temperature is expressed in degrees centigrade (with T_c being about 37°C), and A_a is the total contact area, in meters^2 between the inspired air and the surface lining the respiratory tract, down to the level of the alveoli.

Since the transport mechanism involved in this mode of heat dissipation is convective, we can relate equation [4-78] to equation [4-63], considering h in the latter to be the unit surface conductance of the respiratory tract, and $\Delta T = T_c - T_a$. Then,

$$\Phi_h = h\Delta T = \Phi_L = \rho_a V_a \dot{N} c_p \Delta T \frac{1}{A_a} \text{, and so,}$$

$$\frac{h}{\rho_a c_p} = \frac{V_a \dot{N}}{A_a} \qquad\qquad \text{[4-79]}$$

Going one step further, suppose each of the 300 million or so alveoli of the lungs has a diameter d_a, on the average. Then one may define a characteristic length for the process defined in equation [4-79] as $\ell = \dfrac{A_a}{d_a}$, for example (or, $A_a = \ell d_a$), from which we can thus define a respiratory convection kinematic transport coefficient, α_L, as:

$$\alpha_L = \frac{h\ell}{\rho_a c_p} = \frac{V_a \dot{N}}{d_a} \qquad\qquad \text{[4-80]}$$

Note that α_L goes up with ventilation rate, $\dot{N}$ (e.g., "panting" or hyperventilating), and with the tidal volume, V_a (e.g., "deep breathing"), for a given long anatomy, d_a; and that α_L goes down with increasing alveolar diameter, d_a, as occurs, for example in the debilitating lung disease called pulmonary emphysema (from the Greek, "emphysan," which means, "to inflate").

In addition to the above-defined "sensible" (i.e., controlled panting, deep breathing, and so on) respiratory diffusivity, there is also an *insensible* respiratory diffusivity associated with the conversion of water into water vapor in the lungs, and the subsequent loss of heat by evaporation during exhaling. This rate of "latent" heat loss can be described approximately by the equation (Cooney, 1976):

$$\Phi_\ell = \rho_a V_a \dot{N} c_L (W_E - W_I) \frac{1}{A_a} \qquad [4\text{-}81]$$

where W_E and W_I are, respectively, expired and inspired air water contents, expressed in nondimensional form as kilograms of water per kilogram of dry air (almost like a humidity), and c_L is the latent heat of vaporization at the expired air temperature. The pulmonary mass ventilation rate, $\rho_a V_a \dot{N}$, is primarily dependent on the rate of energy expenditure (i.e., the metabolic rate), $\dot{W}$, and follows pretty closely the formula:

$$\rho_a V_a \dot{N} = 0.006 \dot{W} \qquad [4\text{-}82]$$

when $\dot{W}$ is expressed in kilocalories per hour. Furthermore, W_E and W_I are empirically related to one another under "normal" conditions by the approximate formula:

$$W_E = 0.029 + 0.20 W_I \qquad [4\text{-}83]$$

Thus, equation [4-81] may be written:

$$\Phi_\ell = 0.006 \dot{W} c_L (0.029 - 0.80 W_I) \frac{1}{A_a} \qquad [4\text{-}84]$$

For latent respiratory heat loss due to ventilation, one may define a kinematic transport coefficient, or diffusivity, α_ℓ, as the ratio of c_L to $\dot{N}$:

$$\alpha_\ell = \frac{c_L}{\dot{N}} \qquad [4\text{-}85]$$

and an intensive energy potential, $\phi = c_L$.

Although not considered as important as heat loss mechanisms associated with radiation and perspiration, respiratory losses such as those described herein can contribute on the order of 8 kcal/hr to the organism's efforts to dispose of heat, under relatively routine conditions. This amounts to 192 Cal per day, or, about 10-11% of the basal metabolic rate, with 3-4% being attributed to warming of the inspired air and 7% occurring as a result of evaporation from the lungs. Of course, under conditions of exercise or other strenuous exertion, these rates -- as do *all* of those attempting to get rid of body heat -- go up correspondingly. An individual perspires profusely; (s)he breathes heavily (pants); one radiates more heat to the surroundings; one "fans" oneself to increase convective and evaporative losses; one does everything possible to prevent overheating. If successful, everything is fine. If not, then all of the associated consequences of overheating prevail -- fatigue, heat-exhaustion, collapse, coma, and, if serious enough, death! But what about the other extreme?

Physiologic Response to Cold Stress

We have said a great deal about the fact that the human organism is not a heat engine, and that it must find ways to dispose of the heat energy generated by metabolic processes. But the disposition of such heat depends upon the gradient in temperature between the body and its environment. Suppose T_a is *so* low that this temperature gradient causes heat to be *lost* at a rate potentially *greater* than that at which it is being produced by metabolism? This would create a situation wherein heat loss > heat production, such that T_c could start to decline -- an equally undesirable condition for an isothermal electrochemical engine. In fact, a two degree centigrade drop in T_c from 37°C to 35°C already leads to impairment of enzyme function, which causes metabolic complications associated with a condition of *hypothermia* (muscle weakness, decreased cardiac output, hypotension, slowing of the heart; and so on).

Drop T_c three degrees further to 32°C and the individual becomes unconscious. At 30°C hospitalization is required, and one-half-a-degree lower (29.5°C, or 85°F) generally results in death from heart failure. Thus, just as the organism has developed sophisticated mechanisms for *disposing* of heat, it has evolved equally sophisticated mechanisms for *conserving* and *generating* heat when it needs to do so to maintain its isothermal configuration. Let us take a brief look, then, at how the body responds to cold stress.

In principle, the physiologic response to cold stress is rather straightforward: by means of a variety of feedback control regulatory mechanisms, the human organism attempts to *decrease heat loss* to the environment, while at the same time *increasing the heat produced* by its various subsystems. These might be termed first-order effects, or responses that transpire as a direct consequence of, and in an effort to protect the organism from excessive heat loss due to exposure to extremely cold temperatures. While intended to stabilize body core temperature with the ultimate intent of keeping the organism alive, these first-order responses do, unfortunately, have associated with them, second-order consequences, or, "spin-offs", if you will, that significantly affect man's ability to perform or function effectively in a cold environment. So dramatic are some of these consequences that the human body -- in an attempt to keep itself from freezing to death -- may, ironically, very well kill itself by another mechanism, such as uncontrolled dehydration, frost-bite, gangrene, cardiac arrest, renal failure, liver failure, and a variety of other ailments or afflictions that may be precipitated by the human response to cold stress. We shall address these briefly in passing as we take a look at the mechanisms concerned with decreasing heat loss, and those concerned with increasing heat production.

Decreased Heat Loss to the Environment

Because the primary mechanism for heat loss involves cardiovascular convection to a relatively large skin surface area (up to 1.8 m^2 or more), the first line of defense in the organism's efforts to conserve heat is to *prevent* blood from getting to the skin. That is to say, the body attempts to discourage heat loss via this route by arranging it such that less total fluid, at a reduced temperature, spends as little time as possible, as far away from the periphery as possible -- and, that the heat transferred has to cross a minimum amount of external surface area, and a maximum amount of insulation thickness in order to reach the environment. All of this is accomplished by a cold-induced, sustained peripheral venous, capillary and arteriolar vasoconstriction, which decreases peripheral limb flow on the order of 50%, and even reduces blood flow to the brain by as much as 60%. The vascular constrictions result directly from sympathetic-adrenergic-receptor control of both venomotor and arteriomotor tone, and the resulting pooling of blood in the core regions of the body significantly reduces the peripheral vascular surface area made available for heat transfer to the environment. That's the good news.

The bad news is that peripheral vasoconstriction is not without serious consequences, like significantly impaired mental function due to substantial deprivation of a reasonable blood supply to the brain. There are others that we shall touch upon as we examine some of the more specific details of this heat-conserving mechanism.

1) Peripheral venous constriction channels blood back to the heart selectively through *deep* veins, rather than through more *superficial* veins that would normally promote heat loss. Known as the *Venous Shunt Response,* this mechanism also forces the venous return of cooler blood to take place in association with, or in closer proximity to, central arteries carrying warmer blood. The ratio is usually one deeper artery to two adjacent deep veins. What happens, then, is that warm blood travelling towards the periphery in arteries gives off some of its heat to the adjacent cooler blood heading back towards the heart in veins. The net effect of this *Arterio-Venous Countercurrent Heat Exchange* is two-fold:

First, *arterial* blood temperature falls, as it gives off heat to the venous blood while it (i.e., the arterial blood) is en route to the extremities. The *arterial* blood thus arrives near the body surface at a lower temperature than it would otherwise, which significantly reduces temperature differences such as $T_B - T_E$ in equations [4-61] and [4-63], $T_s - T_a$ in equations [4-74] and [4-76], and $T_c - T_a$ in equation [4-78]. The smaller ΔT's at the surface, in turn, decrease the gradients driving heat transport from the skin, and thus heat loss to the environment is discouraged at the outer boundaries.

Second, *venous* blood temperature rises, as it picks up heat from the arterial blood while it (i.e., the venous blood) is en route back to the heart. This helps to maintain body core temperature at its desired level by promoting heat transport at the internal boundaries between the cardiovascular system and the visceral organs. The arterio-venous countercurrent heat exchange mechanism, then, acts to keep as much heat as possible from getting *to* the extremeties (by cooling the blood that *does* reach the cutaneous capillary beds), and to conserve heat centrally by taking it *from* the arteries going towards the periphery and transferring it back *into* veins going towards the heart. That's the good news.

The bad news is that the resulting lack of heat in *peripheral* tissues can lead to the formation of ice crystals in the extracellular fluid, causing the latter to become hypertonic (more concentrated) with respect to the intracellular fluid. This can cause the cells to shrink, shrivel-up, suffer impaired cellular metabolism (such as destruction of the mitochondria, the power-plants of the cell), become gangrenous, and undergo necrosis (a degenerative death). Coupled with the extracellular edema that results from its hypertonicity, this condition is known as frostbite, and occurs most commonly in exposed areas such as the ears, cheeks, nose, fingers, and toes. Moreover, the lack of sufficient blood flow to peripheral muscles leads to significant impairment of motor function and other problems secondary to ischaemia. We shall have more to say about that later.

2) Peripheral arteriolar and capillary vasoconstriction in the skin, muscles, ears and nasal mucosa causes a significant build-up in backpressure. This "pops open" normally-closed collateral blood vessels that allow the fluid to by-pass superficial flow pathways. Known as arteriovenous anastomoses (or, "AVA's" in the trade), these collateral vessels

also shunt blood away from the closed-off peripheral vascular beds and into deeper veins, where the countercurrent exchange processes described earlier can come into play, thereby reducing heat loss at the expense of intermittent peripheral ischaemia. These AVA's also keep arterial blood warmer by not letting it get to the skin surface to lose heat.

Peripheral arteriolar vasoconstriction can completely shut off the supply of blood to many peripheral capillary beds at once. Such diverting of blood away from the periphery also effectively increases the total *insulation* thickness of the body of making the heat have to travel further to reach the skin surface before being dissipated. Insulation thickness is further enhanced by such mechanisms as piloerection -- the raising of hair follicles as a result of stimulation and contraction of the arrector pili muscles -- and the generation of "goose bumps" or transient skin roughness known as *cutis anserina*. That's the good news.

The bad news is that such peripheral vasoconstriction also acts to increase total peripheral resistance (see Chapter 2, discussion of equation [2-10]), which puts a severe overload on the heart in its efforts to pump blood through the system. Furthermore, as already mentioned, perceived extracellular fluid overloads that result from the physiologic response to cold stress lead to excessive urination (Diuresis), which brings with it the consequent danger of dehydration.

3) Peripheral vasoconstriction, arterio-venous anastomoses, and venous shunts collectively succeed in reducing the heat transfer area between the body and its environment, while simultaneously increasing the area of contact among internal heat transfer surfaces. This, too, discourages or impedes the loss of heat to the environment, at the expense of storing more heat within the organism. Furthermore, by short-circuiting the peripheral flow pathways, these same mechanisms also allow less *time* for heat transfer to take place at the surface of the body because they act to minimize *both* the quantity of blood that reaches the periphery *and* the length of time it spends there en route back to the heart. Conversely, the fluid (and more *of* it) spends more time in the in-

terior regions of the body, and thus has ample opportunity to give off more heat to the core. That's the good news.

The bad news is that the combination of peripheral ischaemia and very *intense* cold (below $-20°C$ or $-4°F$) may cause intermittent paralysis of both the vasoconstricting sympathetic nerve fibers that control arteriovenous sphincters, and, the smooth muscles that control the arterioles, themselves. The latter thus become helpless in terms of their role in redistributing blood flow, which leads to a syndrome known as *Cold-Induced Vaso-Dilatation,* or, CIVD. In CIVD, the peripheral blood vessels actually *dilate* (because their motor-controlled tone is nonfunctional), so that instead of *less* blood reaching the periphery, *more* blood does so at a time when you least want it to. This aggravates the heat loss situation, with very detrimental effects.

So we see that while the organism has developed some rather elaborate and ingenious methods for decreasing heat loss to the environment during periods of cold stress, these are not without some serious and potentially fatal consequences. In fact, the same can be said about those mechanisms that are involved in increasing the amount of heat produced by the body's various subsystems.

Increased Production of Heat Within the Organism

Akin to "throwing as many logs on the fire as possible," the philosophy here is to supply as much fuel as possible for combustion, and to insure that the major product of metabolism is the generation of heat, not work. Central to the mechanisms involved is the hormone Thyroxin, which is secreted in voluminous quantities by the Thyroid gland in response to cold stress. Some of the details associated with the action of this hormone are as follows:

1) The most striking effect of Thyroxin in intact homeothermic animals is the stimulation of oxygen consumption and the production of heat for the maintenance of body temperature. This so-called *Calorigenic Effect* reflects the role of Thyroxin in body metabolism, or, more accurately,

its role in making metabolism *less* efficient. That is, one of the ways in which this hormone works is to *block* the symbiotic coupling discussed earlier (pg. 269) between exergonic and endergonic biochemical reactions. As an example, consider what happens specifically in muscle tissue.

Recall (Chapter 1, pg. 17) that the air we breathe (oxygen) and the food we eat (in particular, glucose and fatty acids) eventually find their way into the mitochondria of cells, in general, and muscle cells, in particular. These tiny intracellular organelles carry out many reactions essential to the survival of a cell, but the most important ones involve oxidative phosphorylation (see Chapter 5 which follows). Here, exergonic oxidation of glucose and fatty acids is coupled with endergonic phosphorylation of the adenosine substrate, in such a way that the chemical energy inherent in the food we eat is stored in retrievable form in the high-energy bonds of Adenosine Triphosphate (ATP). This potential energy in ATP can subsequently be utilized for muscular work.

Adding Thyroxin to intact mitochondria causes it to swell, leading to structural changes that impede its ability to function normally. Ultimately, there results an uncoupling of oxidative phosphorylation, such that the mitochondria *continue* to take up oxygen (in fact, *increase* their uptake of oxygen), but do not continue to synthesize the high-energy phosphate bonds of ATP. Under these circumstances, oxidation becomes more rapid, but less efficient, and more of the energy gets released as heat, i.e., is "lost" rather than "stored". Furthermore, because of the decline in ATP production, the muscle tissue also starts depleting its glycogen stores in an effort to provide more fuel for contraction. This increased glycogenolysis also generates more heat from stored fuel reserves. That's the good news.

The bad news is that the failure of the organism to generate ATP soon runs down the muscle "battery" and severe impairment of motor function ensues. Indeed, Napolean found out the hard way, in Russia, in 1812, that combat effectiveness drops to near-zero under conditions of cold stress, due to musculo-skeletal dysfunction. Hitler was (fortunately) destined to make the same discovery as he was miserably defeated at Stalingrad in 1941.

2) In addition to increasing the rate of oxygen consumption and glycogenolysis (particularly in striated skeletal muscles, heart muscle and liver), Thyroxin also enhances the *absorption* of glucose from the intestines, and the *manufacture* of glucose from amino acids (gluco-neogenesis). In other words, it does everything it can to *supply* as much fuel as possible for combustion, while triggering the depletion of fuel reserves. It increases nitrogen retention and protein synthesis in peripheral tissues; it increases the breakdown of glycogen into glucose; it increases the manufacture of glucose from proteins; it increases the rate of absorption of glucose; it does just about everything that *can* be done to get a maximum amount of sugar into the blood stream and to the "furnace". That's the good news.

The bad news is that all of these processes tend to cause blood sugar levels to rise -- which overloads the Pancreas in terms of its insulin-secreting response to hyperglycemia (excess blood sugar). Thus, in severe circumstances, exposure to cold stress can lead to secondary symptoms of Diabetes, with its detrimental consequences.

3) Because the breakdown of ATP into ADP is a highly *exergonic* reaction, muscles serve as a very convenient source of heat (see Chapter 1, pg. 43). This fact is exploited by the hypothalamus, which functions as the temperature control center of the human body (ibid., pgs. 77-79). If it senses that T_c is too low, the hypothalamus (or, more specifically, the "shivering center" in the posterior hypothalamus) discharges caudal (towards the tail) efferent impulses that travel through the mid-brain tegmentum and Pons to leave the brain and enter the spinal column. These action potentials then course down the spinal column via the Tectospinal and Rubrospinal Tracts, ultimately supplying excitatory synapses to alpha-motoneurons that innervate the body's striated skeletal musculature. This initiates the *Shivering Reflex*, in the hope that the heat released through muscular contraction will bring the body temperature back up to acceptable levels. That's the good news.

The bad news, again, is that shivering causes *all* muscles to contract, even those flexors and extensors that exist in opposing pairs. Thus, no useful work results from these contractions, the energy available for *doing* work is soon depleted, and severe motor function impairment follows, as already discussed.

To summarize, then, the body does a pretty good job of sensing and maintaining body temperature at 37°C, but ambient temperatures at the extremes of hot and cold can and do lead to serious consequences. In particular, the prevailing belief today is that vascular changes and tissue hypoxia are responsible for *all* types of local cold injury, and that variation in the clinical features or manifestations reflects variation in the nature of the insult and the host response (NRC Ad Hoc Committee on Polar Biomedical Research, 1982). Problems with dehydration, impaired musculo-skeletal performance, metabolic disturbances, tissue necrosis, decreased mental proficiency, kidney and liver malfunction, and many other pathologic "correlates" of *both* hot and cold stress can be traced directly to consequences resulting from cardiovascular responses to increased or decreased environmental temperatures. Of course, when all else fails, there are air-conditioning, fans, cold showers, travel to colder climates, and other man-made methods for handling heat stress; together with sweaters, blankets, furnaces, thermal clothing, travel to warmer climates, and other man-made methods for handling cold stress. The discussion of these is left for the reader to pursue in the literature.

Concluding Remarks

In this chapter, the human body has been defined in a thermodynamic sense to be an isothermal, isobaric, open system that utilizes the free energy inherently available in chemical bonds to do useful work. This energy is released gradually to avoid excessive generation of heat, and as much of it as is possible is stored in usable form by coupling degradative, exergonic reactions with synthetic, endergonic reactions, and by allowing the reactions to proceed in smaller steps through more complicated routes, rather than in one big step from substrates to products. This also allows the reactions to be carefully controlled. Because of all of these complexities associated with metabolism, and the need for chemical reactions to proceed at a rate sufficient to maintain life, they must be catalyzed by substances that accelerate biochemical reactions without, themselves, being consumed in the over-all process. That is precisely what enzymes do. In addition to the attributes already cited for these protein catalysts, we point out further that:

1) Individual enzymes catalyze a small number of reactions (most often, only one), but have the ability to catalyze the same *type* of reaction with several structurally related substrates. According to the specificity of the reactions they catalyze, they are broken down into six major classes:

(a) Oxidoreductases: These are enzymes that catalyze oxidation/reduction-type reactions, and they include *dehydrogenases* and *oxidases* that participate in generic reactions defined by:

$$S_{\text{reduced}} + S_{\text{oxidized}} \xrightarrow{\text{Enzyme}} S'_{\text{oxidized}} + S'_{\text{reduced}} \qquad \textbf{[4-86]}$$

(b) Transferases: These are enzymes that catalyze the transfer of a group *other than hydrogen* between a pair of reacting substrates, and they include carboxylases, transaminases, aldolases, acyltransferases, glycosyltransferases, phosphorylases, and so on, that participate in generic reactions defined by:

$$S_1 - D + S_2 \xrightarrow{\text{Enzyme}} S_1 + S_2 - D \qquad \textbf{[4-87]}$$

(c) Hydrolases: These are enzymes that catalyze the hydrolysis of an ester, an ether, a peptide, a glycosyl, an acid-anhydride, a carbon-carbon, a carbon-halide, or a phosphorus-nitrogen bond; in each case, involving the combination of water with a salt (containing the bond to be hydrolyzed) to produce an acid and a base, one of which is more dissociated than the other, i.e.,

$$\text{Salt} + \text{Water} (H_2O) \xrightarrow{\text{Enzyme}} \text{Acid} (-H) + \text{Base} (-OH) \quad \text{[4-88]}$$

(d) Lyases: These are enzymes that catalyze the removal of groups from substrates by mechanisms *other than hydrolysis,* leaving behind double bonds. They act on carbon-carbon bonds, carbon-oxygen bonds, carbon-nitrogen bonds, carbon-sulfur bonds, and carbon-halide bonds, leaving behind carbon-carbon double bonds, i.e.,

$$\begin{matrix} S_1 & S_2 \\ | & | \\ C\!\!-\!\!-\!\!-\!\!C \end{matrix} \xrightarrow{\text{Enzyme}} S_1\!\!-\!\!S_2 + C = C \qquad \text{[4-89]}$$

(e) Isomerases: These enzymes catalyze the conversion from one state into another of optical, geometric, or positional isomers. They include, for example, cis-trans isomerases, racemases, and epimerases.

(f) Ligases: These enzymes catalyze the linking together of two compounds, coupled with the breaking of a pyrophosphate bond in ATP or a similar compound. Included are enzymes catalyzing reactions that *form* carbon-oxygen bonds, carbon-sulphur bonds, carbon-nitrogen bonds, and carbon-carbon bonds. Also included are enzymes such as those involved in active transport mechanisms, as defined, for example, by equations [3-63] to [3-66] and [3-88] to [3-91] in Chapter 3.

The specificity of enzymes further gives the body the advantage of carefully being able to *control* reactions at several points along the way, by making an enzyme available (or not available) to catalyze the reaction. However, this also introduces the disadvantage of having critical reactions *blocked* by "poisons" (such as carbon monoxide) that have stereo-chemically related structures. This is termed *Enzyme Inhibition.*

2) Enzymes improve the *effectiveness* of biochemical reactions in the sense that they increase the yield of *desired* end-products, while minimiz-ing (or reducing the likelihood for) the formation of *undesired* by-

products. This is due to the fact that they remove as much as possible the element of chance in these reactions, thus minimizing the probability that substrates will react in the wrong order, or with the wrong chemical species, or at the wrong time, or in the wrong place, while at the same time maximizing the probability that they will react the right way to yield the right products, at the right time, and in the right place.

Furthermore, by carefully *choreographing* the various steps involved in a series of biochemical reactions, enzymes allow the organism to *economize* on the *energy expenditure* associated with these reactions. They introduce a sense of order to the system, in assembly-line fashion, which also increases the *efficiency* of the reactions they catalyze. That is, the assembly-line concept, wherein reacting substrates are carefully lined up next to each other in the exact sequence in which they are expected to react, greatly enhances both the efficiency of the reaction and the ease with which it can occur. Much more so, for example, than the alternative of requiring a substrate to "go find" its reacting mate, which could be wandering around aimlessly in some obscure location far removed from where the reaction is expected to take place.

3) A summary of enzyme function in metabolism would not be complete without mentioning the role of *Coenzymes* in activating enzymes to catalyze reactions. Many types of enzyme reactions -- in particular, those involving group transfers (c.f., equation [4-87]) -- require an enzyme *system* in order to function properly. The enzyme system is called a *holoenzyme,* and includes a protein part (the *apoenzyme*) covalently bonded to a non-protein, prosthetic part (the *coenzyme,* or co-substrate).

Coenzymes, many of which are members of the Vitamin-B-Complex family, are basically derivatives of Adenosine Monophosphate (see next Chapter). They play several roles in enzyme reactions. By attaching themselves covalently to specific sites on the apoenzyme proteins, they can alter the stereochemistry of the molecule to "activate" it. That is, in the absence of the coenzyme, the apoenzyme has the wrong geometry relative to the substrate upon which it is supposed to act, thus rendering it inactive as far as its catalytic capability is concerned. But in the pres-

ence of the coenzyme, the molecular structure of the apoenzyme is re-
configured, such that it is now in its "active" catalytic state, meaning that
it possesses geometric congruence with its respective substrate.

Secondly, coenzymes can also act as substrates (hence the name,
co-substrate) to receive one of the end-products of an associated chemical
reaction, and, to carry it away from the reaction site. By doing so, they
prevent the end-products of the reaction from building up to either force
the reaction back the other way, or otherwise limit the rate of the reaction. And third, coenzymes can act as donors (as opposed to acceptors)
of groups of atoms required for a reaction to take place. This, too, is part
of their function as co-substrates for biochemical reactions.

In general, coenzymes are *not* required for lytic-type reactions such
as hydrolysis and those involved in digestion. Moreover, when their mode
of action is through *allosteric regulation* (i.e., through stereochemical activation of the enzyme geometry), the coenzymes involved are often trace
elements and inorganic minerals, rather than organic vitamin derivatives.
To distinguish the *mechanism* of allosteric regulation from the *function* of
a co-substrate, vitamin-derivative coenzymes serving as co-substrates are
called *cofactors;* and the term *coenzyme,* itself, is frequently reserved for
trace elements, inorganic metals and/or minerals that activate enzyme
systems through allosteric regulation.

Since, typically, a given enzyme will catalyze only one very specific
type of reaction, and since there are a tremendous number of reactions
associated with metabolism, a further dilemma appears on the horizon.
How can the organism possibly *store* the myriad of catalysts that are required for the myriad of biochemical reactions that are necessary to sustain life? The answer is, it doesn't! It has, instead, developed an
ingenious scheme that allows it to store only the necessary *ingredients* for
each enzyme, and to manufacture it as needed from "recipes" stored in the
organism's DNA code. In this way, economy of *mass* utilization is accomplished, along with the economization associated with the utilization
of energy. These concepts are developed further in the next chapter.

Chapter 5

Basic Principles of Mass, Energy, and Momentum Utilization in Physiologic Systems

Introduction

Thus far in this text, our attention has focussed primarily on the *transport* ($\nabla\phi, \alpha, \Phi$) of mass, energy, and momentum in physiologic systems. In this chapter, we shall concentrate more on some considerations related to the *utilization* of such quantities by the human organism. In particular, thinking of the body as an engine, we shall be concerned about its efficiency in converting potential energy into useful work, and its ability to economize on the utilization of mass, energy, and momentum. Thus, we will address the mechanisms by which raw materials (air, food, and water) are converted into usable fuel (ATP) for metabolism, and, how this fuel is then utilized most efficiently for both cellular processes concerned with the maintenance of life, and musculoskeletal processes concerned with human posture and locomotion. Bear in mind throughout, that *all* physiologic processes ultimately derive from enzyme-catalyzed biochemical reactions; that these reactions usually come in symbiotic, coupled (exergonic-endergonic) pairs; and that economization of mass utilization is accomplished by storing only the *raw materials* required to make enzymes, rather than the enzymes, themselves. This is illustrated in Figure 5-1, which shows schematically one type of mechanism that may ex-

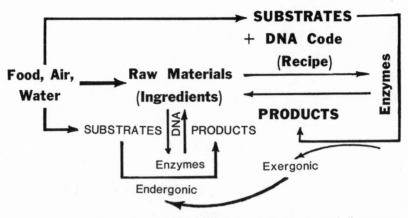

Figure 5-1 Schematic representation of the philosophy of physiologic metabolic processes, including a mechanism for economizing on the utilization of mass.

plain how the enzyme-manufacturing process works. It goes something like this:

When a specific substrate (bold-face in the figure) appears on the scene, the *substrate,* itself, catalyzes the manufacture from stored raw materials (ingredients) of the specific enzyme required to metabolize it. The "recipe" for this manufacturing process is carefully stored in the organism's DNA code, which contains, if you will, separate "subroutines" for every single enzyme that it needs on a regular basis. The raw materials (ingredients) required by the recipe are ingested daily as food, air, and water (see previous Chapter 4). The organism, then, stores only the *ingredients* and the *recipes* (or formulas), not the enzymes, themselves.

This concept is referred to as *enzyme induction,* and gains credence from the observation that enzymes tend to be formed by an organism for the metabolism of a particular substrate only when that substrate is actually present. Going one step further, then, the *products* of the subsequent biochemical reactions, in turn, themselves catalyze the breakdown of the enzyme back to raw materials, while, at the same time *blocking* the manufacture of said enzyme in the forward direction. This is called *enzyme repression,* and has the added advantage of blocking the build-up of too much product. That is, the product, itself, is now responsible for destroying or preventing the manufacture of the very enzyme that catalyzes its formation; and it prevents the cell from making excessive quantities of a potentially toxic substance, when that substance has no substrates upon which to act.

It is possible that enzyme induction and enzyme repression operate through a mechanism of *reciprocal inhibition.* That is, when the substrate appears, it may stimulate (induce) enzyme production by *inhibiting* enzyme repression (perhaps "deactivating", in some sense, the mechanism causing repression). Likewise, when the product appears, it may interfere with (repress) enzyme production by *inhibiting* enzyme induction. Exactly how all of this may be occurring is still not entirely clear, but it is a feasible explanation for the growing arsenal of experimental observations.

Furthermore, in some cases, if the substrates-to-products reaction is exergonic, some of that energy may actually be used to drive the reverse (lighter type in Figure 5-1) enzyme-to-raw-materials reaction if the latter is endergonic. And, conversely, if the raw-materials-to-enzyme reaction happens to be exergonic, some of *that* energy may actually drive the substrates-to-products reaction if *it* happens to be endergonic. This is an ingenious scheme encompassing the very essence of feedback control, utilizing loops-within-loops to maximize the efficiency of the operation and minimize waste and unnecessary expenditures of mass or energy. Ingenious, that is, when it works *right*. One type of breakdown may occur if a substrate appears for which no DNA-sub-routine enzyme recipe exists (either because of an individual genetic defect or because the substrate is manmade and not familiar to the system). We call such substrates "allergens" -- which may trigger food allergies and which may include strange chemical food "additives" such as artificial flavors, preservatives, colors, "flavor-enhancers", and so on, which the organism is not equipped to handle.

Be that as it may, within the framework of the guiding philosophies developed to this point, let us take a closer look at how the body goes about utilizing air, food and water to keep itself functioning.

Cellular Processes

As mentioned in Chapter 4, once inside the cell, biochemical substrates undergo a series of transformations that result in end products which serve basically three purposes: (i) they satisfy both the immediate and long-term energy requirements of the organism; (ii) they repair, maintain or generate structural components, enzymes, hormones, antibodies, and proteins; and, (iii) they provide repositories for the genetic information that characterizes the species (programmed reproduction). All of these are accomplished by specialized structures ("little organs") within the cell, which is organized as follows.

Basic Structure and Organization of the Human Cell

Figure 5-2 illustrates the important features of a typical human cell. This "typical" cell is composed of about 80% water, 15% protein (including most of the nitrogen and sulfur content of the body and some 20 amino acids to be described later), 3% fat and fatty acids, 1% carbohydrate (within which are contained much of the carbon, hydrogen and oxygen content of the cell), and 1% electrolytes (the most important of which are calcium, potassium, sodium, chlorine, and some of the trace elements).

Basically, the cell structure can be broken down into three broad categories: The cell nucleus (together with its contents); the cytoplasm (including various "little organs," or *organelles*); and the cell membrane (see Figure 2-1 and associated discussion in Chapter 2). The nucleus is the *control center* (CPU) of the cell. Herein is contained the DNA source code responsible for all of the physical and functional characteristics of the species. Regulation originates here. Reproduction originates here. And here one finds numerous smaller structures, such as nucleoli (sites of RNA synthesis), chromosomes (large molecules of DNA, 46 to a cell), nucleosomes (smaller "beads" of DNA), chromatin (nucleosomes strung together like a pearl necklace to form genes), and other hereditary material suspended in a transparent, colorless nuclear protoplasm called *nucleoplasm*. Separating the nucleus and nucleoplasm from the surrounding cytoplasm is a thin nuclear membrane.

The cell nucleus varies greatly in size, occupying from as little as 20% to as much as 75% of the total volume of a typical 5 to 10 micron-diameter cell. Some cells of the body are multi-nucleated (as are, for example, striated skeletal muscle cells). Others lose their nucleus as they mature (as do, for example, the 25 trillion or so red blood cells that account for approximately one-quarter of the total cells of the body) -- thus precluding them from having the ability to undergo cell division and reproduction.

Outside the nucleus of a cell, one finds tiny well-defined structural elements called *organelles*, embedded in a fluid (chiefly water) matrix

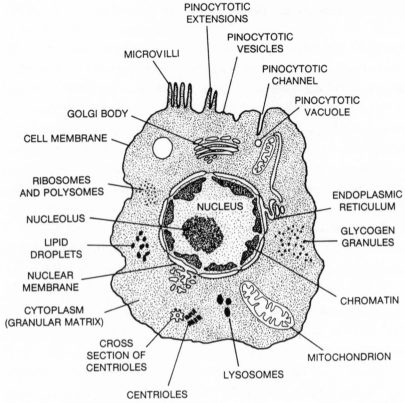

Figure 5-2 Schematic representation of the essential features of a "typical" nucleated human cell containing cytoplasmic organelles.

called *cytoplasm,* which also contains other solutes. Prior to the invention of the electron microscope, there was much controversy over the exact function of these tiny (on the order of 1 micrometer) structures. Cells were simply described as being "granular" because that is how the cytoplasm looked under the light microscope, and no further definition of structure or function could be discerned. Today, we know that these "little organs" are to the individual cells what the cells are to the organs of the body, and what the organs of the body are to the body itself (see Chapter 1). That is to say, organelles are organs inside a cell that are the sites of specific biochemical activities related to all of the functions of the cell. The ones of particular interest to us here have already been described in Chapter 1. Briefly, they include:

1) Mitochondria: Being among the largest (up to 3-4 microns in size) of the cytoplasmic structures, these organelles convert the products of digestion into a usable energy-source for the cell (see later).

2) Ribosomes: Being rich in RNA, these are the sites of protein synthesis within the cell. The proteins are put together from amino acid substrates and released into a network of interconnecting canals (canaliculi) that take them ultimately to the Golgi Apparatus.

3) Endoplasmic Reticulum: This is the network of canals that forms the intracellular transport system in general, and the pathway for the movement of proteins from the attached ribosomes to the terminal Golgi bodies, in particular. In some instances, the endoplasmic reticulum may itself be continuous with the external cell membrane, leading directly to the extracellular fluid, but rarely will a protein molecule leave a ribosome and travel through the endoplasmic reticulum to the extracellular fluid space without first encountering a Golgi body.

4) Golgi Apparatus: The Golgi bodies that comprise this structure have two functions. First, they serve as a means of producing and maintaining the endoplasmic reticulum. Second, protein molecules manufactured in the ribosomes are condensed here and made "ready" to be used by the cell, or transported further to the ECF space as necessary. The endoplasmic reticular system and associated ribosomes and golgi appara-

tus are most highly developed in cells (such as those of the liver and pancreas) which are actively engaged in the production of proteins.

5) Lysosomes: Being rich in digestive enzymes, these organelles are responsible for the digestion of protein (especially *foreign* protein). And,

6) Centrioles, or Centrosomes: These are most visible during cell reproduction, since they form the poles of the spindle apparatus involved in chromosomal replication during mitosis.

Other granular elements shown in Figure 5-2 are glycogen particles (stored forms of glucose), lipid droplets (stored fat), microtubular structures, and filaments that provide some structural integrity to the cell mass. The cell membrane encapsulates all of these structures while dividing the intracellular fluid space from the ECF, thus acting as a barrier to control all mass transport in and out of the cell (c.f., Chapter 2).

Since all cellular processes require energy at some point, the cell must first address purpose number (i) above -- i.e., the generation and storage of energy in a usable or retrievable form -- before it can really do anything else. Thus, we examine first energy-conversion processes in the cell. These take place primarily in the "powerplant," or mitochondrion of the cell, starting with raw materials that include, in order of priority, carbohydrates (preferably, glucose, or, in its stored form, glycogen), lipids (neutral fats or fatty acids), and proteins (or amino acids). Although the cell can generate energy from proteins -- and does so, for example, in response to cold stress or starvation -- it prefers not to, and uses these only as a last resort, since they are primarily intended to be used as structural components, enzymes, hormones, and antibodies.

Energy Conversion Processes in the Cell

KREBS CYCLE

From the point of view of the generation of energy, carbohydrates, lipids and proteins all converge on a final common pathway that causes them to be oxidized in a degradative, exergonic fashion. The energy so released is picked up in a corresponding series of synthetic, endergonic reactions that manufacture the high-energy phosphate compound called Adenosine Triphosphate, or, ATP. Thus, the high-energy chemical bonds in ATP serve as the "battery" of our physiologic engine, storing potential energy in a retrievable form. To talk about "charging" our engine "batteries" to do useful work is thus to talk about the manufacture of ATP in the mitochondria of human cells. Indeed, ATP is the energy "currency" of living cells.

The pathway for the manufacture of ATP begins with the substrate Acetyl-Coenzyme-A, which is one common end-product in the breakdown of proteins, carbohydrates and neutral fats -- the three basic food groups (see Figure 5-3). Acetyl-CoA enters into a sequence of reactions that have come to be known as the *Citric Acid Cycle*, or *Krebs Cycle*, or the *Tricarboxylic Acid Cycle*, one turn of which produces one molecule of ATP and eight atoms of hydrogen.

The Citric Acid Cycle begins with Acetyl-Coenzyme-A combining with Oxaloacetic Acid and water to produce Citric Acid (hence the name of the cycle), in the presence of the condensing enzyme citrate synthase:

$$\overset{\displaystyle S}{\underset{\displaystyle CoA}{\overset{\displaystyle |}{\underset{\displaystyle |}{C_2H_3O}}}} + C_4H_4O_5 + H_2O \xrightarrow[\text{Synthase}]{\text{Citrate}} C_6H_8O_7 + \overset{\displaystyle H}{\underset{\displaystyle CoA-S}{\overset{\displaystyle |}{\underset{\displaystyle |}{}}}} \qquad \textbf{[5-1]}$$

The extra hydrogen is picked up by the thioester bond of CoA, which is accompanied by considerable loss of free energy as heat (see later), ensuring that the reaction goes to completion towards the right. Citric acid is then isomerized to Isocitric Acid in the presence of the enzyme aconitase (aconitate hydratase), following which Isocitric Acid undergoes dehydrogenation in the presence of isocitrate dehydrogenase to form Oxalosuccinic Acid:

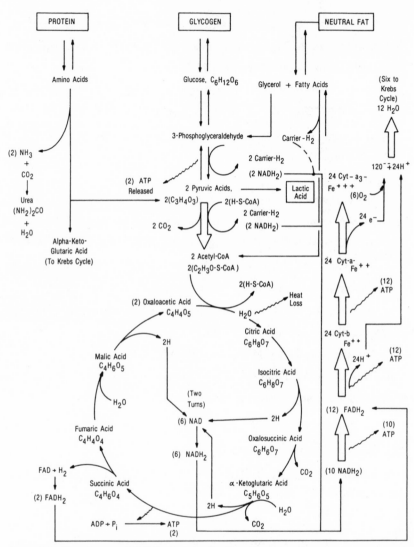

Figure 5-3 Manufacture of ATP in the mitochondria of human cells.

$$\text{isocitrate}$$
$$C_6H_8O_7 \xrightarrow{\hspace{2.5cm}} C_6H_6O_7 + 2H^+ + 2e^- \qquad [5\text{-}2]$$
$$\text{dehydrogenase}$$

At this point, the coenzyme Nicotinamide Adenine Dinucleotide (NAD), acting as a co-substrate, picks up the Hydrogen atoms released above, to form $NADH_2$. There follows a decarboxylation of oxalosuccinic acid to Alpha-Ketoglutaric Acid, also catalyzed by isocitrate dehydrogenase:

$$\text{isocitrate}$$
$$C_6H_6O_7 \xrightarrow{\hspace{2.5cm}} C_5H_6O_5 + CO_2 \qquad [5\text{-}3]$$
$$\text{dehydrogenase}$$

Next, Alpha-Ketoglutaric Acid undergoes oxidative decarboxylation in a hydrolysis reaction catalyzed by an alpha-ketoglutarate dehydrogenase complex, and again NAD picks up the Hydrogen atoms released:

$$\text{Alpha-Ketoglutarate}$$
$$C_5H_6O_5 + H_2O \xrightarrow{\hspace{2.5cm}} C_4H_6O_4 + CO_2 + NADH_2 \quad [5\text{-}4]$$
$$\text{Dehydrogenase} + \text{NAD}$$

The Succinic Acid thus formed is metabolized further by undergoing a dehydrogenation catalyzed by Succinate Dehydrogenase, and this time the hydrogen released is picked up by Flavin Adenine Dinucleotide (FAD):

$$\text{Succinate}$$
$$C_4H_6O_4 \xrightarrow{\hspace{2.5cm}} C_4H_4O_4 + FADH_2 \qquad [5\text{-}5]$$
$$\text{Dehydrogenase} + \text{FAD}$$

The Fumaric Acid formed as a result of this dehydrogenation takes on a water molecule, under the influence of fumarate (fumarate hydratase), to yield Malic Acid:

$$\text{Fumarate}$$
$$C_4H_4O_4 + H_2O \xrightarrow{\hspace{3cm}} C_4H_6O_5 \qquad \text{[5-6]}$$
$$\text{Hydratase}$$

Finally, catalyzed by the enzyme malate dehydrogenase, Malic Acid loses two more hydrogen atoms to NAD, yielding once again Oxaloacetic Acid to start the cycle all over again:

$$\text{Malate}$$
$$C_4H_6O_5 \xrightarrow{\hspace{3.5cm}} C_4H_4O_5 + \text{NADH}_2 \qquad \text{[5-7]}$$
$$\text{Dehydrogenase} + \text{NAD}$$

In all, 8 hydrogen atoms are released in one turn of the cycle, yielding 4 carrier-co-substrates -- 3 in the form of NADH_2 and one in the form of FADH_2. However, since the breakdown of glucose yields *two* molecules of Acetyl-Coenzyme-A for each molecule of sugar metabolized, it takes two complete "turns" of the Krebs cycle, releasing a total of 16 hydrogen atoms (8 carrier-H_2-co-substrates in the form of 6 NADH_2 and 2 FADH_2) to take care of the end-products formed from the breakdown of every molecule of glucose. In fact, 8 *more* molecules of hydrogen (4 carrier-H_2-co-substrates NADH_2) are released *prior* to the Krebs cycle, in the process of breaking glucose down into the two molecules of Acetyl-Coenzyme-A (see Figure 5-3), so we wind up with 10 NADH_2 and 2 FADH_2 co-substrates from each molecule of glucose metabolized completely through two turns of the Krebs cycle.

The point of all of this, is that the oxidation of $C_6H_{12}O_6$, with the corresponding release of anywhere from 686,000 to 738,000 calories of chemical bond energy (see Chapter 4), is an *exergonic* reaction that is coupled to a series of enzyme-catalyzed *endergonic* reactions, in an attempt to store as much as possible of this energy in a retrievable form. To this point in the metabolic process, i.e., to the end of the Krebs cycle, much of this energy is now contained in the coenzyme carriers NADH_2 (10 per molecule of glucose metabolized) and FADH_2 (2 per molecule of glucose metabolized). Let us then see how this energy is transferred to the high-energy bonds of ATP.

OXIDATIVE PHOSPHORYLATION

The 24 hydrogen atoms carried by the 12 coenzymes generated as described above are now transferred to the respiratory chain in the inner mitochondrial membrane of the cell. As they pass along this chain (see Harper, et al., 1977, for example), a series of cytochrome enzymes catalyze reactions that transfer bond energy from the coenzymes to the high-energy phosphate compound ATP as follows: First of all, the 20 hydrogens carried by $NADH_2$ are transferred to FAD to produce 10 more co-substrates $FADH_2$ (we thus have a total of 12 $FADH_2$'s at this point). The energy released in this process is enough to generate one molecule of ATP for every $NADH_2$ thus oxidized. Therefore, per molecule of glucose metabolized:

$$10\,NADH_2 \xrightarrow[\text{Exergonic}]{} 10\,FADH_2 + Energy$$

And,

[5-8]

$$10\,(ADP) + 10\,P_i + Energy \xrightarrow[\text{Endergonic}]{} 10\,ATP$$

As is illustrated in Figure 5-4, the compound Adenosine Triphosphate (ATP) is composed of Adenine (the Nitrogen-containing double-ring Purine system), attached to the number 1 carbon of a 5-carbon sugar (pentose) called D-Ribose -- the combination of the two comprising the nucleoside called *Adenosine*; and a side chain of three inorganic phosphate groups ($H_2PO_4^-$) attached at the number 5 carbon of the pentose. The $\sim$-bonds shown between the three oxygens and the three phosphorus atoms indicate "high-energy" bonds within which the glucose-derived, metabolic potential energy is stored. Hydrolysis of these

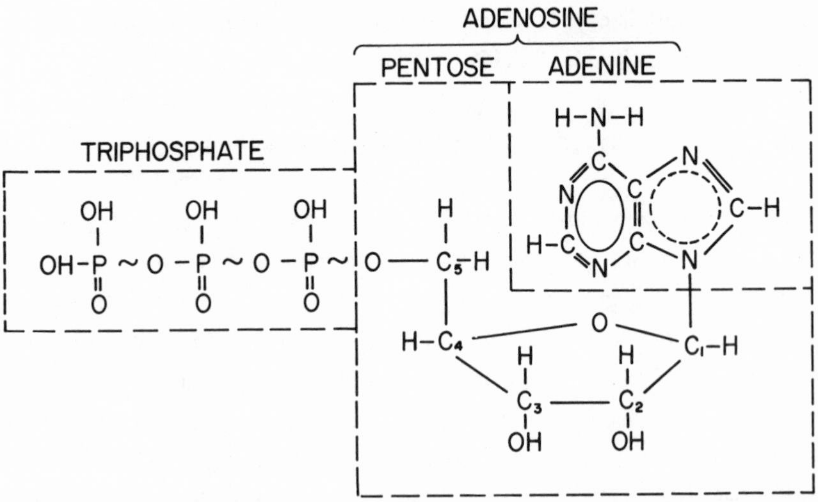

Figure 5-4 The adenosine triphosphate (ATP) complex composed of inorganic phosphate, the pentose sugar, and the base adenine, containing the high-energy bonds that provide potential energy to drive biochemical reactions.

bonds subsequently serves as the energy source (or "battery") for driving biochemical reactions.

In equation [5-8], Adenosine Diphosphate, ADP, is the substrate for the endergonic reaction, and P_i represents inorganic phosphate. At this point in the mitochondrial "assembly line", the 10 $FADH_2$ are joined by the 2 $FADH_2$ released after the dehydrogenation of Succinic Acid (equation [5-5]), resulting in 12 $FADH_2$. These 12 $FADH_2$ are further oxidized, giving up their electrons to convert Cytochrome-b-Fe^{+++} into Cytochrome-b-Fe^{++}, and releasing their hydrogen into solution as hydrogen ions, $24H^+$. The energy released in this process is again enough to generate one molecule of ATP for every $FADH_2$ oxidized:

$$12\,FADH_2 + 24\,Cyt\text{-}b\text{-}Fe^{+++} \longrightarrow 24\,Cyt\text{-}b\text{-}Fe^{++} + 24\,H^+$$
$$\text{Exergonic} \qquad + 12\,FAD + Energy$$

$$[5\text{-}9]$$

$$12\,(ADP) + 12P_i + Energy \longrightarrow 12\,ATP,\ \ \text{per molecule}$$
$$\text{Endergonic} \qquad \text{of glucose metabolized}$$

The FAD is returned to the Krebs cycle to be used again. The 24 Cytochrome-b-Fe^{++}, in turn, get oxidized *back* to 24 Cytochrome-b-Fe^{+++} at the expense of 24 Cytochrome-c-Fe^{+++} getting reduced to Cytochrome-c-Fe^{++}, and these 24 Cytochrome-c-Fe^{++}, also in assembly-line fashion, get oxidized back to Cytochrome-c-Fe^{+++} at the expense of Cytochrome-a-Fe^{+++} getting reduced to Cytochrome-a-Fe^{++} (Harper, et al., 1977). The last set of reactions liberates enough energy to generate still 12 more molecules of ATP, i.e.:

$$24\,Cyt\text{-}c\text{-}Fe^{++} + 24\,Cyt\text{-}a\text{-}Fe^{+++} \longrightarrow 24\text{-}Cyt\text{-}c\text{-}Fe^{+++}$$
$$\text{Exergonic} \qquad + 24\text{-}Cyt\text{-}a\text{-}Fe^{++}$$
$$+ Energy$$

$$[5\text{-}10]$$

$$12\,(ADP) + 12P_i + Energy \longrightarrow 12\,ATP,\ \text{per moledule of}$$
$$\text{Endergonic} \qquad \text{glucose metabolized}$$

Finally, the 24 Cyt-a-Fe^{++} get oxidized back to 24 Cyt-a-Fe^{+++} as 24 Cyt-a$_3$-Fe^{+++} get reduced to 24 Cyt-a$_3$-Fe^{++}. The 24 Cyt-a$_3$-Fe^{+++} get regenerated at the end of the mitochondrial assembly-line, ultimately releasing 24 electrons to 12 molecules of oxygen, if it is present in sufficient quantity (i.e., under conditions of *aerobic metabolism*), to generate 12 oxygen ions:

$$24 \text{ Cyt-a}_3\text{-Fe}^{++} \longrightarrow 24 \text{ Cyt-a}_3\text{-Fe}^{+++} + 24 \, e^-$$

[5-11]

$$\text{(Aerobic Metabolism) } 6 \, O_2 + 24e^- \longrightarrow 12 \, O^{--}$$

The 12 oxygen ions then combine with the previously released 24 hydrogen ions (c.f., equation [5-9]) to form 12 molecules of Water:

$$24 \, H^+ + 12 \, O^{--} \longrightarrow 12 \, H_2O \qquad \text{[5-12]}$$

Six of these water molecules go back into the Krebs cycle to form (in two turns of the cycle) Citric Acid, Succinic Acid, and Malic Acid. The remaining six are net products of the metabolism of Glucose.

Summarizing, then, all of the chemical reactions illustrated in Figure 5-3, incorporating equations [5-1] to [5-12], we have the net result under aerobic conditions that one molecule of glucose combines with six molecules of oxygen to yield a net of six molecules of water and six molecules of carbon dioxide, plus a total of 712 kilocalories (on the average) of energy:

$$C_6H_{12}O_6 + 6O_2 \longrightarrow 6 \, CO_2 + 6 \, H_2O + \sim\!712{,}000 \text{ cal per}$$
$$\text{mole of Glucose oxidized} \qquad \text{[5-13]}$$

The energy thus released is "handed-off" in a series of assembly-line type reactions during which the coenzyme carriers NADH$_2$ and FADH$_2$ get oxidized to yield, ultimately, a *net* of 34 molecules of ATP per molecule of glucose metabolized. This series of reactions occurring in the mitochondria of the cell are collectively referred to as *oxidative phosphorylation* (equations [5-8] to [5-10]), and represent a respiratory chain at the end of which oxygen enters to pick up the 24 electrons released by $Cyt\text{-}a_3\text{-}Fe^{++}$. Keep in mind that oxygen enters *only* at the very end of this sequence of reactions -- ultimately getting reduced to regenerate the co-substrates required in the respiratory mitochondrial chain.

AEROBIC METABOLISM

If *enough* oxygen is present, our engine "generator", operating through the cytoplasmic Krebs cycle and the mitochondrial oxidative phosphorylation chain, can manufacture a *total* of 38 molecules of ATP for every molecule of glucose metabolized: 2 resulting from the breakdown of glucose into pyruvic acid via the glycolysis series of reactions known as the Embden-Meyerhof Pathway (see Figure 5-3); 2 resulting from two turns of the citric acid cycle; and 34 resulting from the oxidative phosphorylation reactions already described. However, 4 of these ATP molecules are "recycled" back into the reaction pathways to drive intermediate steps that are endergonic -- such as the generation of Glucose-1-Phosphate and Fructose-1,6-Diphosphate as Glycogen or Glucose undergo conversion to Pyruvic Acid, and a couple of in-between steps in the conversion of alpha-ketoglutaric acid into succinic acid.

The inner mitochondrial membrane of a human cell illustrates perfectly the assembly-line concept discussed earlier with respect to the ability of enzymes to increase both the effectiveness and the efficiency of biochemical reactions. All of the cytochromes required for oxidative phosphorylation are arranged in a well-ordered fashion, in precise sequence, along the tortuous, multi-canaliculated folds of the membrane, and the products of the Krebs cycle are handed along from fold to fold as they undergo each oxidation. Thus, they enter at one point in the mitochondrion as carrier-H_2 molecules, and leave at a well-defined exit as the products of oxidation, i.e., carbon dioxide, water, and, if amino acids should happen to get involved, urea. Left behind are high-energy ATP molecules, ready to drive other cellular processes (see later). This well-ordered, well-controlled, assembly-line anatomical configuration, coupled with an equally well-ordered, well-controlled, stereochemical enzyme system, insures that reactions will proceed quickly, and towards completion in the desired direction.

By providing special *sites* for specific chemical reactions, and by *localizing* the enzymes at these sites, the cell also manages to keep various reactions from getting in each others' way, and it establishes a tight "hold" not only on *whether* or not a reaction will take place, but also *where* it will take place. Thus, *all* of the enzymes required for ATP synthesis are manufactured and localized in the mitochondria of the cell. They are not found elsewhere. Likewise, *all* of the enzymes required for protein synthesis are localized in the ribosome-endoplasmic-reticulum-Golgi-apparatus portions of the cell; *all* of the enzymes required for the proper manufacture of mRNA, DNA and tRNA (see later) are in the cell nucleus; and *all* digestive enzymes for the catabolism of unwanted cell wastes are concentrated in the lysosomes of the cell. The cell runs a tight ship. It keeps its house in order and is in complete control at all times. Little is left to chance. Everything is orderly, carefully calculated, carefully inventoried, meticulously well-defined and organized to the n'th degree. How this is accomplished is addressed further below, when we examine some aspects of protein synthesis (of which enzyme synthesis is an example) and the synthesis of nucleic acids.

ANAEROBIC METABOLISM

If insufficient oxygen is available for oxidative phosphorylation to be completed, as shown in Figure 5-3, then the products of glycolysis begin to pile up and the whole system backs up. In this *anaerobic* case, the Krebs cycle is by-passed and pyruvic acid gets converted, instead, into lactic acid, $C_3H_6O_3$. At this point, note that only two molecules of ATP have been generated. Of the lactic acid produced, however, 80% gets re-synthesized into glycogen, while ATP is derived from the oxidation of the remaining 20%. The resynthesized glycogen is then available, again, to generate more ATP at such time as more oxygen becomes available. If it never does, then the system "fatigues", or breaks down completely, and must stop, as is the case commonly when muscular exertion causes the build-up of too much of an "oxygen debt" in physiologic systems. In this case, all of the hydrogen carrier molecules become saturated with H_2 and there are none available to pick up and store the energy released in glucose metabolism. The saturation of the carrier molecules is what actually backs the system up, preventing the *manufacture* of ATP. The build-up of lactic acid in striated skeletal muscles also prevents them from functioning by preventing the *breakdown* of ATP, as well. This is done in an effort to prevent the *further* accumulation of lactic acid from the breakdown of fructose diphosphate during anaerobic muscular contractions. It also illustrates a type of product-control over biochemical reactions known as *feedback inhibition*, which, akin to enzyme repression, works something like this:

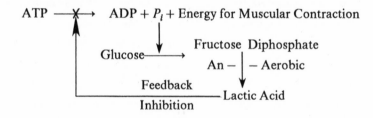

One last point needs to be made in discussing the conversion of the products of digestion into retrievable energy sources for human cells. The high-energy phosphates represent *immediate* sources of energy for short-term needs. They are being continuously generated and continuously broken down by the cell for synthetic reactions, muscular contraction, active mass transport, and so on. Medium-term energy needs are provided from the small stores of *glycogen* that the cell is able to keep in granules in its cytoplasm. This stored form of glucose is derived from the cell's ability to put together 8 to 12 glucose residues in an energy-consuming process that generates a high molecular weight (on the order of 270,000 to 100,000,000) branched-chain polysaccharide ("many sugars") granule. The process takes place in the intracellular fluid, where blood glucose gets converted first to Glucose-6-Phosphate -- in an energy-consuming reaction that uses one molecule of ATP -- then to Glucose-1-Phosphate, and finally to Glycogen, in a dephosphorylation reaction that attaches it to another glucose molecule.

Glycogen granules can be seen floating around in the cell (see Figure 5-2), especially in striated skeletal muscle and liver tissue. This immediately-retrievable stored form of glucose ranges from 0% during fasting, up to 8% of the wet weight of liver following a meal. On the average, it accounts for 4-5% by weight of liver tissue, and 0.5% to 1.0% of muscle tissue. However, because total body muscle mass is so much larger than the mass of the liver, *most* of the *total* glycogen in the organism normally resides in muscle tissue.

Still longer-term energy needs are met by the fats stored in adipose tissue. These are derived either from direct ingestion with subsequent conversion to glycerol and fatty acids, followed by storage as neutral fats; or, by processes that convert excess carbohydrate into fats for long-term storage, when the former are not immediately required to meet the energy needs of the organism. Fats also serve as an insulating material in the subcutaneous tissues and around certain organs. In combination with proteins, they further constitute the major structural component of the cell membrane (see Chapter 2), and they occur as well in the mitochondria within the cytoplasm.

In real emergencies, and in situations that border on starvation, the body can, as a last resort, derive energy from the breakdown of proteins. To do this, it shunts the normal metabolic pathways for amino acids into a series of reactions that generate alpha-keto-glutaric acid instead. The latter is then fed into the Krebs cycle as shown in Figure 5-3. This, however, is an extreme that is not considered routine in the course of normal events. The normal hierarchy established to provide an energy pool for metabolic function involves the ingestion of carbohydrates that are ultimately:

(i) Excreted (as a certain percentage of them inevitably are);

(ii) Oxidized to CO_2 and H_2O while generating ATP as an *immediate* source of energy to drive other metabolic processes;

(iii) Synthesized into the polysaccharide Glycogen for medium-term storage (mostly in the liver and in muscle), in a process called *Glycogenesis* (the reverse of which is called *Glycogenolysis*); and/or,

(iv) Synthesized into fats for long-term storage in Adipose Tissue.

EFFICIENCY OF THE ENERGY-CONVERSION PROCESSES IN THE CELL

When ATP is broken down as an energy source to yield ADP and inorganic phosphate, the *total* enthalpy change (heat of reaction, $h_E = g(T, p) + Ts$) is approximately 11,000 calories per mole, of which the free-energy portion (usable Gibbs Free Energy, $g(T, p)$) is only about 7,800 calories (plus or minus 500, see Chapter 3 and Harper, et al., 1977). Thus, of the original 712,000 calories released in the oxidation of one mole of glucose (c.f., equation [5-13]), 374,000 (i.e., $11,000 \times 34$) are "captured" in the high-energy phosphate bonds of the net 34 moles of ATP generated following oxidative phosphorylation. The efficiency of this energy-conversion process can therefore be calculated to be $\frac{374,000}{712,000} = 0.5253$, or, on the order of 50%. Of course, this is efficiency based on the total heat content of one mole of ATP. If we consider that only about 7,800 calories are ultimately available to do actual work, i.e., that 3,200 calories per mole of ATP subsequently hydrolyzed appears as entropy (Ts), then the real *net* efficiency of this energy conversion process -- going from the total energy available in the chemical bonds of glucose, to the total energy actually utilized to do work following the hydrolysis of ATP -- is, $\frac{(34)(7,800)}{712,000} = 0.37$, or 37%. This is still pretty good when compared with engineering efficiencies of most internal combustion engines. The *aerobic efficiency*, i.e., the efficiency in the presence of a sufficient supply of oxygen, of the ability of a human cell to charge its batteries, starting with raw materials consisting of carbohydrates, oxygen, and water, is thus around 50% gross and 35% net. The energy lost appears as heat and is dissipated by mechanisms such as those described in the previous chapter.

Now that we have described how the cell converts the end products of digestion, using oxygen, into usable sources of energy, let us turn our attention briefly to how this energy is utilized in other cellular processes, the most important of which involve the manufacture of proteins. In Chapter 3, the subject of energy expenditure for active transport processes and facilitated diffusion was already discussed. Later in this chapter, we shall examine how energy is expended for the generation of muscular power. Here, we now address the subject of protein synthesis, which originates basically in the ribosomes of the cell from "patterns" or "templates" contained in the cell nucleus.

Protein Synthesis in the Cell

THE DNA SOURCE CODE

Proteins are synthesized by the cell according to instructions which are coded into the chromosomes contained in the cell nucleus. Chromosomes are simply large molecules of the nucleic acid, DNA. To understand the structure of DNA, refer back to Figure 5-4, which illustrates the ATP complex. Replacing the hydroxyl group, -OH, on the number 2 carbon by a hydrogen bond, -H, generates the pentose sugar, D-2-Deoxyribose. As shown in the figure, this sugar has the base Adenine attached at the number 1 carbon. In the case of DNA, the heterocyclic base attached here may actually be Adenine, Thymine, Cytosine, or Guanine. As also shown in the figure, the ATP complex is a self-contained molecule, whereas DNA is a *strand* of molecules joined in phosphate linkage from the number 3 carbon of one sugar to the number 5 carbon of the adjoining sugar, to get a polynucleotide chain that looks something like this:

$$H \quad\quad O^- \quad\quad H \quad Base \quad H \quad\quad O^- \quad\quad H \quad Base \quad H$$
$$\mid \quad\quad\quad \mid \quad\quad\quad \mid \quad\quad \mid \quad\quad \mid \quad\quad\quad \mid \quad\quad\quad \mid \quad\quad \mid \quad\quad \mid$$
$$- C^3 - O - P - O - C^5 - Sugar - C^3 - O - P - O - C^5 - Sugar - C^3 - \ldots$$
$$\mid \quad\quad\quad \mid \quad\quad\quad \mid \quad\quad\quad\quad\quad\quad\quad \mid \quad\quad\quad \mid \quad\quad\quad \mid$$
$$O^- \quad\quad H \quad (C^1) \quad\quad\quad\quad O^- \quad\quad H \quad (C^1)$$

These polynucleotide chains reside in the nucleus of the cell and *never* leave. The nucleus thus acts as a kind of "safety deposit box," carefully storing the "recipes" handed down from generation to generation that control cell division, the duplication of genetic material, and the synthesis of the metabolic units of the cell (most importantly, essential enzymes and structural proteins).

Deoxyribonucleic acid (DNA) chains are structured as a double helix, two strands of nucleic acid being wound around one another and bound together through hydrogen bonding of the base pairs Adenine-Thymine and Cytosine-Guanine. Each half turn of the double helix contains five base pairs. A single gene is a chain of about 1,000 nucleotides; a single chromosome contains up to 9000 genes, plus Ribonucleic Acid (RNA, see later), a low-molecular-weight protein called histone, and a more complex protein.

THE MESSENGER RNA TEMPLATE

In the nucleus -- probably in the nucleolus -- the double helix of DNA undergoes a breaking of the hydrogen bonds between base pairs, and the two strands unwind. Each strand then has exposed hydrogen bonds which are capable of attracting bases in the synthesis of a strand of RNA. The *sequence* of the bases, Cytosine, Guanine, Thymine, and Adenine, determine the sequence of Cytosine, Guanine, *Uracil,* and Adenine in the new RNA molecule. Therein lies the code -- in the sequencing of these four bases, and in the virtually infinite number of possibilities that exist for such sequencing when thousands of bases are required to decipher one tiny bit of information. Every possibility corresponds to a particular enzyme, a particular protein, a particular hereditary characteristic, a particular antibody, perhaps a particular mutation, and so on, and so forth. The order and spacing of the genes in a single chromosome can be correlated with the structural configuration of synthesized proteins, and the absence of a specific small segment of chromosome can sometimes be likewise related to the absence of a specific gene that controls a known biochemical step.

Ribonucleic Acid (RNA) differs from DNA in three ways: (i) It contains the sugar ribose instead of deoxyribose; (ii) It contains the base Uracil instead of Thymine; and (iii) It contains a single strand, rather than a double helix. This single strand, called *messenger* RNA, or, more simply, mRNA carries the DNA source code from the cell nucleus out into the cytoplasm. Note, again, that the "source program", DNA, *never* leaves the safety of the cell nucleus. It is the "standard" by which all protein synthesis is patterned. It makes "carbon" copies (clones) of itself in the form of mRNA and sends the clones out to do the work, keeping the original safely behind.

THE ROLE OF TRANSFER RNA AND RIBOSOMES

In the cytoplasm, protein is synthesized, the amino acid sequence of the protein being determined by the base sequence of the messenger RNA. As listed in Chapter 1, page 87, some 20 amino acids are involved in protein synthesis: 7 neutral Aliphatic ones, 3 neutral Aromatic ones, 3 neutral ones containing Sulphur, 4 being acidic, and 3 being alkaline. Ten of these amino acids, i.e., Arginine, Histidine, Isoleucine, Lysine, Methionine, Phenylalanine, Threonine, Tryptophan, Leucine, and Valine, have been termed *essential amino acids* in the sense that the animal organism cannot synthesize them *de novo* and so must depend on ingesting them through dietary sources.

The amino acids required for protein synthesis are brought to their proper site on the messenger-RNA-chain by specialized carrier molecules, or "taxi-cabs", called *transfer-RNA*, or tRNA. Each tRNA active transport molecule is specific for a particular amino acid. The tRNA brings its respective amino acid to within a certain proximity of the mRNA-template, and basically "holds" it there until *all* of the other tRNA molecules have arrived in their proper sequence. This having been accomplished, the ribosomes now move along the strand of mRNA, in zipper-like fashion, "sewing-up" the series of amino acids, in sequence, according to the base sequence determined by the corresponding mRNA strand. The resultant protein molecule then has a specific amino acid sequence, which determines its characteristics. If it is an enzyme to be used within the cell, it gets released from the mRNA matrix once formed and diffuses into the ICF to catalyze the biochemical reaction for which it is specific. In this sense, it has been hypothesized that end products of a biochemical reaction can repress enzyme-synthesis by acting *directly* on the portion of the DNA molecule in the nucleus that codes for the manufacture of this particular enzyme. By thus *blocking* the manufacture of the appropriate mRNA chain for that enzyme, the product, itself, interferes with the manufacture of the catalysts necessary to make more end product. Substrates, then, do exactly the opposite, to *induce* enzyme-synthesis in response to their presence.

If the protein molecules are being manufactured for use elsewhere, as is the case of antibodies or hormones, for example, they get released into the endoplasmic reticulum, whereupon they are transported to the Golgi apparatus for ultimate release into the extracellular fluid. Along the way, they get further refined for their ultimate intended purpose.

THE CONCEPT OF A CODON

Since there are about 20 amino acids involved in protein synthesis, but only four bases that code for them, the bases must obviously operate in groups to identify a particular amino acid. That is, four quantities, taken four at a time, can easily be arranged in 24 different ways to code for 24 different items. Furthermore, four quantities, taken three at a time, can *also* be arranged in 24 different ways if no quantity appears more than once in any arrangement, or in $4 \times 4 \times 4 = 4^3 = 64$ ways if quantities are allowed to repeat (i.e., appear more than once) in each arrangement. Most of the evidence thus far available suggests that the latter "triplet-code" is the correct one for protein synthesis. Thus, each tRNA contains three bases arranged in one of 64 possible sequences, each sequence being called a *codon,* and being congruent to the amino acid for which it is coded. This congruence, then, determines which amino acid the tRNA will transfer to the strand of mRNA, since the sequence of bases on tRNA is complementary to its corresponding site on the mRNA strand.

If the tRNA has a sequence Uracil-Adenine-Guanine (UAG), for example, it will collect the amino acid -- say, Arginine -- that is congruent to this codon configuration and transport that amino acid to the part of the mRNA strand that has the corresponding codon sequence Adenine-Uracil-Cytosine (AUC). This is because Adenine bonds or pairs itself with Uracil, Uracil bonds with Adenine, and Cytosine pairs with Guanine in the hierarchy of affinities between base pairs. Again, in "zipper-like" fashion, the ribosomes travel up the mRNA chain, depositing amino acids brought by tRNA, in the sequence determined by mRNA, as originally coded into the DNA source, along the protein chain being manufactured.

As an example of the versatility that such a coding system offers, let us get back to the organism's immune system response mechanisms as discussed briefly in Chapter 2. Recall that when Lymphocytes contact a foreign antigen, they become "activated" and trigger a response that calls for the manufacture of a specific antibody. Now, *each* kind of antigen stimulates its *own* corresponding antibody (or, at least it *should* for the system to function effectively). Estimates of how many different kinds of lymphocytes would be required to endow the body with the capability of synthesizing antibodies against virtually *any* antigen run as high as 10 billion or more. At first, this may sound preposterous. But, since each antibody is composed of four polypeptide chains -- two identical "heavy" chains and two identical shorter "light" chains, linked by disulfide bonds -- only 100,000 genes for light chains and 100,000 genes for heavy chains are necessary to code for 10 billion different antibodies. Thus, 200,000 genes can code for 10 billion antibodies. Again, this may *sound* high, but remember, the human genetic code is programmed into *twice* that many genes, not all the same size, and various calculations suggest that this *particular* 200,000 genes would occupy less than 10 percent of the total DNA content of a typical human cell. Hence, the hypothesis now most widely accepted postulates the existence in the human body of an enormous number of *slightly* different lymphocytes, each specialized for the potential production of a different antibody. Antibodies are globular proteins, with about three fourths of each heavy chain and half of each light chain showing great constancy of amino acid sequencing. The variability among them lies mostly in the remaining portions of each chain, at the free amino ends, and it is believed that the binding sites for antigens are at these variable end-portions of the protein molecule.

It is also believed that the binding sites work primarily through their geometric configuration, as already addressed several times earlier. That is to say, at the end of the protein synthesis process, we are left with a very *long, thin* (i.e., structurally unstable) protein molecule composed of some polar (Arginine, Aspartic Acid, Asparagine, Cysteine, Glutamic Acid, Glutamine, Glycine, Histidine, Lysine, Serine, Threonine and Tyrosine) and some non-polar (Alanine, Isoleucine, Leucine, Methionine, Phenylalanine, Proline, Tryptophan and Valine) amino acid side chains that are capable of interacting with one another. This structurally unstable configuration collapses into a more stable, complex, coiled-up, secondary and tertiary physical arrangement that is crucial to the stereospecificity of the protein function -- especially if it is to perform as an enzyme in a critically-catalyzed biochemical reaction.

Catabolic Reactions

What we have discussed so far are *anabolic* types of cell reactions, which tend to take place in the nucleus of the cell, in the mitochondria, along ribosomes, in the intracellular fluid, and in the endoplasmic-reticular-Golgi-apparatus-complex. There are other enzyme-catalyzed *catabolic* reactions that occur inside the cell, taking place mostly in the Lysosomes.

Lysosomes can digest proteins, nucleic acids and polysaccharides via basically six types of enzyme-catalyzed reactions that have four primary purposes. The six reactions include: (i) the breakdown of RNA (mRNA, tRNA) by the enzyme Ribonuclease; (ii) the breakdown of "foreign" DNA by the enzyme Deoxyribonuclease; (iii) the lysing of phosphate esters by the enzyme Phosphatase; (iv) the catabolysis of proteins by the enzyme Cathepsin; (v) the decomposition of sugars and polysaccharides by the enzyme Glycosidase; and, (vi) the breakdown of sulfate esters by the enzyme Sulfatase. These six types of reactions are accomplished for the purpose of: (i) digesting large particles of foreign or undesirable matter that enter the cell; (ii) digesting wastes generated from intracellular substances; (iii) digesting substances external to the cell; and/or, (iv) digesting the cell, itself, when it has outlived its usefulness.

Control of Cellular Processes

Although we have touched upon several metabolic control mechanisms throughout this text, it would be useful at this point to review some of them in a more generalized sense. First of all, when we speak of physiologic control, we distinguish between *intrinsic* control mechanisms, and those that are *extrinsic*. By intrinsic, we mean, specifically, those control mechanisms that are inherent to a system, and become manifest as a result of the very nature or characteristics of the system. Thus, for example, the ability of a cell membrane to keep undesirable substrates out of the cell by virtue of the geometric properties of the membrane, itself (pore size, pore configuration, molecular structure, and so on, see Chapters 2 and 3) would be considered intrinsic control of cellular processes. In this sense, the cell can control the *access* of substrates, S, to the intracellular fluid spaces by all of those intrinsic mechanisms described in Chapters 2 and 3 with respect to cell membrane permeability to S. Moreover, because of the intrinsic nature of lysosomes, the cell has the ability to dispose of, or otherwise denature undesirable substrates if they somehow *do* manage to gain access to the ICF.

By far, the primary intrinsic control of cell metabolism is accomplished through the regulation of enzyme synthesis. If the enzyme necessary to catalyze a particular biochemical reaction is absent or otherwise denatured or inhibited, the reaction will simply not go at a rate sufficient to maintain life. For the most part, the presence or absence of a specific enzyme required to catalyze a specific reaction is determined by the genetic code that resides in the cell nucleus. And because of the very specificity, itself, a substitute enzyme cannot generally step in for a missing one. The need to have the right enzyme present, at the right time, in the right place, in an activated, functioning condition gives the cell a tremendous amount of control over which reactions will go, and when, and to what extent.

Enzyme induction by substrates, and enzyme repression by the products of a reaction are further examples of intrinsic control of cell metabolism. These have the advantage of allowing the cell to make specific enzymes only under certain conditions, thus economizing on both effort (energy efficiency) and space (metabolite or mass efficiency). Furthermore, as a consequence of this regulated type of enzyme synthesis, cells are prevented from making excess concentrations of enzymes that may, in themselves, either be toxic to the cell, or inhibit other reactions from taking place.

Other types of intrinsic control include enzyme-substrate affinity (or the lack thereof in the case of adverse competition for binding sites on the enzyme molecule); the number of binding sites on an enzyme; their degree of saturation (as discussed relative to inhibition of the Krebs cycle); feedback inhibition by the products of a reaction; and feedforward enhancement by the "siphoning off" or accelerated catabolism of the products of a reaction (akin to the way the oxygen-reserve mechanism described in Chapter 3 works). All of these and more are *intrinsic,* in the sense that there is no intervention from external sources -- such as the Endocrine System or the Autonomic Nervous System involving hormonal or neural stimulation. This is as opposed to *extrinsic* control where the latter are intimately involved.

For example, we saw in Chapter 4 that during cold stress, metabolic processes involving the mitochondria can be significantly affected by the hormone Thyroxine. Enzyme synthesis, too, can come under the influence of hormones where they interfere with or enhance either the manufacture of the enzyme, itself, or its subsequent activation following synthesis. Different molecular configurations of the same enzyme are called *isozymes*. Hormones can make isozymes active as catalysts. This is not simply a manipulation of the geometry of the enzyme (as is the case in allosteric regulation), but also a redistribution of the actual chemical constituents of the enzyme. Its actual chemical *formula* is *not* changed, but the location on the enzyme chain of key side groups may be redistributed under the influence of the right hormone, making the enzyme active where it previously was not. In other words, the same building blocks are assembled, but in a different way, due to hormone-induced association or dissociation of specific protein subunits. Furthermore, some *co-enzymes* can be influenced by hormones that activate *them*, so that they, in turn, can activate their respective enzyme.

And the list goes on. Indeed, one might propose a corollary to the theory of evolution that proposes survival of the fittest based on mass transport processes (c.f., Chapter 2, pg. 91). This corollary would suggest further that those species that could affect the most efficient *control* of the most desirable *enzyme systems* -- to catalyze the most desirable reactions, most effectively, and most efficiently -- survived; while those who could not, did not! In other words, those species who could best *utilize* nutrients for the generation of energy and the synthesis of proteins; who could best digest foreign invaders (through immune mechanisms); and who could reproduce most efficiently -- all processes very much dependent upon the right enzyme, being in the right place, at the right time, in proper functioning order -- these were the species who were most likely to survive. Along with this concept, one may develop two others of perhaps equal importance: one, discussed further below, advances the idea of a minimum energy principle in Biomechanics; and the second, addressed further in Chapter 6, examines *adaptation*, as it relates to the concept of Homeostasis.

Principles of Human Posture and Locomotion

Among the most important of the metabolic processes driven by ATP are active transport mechanisms such as those described by equations [3-63] to [3-97] in Chapter 3. Indeed, as mentioned in that chapter, as much as 20% of resting oxygen consumption in mammalian muscle tissue may go entirely for the purpose of actively transporting sodium ions via the sodium pump -- mainly for the purpose of generating a resting membrane potential in this excitable tissue. The *resting* membrane potential subsequently becomes the source of an *action* potential, and the latter allows muscle tissue to *contract*. Thus, upon the command of an action potential, cleavage of the phosphate bonds in ATP drives a series of biochemical reactions that result in muscular contractions that can affect movement of parts or all of the animal body. Let us take a closer look at how this is accomplished.

Generation of an Action Potential in Excitable Tissue

The establishment of a *resting* membrane potential in excitable tissue was described in some detail in Chapter 3, pgs. 231-241. Suppose, now, that we somehow disturb that resting equilibrium situation defined by equation [3-99], viz., $\Delta B_m = -90$ millivolts. The disturbance in the case of excitable tissue takes the form of doing something to alter the cell membrane permeability. This may be accomplished *physically* (e.g., by "stretching" the membrane, as occurs when the heart passively fills with blood returning from the systemic and pulmonary circulations); or, it may be accomplished *chemically* (e.g., as a result of exposure to the neurotransmitters Acetylcholine and Norepinephrine, or the hormone Adrenalin); or, it may be accomplished *pharmaceutically* (e.g., through the action of drugs such as cardiac glycosides, anti-arrhythmics, and antihypertensives); or, the disturbance may be in a number of other electrical and mechanical forms, both normal and pathological, natural and/or synthetic (e.g., see Chapter 3, pgs. 199-201, items 1-6 at end of section entitled, *Gibbs-Donnan Equilibrium and Capillary Mass Transport*). Regardless of the cause, disturbing a resting membrane in such a way that its permeability *increases* leads to a slow process of depolarization that culminates in the generation of an *action potential*. The exact mechanisms involved were identified by two scientists named Hodgkin and Huxley, who received the Nobel price for their work in 1963.

A disturbance that affects the permeability of excitable tissue generally causes it first to become more permeable to the sodium ion. Thus, sodium starts to "leak" back into the ICF, decreasing its negativity relative to the ECF, and starting to depolarize the resting cell membrane potential from -90 millivolts back towards zero. From the Goldman equation [3-98], it can be seen that if P_{Na} increases, the steady-state potential of the membrane will change in the direction of further depolarization. In other words, the more the cell is depolarized, the more P_{Na} increases, and the more P_{Na} increases, the more the cell is depolarized. If the membrane becomes depolarized enough, then P_{Na} will have increased enough so that the cell will try to move towards a Goldman potential which is even more depolarized, so the process is regenerative.

At a certain point in this regenerative process (called the *Threshold Potential*) -- typically, at a 20 to 40 mv depolarization point relative to the resting potential (although it can be as high as 50 mv) -- the process now becomes not only regenerative, but *irreversible,* as well. Indeed, the major characteristic of *excitability* or irritability) is that if a polarized cell membrane is depolarized a certain amount from a resting equilibrium level to a specific threshold value, the process of depolarization becomes irreversible and self-generating, and the propagation of a depolarization wave (action potential) ensues.

At the threshold potential P_{Na} increases dramatically several hundred fold, so that it is many times greater than P_K, and a sudden, very rapid stream of sodium ions penetrates the cell membrane. This, "Fast Inward Sodium Current," as it is called, leads to a rapid rise in the rate of membrane depolarization, causing a "spike" in the depolarization (mv) vs. time curve for the generation of an action potential (see Figure 5-5, Phase "O"). The rising phase "O", or *upstroke* phase of the action potential curve is due to the regenerative, irreversible increase in P_{Na} as the transmembrane potential goes toward the sodium equilibrium potential; and, within about $\frac{1}{2}$ to 1 millisecond, the cell membrane actually *reverses* polarity, (i.e., "overshoots"), becoming inside positive by about 30-40 millivolts, relative to the outside. This change in potential is called the action potential spike. It can occur very fast (e.g., at the rate of 200 - 800 volts per second) in tissues like the conducting pathways of the Atria, His-Purkinje system and Ventricles of the heart; or, it can proceed considerably slower (e.g., at the rate of 5-10 volts pe second) in tissues such as the Sino-Atrial and Atrio-Ventricular Nodes of the heart. Moreover, the amplitude of the spike can be large (on the order of 110 mv or more) for fast upstrokes, and/or small (on the order of 60 mv or less) for slow upstrokes. The faster the upstroke, and the larger the amplitude of the action potential spike, the faster will be the subsequent conduction velocity of the action potential.

In any case, if an even greater stimulus is administered to depolarize the cell, the action potential will be exactly the same, which is to say, if the cell is brought to its threshold, it will "fire" in an all-or-none response. The only thing determined by the *strength* of the stimulus is *whether or not* an action potential will be generated *at all*. The strength of the stimulus has no effect on the strength of the action potential generated, or on its subsequent speed of propagation. The rise time and amplitude of the action potential both depend not on the strength of the stimulus, but on the nature of the excitable *tissue,* and so, for a *given* type of tissue, they will always be the same, regardless of the type or intensity of the original stimulus. However, different excitable tissues in the body have different threshold potentials, so the strength of a stimulus *will* determine *which* tissue (or combination of tissues) fires at any given time. Tissues with a high threshold potential will be unaffected by weak stimuli.

Following stimulation of excitable tissue, the increase in P_{Na} and the rising membrane voltage are both short-lived. A second, slower process called *inactivation* decreases the sodium permeability, while chlorine also starts to get into the act. Note from Table 3-1 that the ECF concentration of chlorine is much higher than is the ICF concentration of this ion, so there is a natural concentration gradient tending to drive it into the cell. Add to that the (i) chemical affinity of chlorine for sodium, and (ii) the inside-positive state of affairs that exists at the peak of the action potential spike (which electrostatically attracts this negatively-charged ion), and the stage is set for a rapid influx of chlorine to follow on the heels of a rapid influx of sodium. Indeed, it does, so that the negativity of the interior of the cell begins to return with only a very small time delay after the cell polarity reverses. The inward chlorine current thus produces the *rapid repolarization* phase 1 shown in Figure 5-5 for the events associated with the generation of an action potential, and causes the cell membrane potential to start to fall back towards the resting potential.

If the excitable tissue involved is nerve tissue or striated skeletal muscle tissue, then following the rapid repolarization phase the sodium-potassium pump quickly takes over, and, within about 2.5 milliseconds, the transport mechanisms in the cell membrane have rapidly brought it back to its resting membrane potential of −90 millivolts. The cell is ready to fire again. In this respect, it is worthy to note that a sudden influx of sodium ions can cause an increase of three-to-four-fold in the activity of the sodium-potassium electrogenic pump (Guyton, 1981).

If, on the other hand, the excitable tissue involved is cardiac muscle tissue, or certain types of smooth muscle tissue, then we enter still another phase of the ionic events associated with the generation of an action potential. In this phase (Figure 5-5), positively charged calcium ions, which are more highly concentrated in the ECF than they are in the ICF (see Table 3-1), start to diffuse slowly across the cell membrane along with the negatively charged chlorine ions. For a brief period of time,

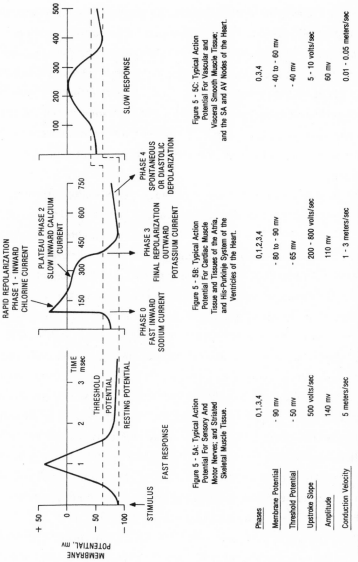

Figure 5-5 Various types of action potential profiles for physiologic excitable tissue.

therefore, there is a momentary "plateau" in the membrane depolarization curve, as the influx of negative charge is just balanced by the *slow inward calcium current*. Due to the important role that calcium plays in helping to generate contraction of muscle fibers (Schneck, D. J., 1984b), this phase 2 of the action potential profile is quite important in the physiology of both cardiac muscle tissue and some types of smooth muscle tissue (especially visceral smooth muscle concerned with peristaltic activity and vascular smooth muscle tissue concerned with the maintenance of blood vessel tone).

As time moves on, the cell membrane permeability returns to its pre-disturbed levels -- inactivation decreases P_{Na} and is accompanied by a delayed increase in P_K -- allowing the sodium-potassium pump to re-establish the resting membrane potential in the *final repolarization phase 3*, or outward potassium current phase of the action potential process. Since the Goldman potential is determined essentially by the ratio of P_{Na} to P_K (ignoring the chloride terms which, for the most part, are passive), the repolarization of the membrane is brought about both by inacti-vation, which decreases P_{Na}, and the increase in P_K, which leads to the fast outward potassium current. During an action potential the cell has a net gain of sodium and a net loss of potassium. These changes in con-centration must be counterbalanced by active ion transport. If the active transport is eliminated, however, the cell can still experience many action potentials before there are significant changes in concentration. The ex-act number will depend on the size of the cell, but even a small cell of 1 micron radius can experience up to 600 consecutive action potentials be-fore the net change in concentrations gets to the order of 10%.

Referring again to Figure 5-5, final repolarization of an excitable tissue membrane is designated as Phase 3 of the events associated with the generation of an action potential. Phase 4, then, starts the process all over again with a disturbance depolarizing the membrane to threshold. In the language of Cardiology, Phase 4 is called the "Diastolic Depolarization" phase of the Cardiac cycle, and partly gives rise to the *automaticity* that allows the heart to beat at a natural rhythm independently of any specific external neural stimulation. Phase 4 associated with visceral smooth muscle is what gives the small intestines, for example, their continuous, rhythmic, peristaltic undulating motion that keeps food moving along in the alimentary canal.

In the process of completing the repolarization Phase 3 of the ionic events associated with an action potential, the membrane may actually experience a momentary hyperpolarization before actually returning to its normal resting potential. When this occurs, the momentary hyperpolarization during Phase 3 is called a *repolarization overshoot,* and the return to normal resting potential from the overshoot potential is called the after-potential of the action potential. These events represent a phase lag between depolarization and repolarization activities, causing the latter to "overcompensate" before establishing once again an equilibrium configuration. This hyperpolarization during the relative refractory period (see later) also contributes to the fact that an action potential can only be generated by a stronger-than-normal stimulus during this period of time.

Figure 5-5, then, summarizes the three basic types of action potential profiles that are associated with various types of excitable tissue: Slow Action Potentials, and Fast Action Potentials with, or without a Plateau Phase 2. All of these may or may not have a spontaneous depolarization Phase 4, depending on the particular tissue involved. If there is *no* spontaneous depolarization, as is the case most commonly in striated skeletal muscles, then Phase 4 simply becomes an iso-electric, constant polarity phase within which the tissue simply awaits an external stimulus (such as from an alpha-motorneuron) to generate the next action potential.

Formulation of an Action Potential Equation

The preceding discussion may be quantified following the method outlined by Fleming (1969). Thus, let ΔB_m be the resting potential of the cell membrane (c.f., equation [3-99]), let ΔB_t be the transmembrane threshold potential for initiation of an action potential (c.f., Figure 5-5), and define $\Delta B_s = |\Delta B_t - \Delta B_m|$ to be the amount of depolarization from a resting state required to generate a self-perpetuating action potential, or, to "trigger" a response. Furthermore, let R^*_m be the membrane resistance (c.f., equation [3-60]), C^*_m be the membrane capacitance (c.f., equation [3-62]), and $\tau^*_m = R^*_m C^*_m$ be the corresponding time constant associated with the parallel resistance-capacitance circuit representing the electrical analogue of one unit area of the membrane. Finally, let I^* represent the "intensity" (expressed as a stimulus current or flow of ions across a depolarizing membrane) of a disturbance applied to an excitable cell membrane, whose conduction resistance is R^*_m, whose membrane capacitance is C^*_m, and whose resting membrane potential is ΔB_m. Then, if I^* is applied (or persists) for at least the time interval, τ^*, necessary to depolarize the excitable membrane by a net voltage change ΔB_s, i.e., to threshold from ΔB_m to ΔB_t, thereby generating an action potential, we may write (Fleming, 1969):

$$\Delta B_s = I^* R^*_m (1 - e^{-\tau^*/\tau^*_m}), \quad \text{or,} \qquad \text{[5-14]}$$

$$I^* = \frac{\Delta B_s}{R^*_m (1 - e^{-\tau^*/\tau^*_m})} = \text{Stimulus "Strength"} \qquad \text{[5-15]}$$

Note from equation [5-15] that if it takes an infinitely long (i.e., $\tau^* \to \infty$) period of time to elicit an action potential response, then obviously the stimulus strength (depolarization current *across* the membrane) is just at a bare minimum in terms of its ability to excite the membrane. This "bare minimum stimulus strength" which will still elicit a response from the excitable membrane is called the *Rheobase Current,* or, simply, the Rheobase of the membrane, and it is given from equation [5-15] by:

$$I^*_r = \frac{\Delta B_s}{R^*_m} \qquad \text{[5-16]}$$

If ΔB_s is, say, typically 40 millivolts (40×10^{-3} volts) and R^*_m is on the order of 1000 ohms/cm^2 of membrane surface area (see Chapter 3), then $I^*_r = 40 \times 10^{-6}$ amps, or, 40 microamps. Thus, a 40 microamp depolarization stimulus applied for a very long period of time to this particular type of excitable tissue will *eventually* get it to respond (i.e., propagate a depolarization wave *along* the membrane).

Note the dependence of I^*_r on $\Delta B_s = |\Delta B_t - \Delta B_m|$, such that raising the threshold potential, ΔB_t, or lowering the resting potential, ΔB_m (e.g., by hyperpolarization), or both, can lead to a significant increase in the stimulus strength required to generate an action potential (and vice versa). This is the *only* way in which the stimulus strength and resulting action potential are related to one another, i.e., whether or not an action potential will be generated at all. The stimulus strength, as mentioned previously, does *not* have anything to do with the subsequent strength, amplitude, or speed of propagation of the generated action potential. It is important, however, to realize that the human organism has established a hierarchy of excitable tissues that have different values of ΔB_m, ΔB_t, R^*_m, C^*_m, and τ^*_m -- such that it can discriminate, based on *which* individual or groups of excitable nerve fibers (for example) are "firing" at any given time, the frequency (proportional to $\frac{1}{\tau^*}$) and intensity (proportional to I^*) of the disturbing signal. Moreover, when necessary, the organism can *control* (see Chapter 6) ΔB_m, ΔB_t, R^*_m, C^*_m, and τ^*_m in order to adjust, adapt or otherwise alter the tissue sensitivity to an excitation stimulus. But, once the tissue fires, it fires, following the all-or-none response principle discussed earlier.

Substituting equation [5-16] into equation [5-15], and rearranging, we can write an action potential response equation of the form:

$$\frac{I^*_r}{I^*} = 1 - e^{-\tau^*/\tau^*_m}, \text{ which, if} \qquad [5\text{-}17]$$

we plot nondimensional stimulus strength, I^*/I^*_r, vs. nondimensional time, τ^*/τ^*_m, yields the hyperbola shown in Figure 5-6. The value of τ^*

for which $I^* = 2I^*_r$ is called the *Chronaxy Time,* τ^*_c, and it corresponds to $(\tau^*/\tau^*_m) = 0.693$. Again, for illustrative purposes, if we let $R^*_m = 1000$ volts-sec/coulombs/cm², and $C^*_m = 1$ microfarad per cm² $= 10^{-6}$ coulombs/volt/cm² (see Chapter 3), then τ^*_m is of order $(10)^3(10)^{-6} = 10^{-3}$ seconds, or, one millisecond per square centimeter of excitable tissue membrane area; and $\tau^*_c = 0.693 \times 10^{-3}$ seconds. For this tissue, then, a stimulus I^* equal to twice the Rheobase Current, $I^*_r = 40$ microamps, would have to be applied for at least 0.693×10^{-3} seconds in order to elicit an action potential response. In other words, it would take an 80 micro-amp local depolarization disturbance, applied at a *maximum* frequency of 1443 cycles per second ($\frac{1}{\tau^*_c}$, where $\tau^*_c \geq 0.693 \times 10^{-3}$), to generate an action potential in this particular tissue. At higher frequencies, *this* tissue *probably* would not be very responsive to an 80 micro-amp stimulus, although *another* membrane, having different values for R^*_m and/or C^*_m, might be. Of course, the tissue, hypothetically, will also respond to a 40 micro-amp (I_r^*) current applied for a very *long* period of time (i.e., frequency approaching zero) -- but the latter time required (called the *utilization* time) is ambiguous and difficult to measure, so the chronaxy time is more commonly taken to be a measure of excitability.

Going one step further, this tissue will *not* respond to a stimulus *weaker* than the equivalent of 40 microamps in intensity (i.e., the Rheobase, I^*_r), although, again, another membrane might. Moreover, note from Figure 5-6 that stimulus intensity drops off exponentially with stimulus period (increasing τ^*, or decreasing stimulation frequency), such that higher frequency disturbances are generally associated with higher stimulation intensities, and vice versa, or, the more intense the stimulus, the shorter the time required to excite the tissue.

The significance of the Chronaxy time may be addressed physically by noting that an excitable membrane can only generate an action potential when it is properly polarized to begin with (i.e., when a resting potential exists), and when its permeability characteristics are in the correct state for appropriate ion transport as described for phases 0 through 4 of the action potential profile. Thus, since just after an excitable

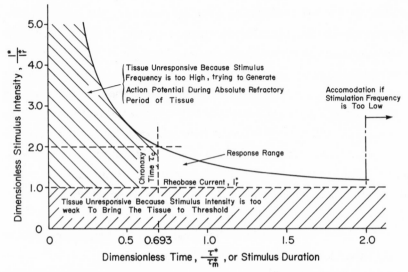

Figure 5-6 Non-dimensional stimulus strength vs. non-dimensional time for the generation of an action potential.

membrane has been stimulated, both the polarity of the membrane and its permeability characteristics are significantly altered, it is impossible to have another action potential initiated at this time. This "time" is called the *absolute refractory period* of the tissue, and represents the absolute minimum period of time required for the membrane to "re-set" itself in order to "fire" again. This absolute refractory period is generally of the order of 0.5 milliseconds for most types of excitable tissue -- so we see that the chronaxy time of 0.693 msec for our illustrative tissue is in some sense related to the absolute refractory period of 0.5 msec for that tissue, just as the rheobase current for the tissue is in some sense related to its threshold potential for firing.

One may thus speak of a "response range" for excitable tissue as being the area bounded in Figure 5-6 by the Rheobase Current, I^*_r along the ordinate axis, the Chronaxy Time, τ^*_c, along the abscissa axis, the action potential response equation [5-17], and some upper abscissa limit determined by the physical characteristics of the tissue involved as it relates to the phonomenon of *accommodation* (see below). Furthermore, for up to 10 milliseconds *after* the absolute refractory period, an excitable membrane has repolarized enough to generate an action potential; but, the threshold is increased (because the permeability to potassium remains somewhat elevated), the depolarization distance to threshold is increased (because of afterpotentials and "overshoots" that lower the resting potential), and the membrane resistance is decreased (i.e., its conductance is momentarily increased). Because of all of these effects, the intensity of signal required in order to generate an action potential during this period of time is much larger than it would be normally, and the time interval involved is called the *relative refractory period*. Shortly after reaching its resting membrane potential once more, the membrane is no longer refractory, and is ready to conduct another impulse normally.

If an excitable membrane is depolarized *slowly*, letting τ^* be spread over at least 5 to 10 milliseconds (i.e., for low frequency disturbances on the order of 100 cycles per second), the threshold potential will gradually rise. This phenomenon is called *accommodation*, and illustrates the fact that not only the *amount* of depolarization, ΔB_s, determines whether or not an action potential will be generated, but the *rate* of depolarization, $\dfrac{\Delta B_s}{\Delta t}$, is a factor, as well. This is because during slow depolarizations, *inactivation* -- which is to say, the sodium-potassium pump and other factors that control membrane permeability -- can keep up with the normally more rapidly occurring factors that increase sodium permeability with depolarization. If we imagine an analogy whereby one is trying to bale water out of a leaking boat (i.e., the sodium-potassium pump trying to *repolarize* a membrane leaking sodium ions into the ICF), then if the leak is slow enough (very slow depolarization stimulus), we can both bale the water out as fast (or faster) than it comes in, *and,* have the time to "plug" the leak. Also, the increase in potassium permeability with depolarization can keep up with the depolarization itself, if it is slow enough, and this is basically what results in accommodation. It prevents a "rush" of sodium ions into the ICF, which is essentially what triggers the action potential spike (i.e., "sinks" the boat).

Going one step further, if an excitable membrane is *hyperpolarized,* say, to a resting potential of −120 mv instead of −90 mv, for at least 5 to 10 msec, the threshold will gradually fall -- sometimes even to the normal *resting* potential -- in fact, so far that when the hyperpolarization is ended and the cell returns (i.e., depolarizes 30 mv) to its resting potential of −90 mv, it actually *fires* at this normal resting potential. This is known as the *anodal break phenomenon,* and is believed to be due both to a decreased membrane permeability to potassium that results from prolonged hyperpolarization, and, to a somewhat depressed, or "sluggish" membrane inactivation mechanism. In other words, the anodal break phenomenon is essentially the exact opposite of accommodation.

In summary, then, stimulation of excitable tissues at too high a disturbance frequency leads to problems related to the absolute and relative refractory periods following depolarization; stimulation at too low a disturbance frequency leads to problems related to accommodation. Stimulation at too low a disturbance amplitude leads to problems related to the resting and threshold potential levels, as well as factors related to the rheobase current; stimulation at too high a disturbance amplitude may damage the tissue involved. Somewhere between these extremes of amplitude and frequency lies the response range of any given (R^*_m, C^*_m) excitable cell membrane, in terms of its ability to *generate* an action potential at the site of stimulation. But the story does not end here, because once generated, the action potential now proceeds to *propagate* along the membrane away from the actual site of stimulation. Let us describe briefly how this is accomplished.

The Propagation of an Action Potential in Excitable Tissue

At a site where an action potential is occurring, the depolarization of the membrane locally influences sites further along the tissue by the passive "cable" properties of the tissue, so that these sites are progressively brought to threshold as well. Recall that during an action potential spike (Figure 5-5, Phase 0), the fast inward sodium current actually *reverses* the membrane polarity, locally making the *inside* positive and the *outside* negative at the actual site of stimulation. Now, if, in the immediate neighborhood of the excitation spot, excitable tissue is in a *resting* state (membrane *inside* negative, *outside* positive), then we have created a situation where there exist on either side of the membrane continuous regions in space, which are in immediate contact with one another, but which are exactly opposite in charge distribution. There being no intervening barrier to impede the *flow* of such charge, positive ions will move to neutralize negative ions and there will result a self-perpetuating "surge" or wave of depolarization that will propagate down both sides (inside and outside) of the membrane with a very specific and relatively constant velocity (see Chapter 6 and Plonsey, 1969). The action potential impulse will thus be conducted away from the site of stimulation.

Moreover, since the area immediately behind a propagating action potential is in an absolute refractory state, as discussed earlier, conduction of an action potential generally proceeds *away from* the stimulation site. For this reason, when a depolarization wave reaches the end of a conducting pathway along the membrane of an excitable tissue, it will stop. It normally cannot, and does not turn around and propagate back in the direction from which it came. The only thing that can alter this state of affairs is somehow to impede, slow down (by disease, malfunction, or extrinsic control mechanisms), or otherwise delay the propagating depolarization wave, thus giving the area behind it a chance to recover enough to fire again. Only then can the conducted impulse propagate both backwards and forwards, in a phenomenon called re-entry of excitation -- or, more simply, *Re-entry* -- but this is not the normal physiologic state of affairs.

The mathematical description of the events associated with propagation of an action potential along excitable membranes is addressed in some detail in the fine work of Plonsey (1969), and others (e.g., Deutsch and Micheli-Tzanakou, 1987). Considering this topic to be "beyond the scope of the present work," we refer the reader to the literature for further details, but point out the fact that the depolarization wave, propagating down a muscle membrane at a speed of about 5 meters per second, alters the electrical potential of the membrane and acts just like an electric current. Thus, the electrical activity associated with the action potential causes hydrolysis of water molecules in the muscle tissue, releasing hydrogen and hydroxyl ions according to the reaction:

$$H_2O \xrightarrow{\text{Electro-lysis}} H^+ + OH^- \qquad \text{[5-18]}$$

The oxidation potential associated with the above reaction involving *two* moles of water (see Chapter 4, equations [4-31] to [4-37]) is -1.23 volts at 25°C. The negative sign means that the reaction "prefers" to go spontaneously to the *left* (c.f., equation [4-34], for which the standard reduction potential was given as $+0.816$ volts for the *formation* of two moles of water from a mole of O_2 and two moles of H_2 at 25°C). Thus, this reaction is *driven* to the right by the energy supplied in the action potential (electromagnetic energy transduced from the work done by the original stimulus). The hydrogen ions released by reaction [5-18] attack the terminal $P{\sim}O$ bond of the Adenosine Triphosphate (ATP) complex illustrated in Figure 5-4, forming a terminal $- OH$ bond on what is to become Adenosine Diphosphate, and splitting off the terminal phosphate group, to combine with the OH^- ion released above and thus form phosphoric acid. The reaction,

$$ATP + H^+ + OH^- \xrightarrow{\text{ATP-ase}} ADP + H_3PO_4 + \text{Energy} \qquad \text{[5-19]}$$

is catalyzed by a calcium-activated form of the protein Myosin (ATP-ase) and releases the energy that is subsequently to be transduced into the generation of a muscular force (Schneck, 1984b). This energy can amount to a maximum useful power output of some 800 calories per minute, per kilogram of an individual's body weight (0.075 Horsepower per kilogram). The reaction [5-19] can be inhibited by any chemical compound (such as the glycoside Ouabain) that de-activates or immobilizes ATP-ase.

Propagation of an action potential along the membrane of polarized muscle cells, then, provides the stimulus for the cleavage of the high-evergy bonds of ATP, which, in turn releases the energy from which muscular contractions are derived. A more complete discussion of the actual transduction mechanism whereby the potential energy released from ATP is actually converted into muscular work can be found elsewhere (e.g., Schneck, D. J., 1984b, Squire, 1981, and Hatze, 1981). Again, it is not our intent here to develop a treatise on the microscopic aspects of musculoskeletal function, but to propose some basic principles that can be used to analyze such function from an engineering point of view. Thus, we turn our attention now to a simple type of muscular activity, in order to develop the principles involved.

Inductive Analysis of the Kinetics of Human Movement

Let us examine a very simple type of human movement: forearm flexion that results from the generation by the biceps musculature of a moment at the elbow joint. The free-body diagram that describes this type of motion is illustrated in Figure 5-7. Basically, the flexor musculature of the upper arm (i.e., the biceps muscle) is attached to (i.e., inserts into) the lower arm, or forearm, at a point $\vec{\ell}$ measured relative to the elbow joint, O. Generation of a force, $\vec{F}$, the arrow indicating a vector quantity, by contraction of the biceps thus produces a counterclockwise (positive) moment, $\vec{M}_o$, about the elbow joint, which moment acts to produce a forearm flexion. That is, the moment produces an angular acceleration, $\ddot{\Theta}$, a corresponding angular velocity, $\dot{\Theta}$, (the dots indicating differentiation with respect to time), and a resulting angular displacement, Θ, of the forearm having mass m_f. Increasing Θ (*closing* the angle between the upper and fore arms at the elbow joint) defines a forearm flexion, which, for simplicity, is assumed to transpire completely in the sagittal plane (the plane of the paper). The latter is a vertical plane through the longitudinal axis of the body that divides it into a right and left half.

FORMULATION OF THE GOVERNING EQUATIONS
OF MOTION

To the situation described above and illustrated in Figure 5-7, we apply Newton's Second Law(s) of Motion, which addresses conservation of linear and angular momentum as it relates to the existence of external forces and moments, viz.,

$$\sum \vec{F}_{\text{External}} = m_f \vec{a}_{CG}, \quad \text{and,} \qquad \text{[5-20]}$$

$$\sum \vec{M}_{\text{External}} = I_{CG} \vec{\alpha}_{\text{forearm}} \qquad \text{[5-21]}$$

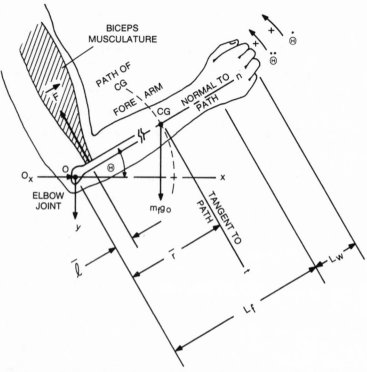

Figure 5-7 Free body diagram of a simple forearm flexion generated by the biceps musculature acting to develop a moment, M_O, about the elbow joint.

where $\vec{a}_{CG}$ represents the linear acceleration vector of the center of gravity (CG) of the forearm (m_f), $\vec{\alpha}$ represents the angular acceleration vector (magnitude, $\ddot{\Theta}$) of the lower limb, and I_{CG} represents the mass moment of inertia of m_f calculated with respect to the center of gravity of the forearm, and equal to $m_f k_r^2$, where k_r is its corresponding radius of gyration relative to CG. If the forces at the elbow joint have resultant components O_x and $-O_y$ as shown in Figure 5-7, and the biceps muscle force has components $-F_x$ and F_y in the sagittal plane as shown, then,

$$\sum F_x = O_x - F_x = m_f a_x, \quad \text{and,} \qquad \text{[5-22]}$$

$$\sum F_y = -O_y + F_y - m_f g_o = m_f a_y \qquad \text{[5-23]}$$

Now, the acceleration vector, $\vec{a}_{CG}$ in equation [5-20] may be written with respect to the elbow joint at O by the relative motion equation:

$$\vec{a}_{CG} = \vec{a}_{CG/O} + \vec{a}_O \qquad \text{[5-24]}$$

where:

$$\vec{a}_{CG} = a_x \hat{i} + a_y \hat{j} \qquad \text{[5-25]}$$

$$\vec{a}_O = a_h \hat{i} + a_v \hat{j} \qquad \text{[5-26]}$$

and $\vec{a}_{CG/O}$ has normal (centripetal) and tangential components:

$$\vec{a}_{CG/O} = -\bar{r}(\dot{\Theta})^2 \hat{i}_n + \bar{r}(\ddot{\Theta})\hat{i}_t \qquad \text{[5-27]}$$

Substituting equations [5-22] to [5-27] into equation [5-20], one can write the following set of equations in the sagittal plane:

$$\sum F_x = O_x - F_x = m_f a_x$$

[5-28]

$$= m_f [- \bar{r}(\dot{\Theta})^2 \cos \Theta - \bar{r}(\ddot{\Theta}) \sin \Theta + a_h]$$

$$\sum F_y = - O_y + F_y - m_f g_o = m_f a_y$$

[5-29]

$$= m_f [- \bar{r}(\dot{\Theta})^2 \sin \Theta + \bar{r}(\ddot{\Theta}) \cos \Theta + a_v]$$

Now, turning to equation [5-21], the sum of the external moments acting around CG may be determined from Figure 5-7 to be:

$$\sum M_{CG} = I_{CG}\ddot{\Theta} = O_x \bar{r} \sin \Theta + O_y \bar{r} \cos \Theta$$

[5-30]

$$- F_x(\bar{r} - \bar{\ell}) \sin \Theta - F_y(\bar{r} - \bar{\ell}) \cos \Theta$$

Note that equation [5-21] has become a scalar equation [5-30], since all of the forces are assumed to lie in the sagittal plane so the moment axis is perpendicular to the paper and represents the *only* axis around which rotation is occurring. Note further that the elbow joint is assumed to be a smooth (no frictional joint couples), pinned (i.e., capable of allowing only pure rotation), connection. Thus, equation [5-30] may be expanded and rearranged to yield:

$$\sum M_{CG} = I_{CG}\ddot{\Theta} = F_x \bar{\ell} \sin \Theta + F_y \bar{\ell} \cos \Theta$$

[5-31]

$$+ \bar{r} \sin \Theta(O_x - F_x) - \bar{r} \cos \Theta(F_y - O_y)$$

But, $F_x \bar{\ell} \sin \Theta + F_y \bar{\ell} \cos \Theta$ is precisely equal to M_o, i.e., the flexion moment generated about the *elbow* joint by the biceps muscle force, $\vec{F} = - F_x \hat{i} + F_y \hat{j}$. Also, we may substitute into equation [5-31] equation [5-28] for $O_x - F_x$ and equation [5-29] for $F_y - O_y$, to get:

$$I_{CG}\ddot{\Theta} = -m_f \bar{r}^2 (\dot{\Theta})^2 \sin\Theta \cos\Theta - m_f \bar{r}^2 (\ddot{\Theta}) \sin^2\Theta + m_f \bar{r} a_h \sin\Theta$$

$$+ m_f \bar{r}^2 (\dot{\Theta})^2 \sin\Theta \cos\Theta - m_f \bar{r}^2 (\ddot{\Theta}) \cos^2\Theta - m_f \bar{r} a_v \cos\Theta \quad \textbf{[5-32]}$$

$$- m_f g_o \bar{r} \cos\Theta + M_o$$

Note that the leading terms lined-up under one another to the right of the equality sign on the first two lines of equation [5-32] cancel one another. Moreover, using the trigonometric identity, $\sin^2\Theta + \cos^2\Theta = 1$, and the parallel axis theorem for moments of inertia, i.e., $I_{CG} + m_f \bar{r}^2 = I_o$ -- the mass moment of inertia of the forearm with respect to the elbow joint, equation [5-32] reduces to:

$$M_o = I_o \ddot{\Theta} + m_f \bar{r} \cos\Theta [a_v + g_o] - m_f \bar{r} a_h \sin\Theta \qquad \textbf{[5-33]}$$

If we impose the additional constraint that the elbow joint remain fixed in space during the entire flexure movement (i.e., no *upper* arm motion is allowed), such that $a_v = a_h = 0$, and the center of gravity of the limb follows the trajectory,

$$x^2 + y^2 = \bar{r}^2 = \text{constant}, \qquad \textbf{[5-34]}$$

then, for this very special case equation [5-33] reduces to:

$$M_o = I_o \ddot{\Theta} + m_f \bar{r} g_o \cos\Theta \qquad \textbf{[5-35]}$$

Equation [5-35], which incorporates both the linear and angular momentum considerations associated with a very simple forearm flexion, is the governing equation of motion that describes such a body movement. Note, however, that in the form given -- even *with* all of the inherent assumptions and constraints -- it is still an indeterminate equation. That is, what we are left with here is one equation, [5-35], containing *two* unknown quantities, $M_o(t)$, and $\Theta(t)$. More specifically, then, it is a *first-*

order indeterminate equation, in that there is one more unknown than there are equations available to solve for them. It is not an accident that the *first*-order indeterminateness corresponds exactly to the fact that there is *one* joint involved in this movement (the elbow joint only). We shall get back to this point later on in this chapter.

More immediately, however, the situation we have posed in Figure 5-7, governed by equation [5-35], in the absence of any additional information, hypothetically allows there to be an infinite number of solutions to the problem. That is, by assigning arbitrary values to *either* of the two unknown quantities, equation [5-35] can be used to solve for the other. But, since the choice of *which* quantity to define, i.e., $M_o(t)$ *or* $\Theta(t)$, as well as the actual *functional form* or *values* one assigns to either, may be purely arbitrary, there are obviously many possibilities, each leading to a totally different solution.

When confronted with indeterminate formulations, investigators attempt to alleviate this dilemma by reducing the total number of unknown quantities in some systematic fashion. This may involve utilizing one, or a combination, of the following strategies:

1) *Making certain simplifying assumptions.* In fact, we have already done that in this problem -- when we confined the movement to the sagittal plane, thereby reducing the formulation to a two-dimensional problem; when we assumed the hand and forearm could be considered to be one continuous limb (i.e., no wrist joint or finger degrees of freedom have been allowed); when we considered the elbow joint to be "frictionless"; and when we tacitly took $\vec{F}$ and $\vec{O}$ to be "point loads" rather than distributed loads at their respective points of application. Simplifying assumptions can often be based on sound and logical physical reasoning. One would not choose, for example, arbitrary values of M_o that might be *mathematically* allowable, but, that *realistically* far exceed any muscular force (or torque) that an individual could possibly achieve. Or, such assumptions could be based on a systematic dimensional analysis of the problem -- as was illustrated in Chapter 4 for the case of thermoregulation; or, they could be based on the imposition of *math-*

ematical continuity requirements -- such as throwing out terms in an equation that "blow up" when some finite limit is reached, like $\dfrac{\text{Constant}}{r} \to \infty$ as $r \to 0$.

2) *Introducing additional boundary conditions or constraints.* In fact, we already did this, too, when we decided to examine the specific problem where $a_h = a_v = 0$, thus uncoupling the forearm motion from the upper arm motion. Choosing specific values for a governing parameter or variable, or otherwise constraining the solutions to well-defined degrees of freedom -- in this case, forcing the *path* of CG to be constrained to the plane circle defined by $x^2 + y^2 = \bar{r}^2$ (so that x and y no longer become independent of one another), allows one to generate an arsenal of solutions for *specific* cases. The hope, of course, is that generalizations will emerge as the arsenal of solutions gets larger and larger.

3) *Utilizing approximation and scaling techniques to examine asymptotic or perturbation solutions.* This is actually a more rigorous generalization -- or, if you will, a very special case of (1) and (2) above -- wherein solutions to a problem are formulated or "expanded" mathematically in terms of a parameter "governing" the problem. The idea here, is that the governing parameter may, itself, become either very large or very small; and, the guiding philosophy is that the *actual* problem being examined is considered to differ only slightly from a known *standard* solution which is that existing under "limiting" conditions (zero, or infinity) of the "perturbation" quantity. The two solutions, in fact, become exactly the same as the limit of the perturbation quantity approaches zero or infinity, as the case may be.

4) *Finding some functional relationship among certain of the variables,* so that they may be *combined* into parameters that are less in number than the original number of unknowns. Here, too, the relationships may be derived from some nondimensionalization scheme, or by similarity methods, or by some other technique that establishes a well-defined dependence of one variable upon another -- so that they always appear in the restructured equations as the same product, or ratio, or whatever. This is where constitutive relationships play a key role, as do so-called

"compatability" conditions for deformable materials. And finally, if and when all else fails from an analytical point of view, one may,

5) *Perform various carefully designed laboratory experiments,* to define some unknown quantities from actual data, rather than by guessing. For example, a goniometer is a specially-designed electrical potentiometer that can be attached to the limbs on either side of a joint in order to measure joint angle, Θ, experimentally (see, for example, Winter, D. A., 1979). Piezoelectric or strain gauge accelerometers attached to a body segment at a specific point (such as CG or O on Figure 5-7) allow one to measure absolute acceleration in one, two, or three orthogonal directions simultaneously (ibid., and Davis, R. B., 1986). Cinematographic and videotape techniques employing strobe lights allow one to image the kinematics of human posture and locomotion, from which images ("stop-action" analysis of multiple exposures) equations for $\Theta(t)$ may be empirically developed (ibid.). These are actually the techniques that we shall exploit here, although later in the chapter we shall address a novel concept that may allow us to avoid the need for experimentation. What follows, then, is the *empirical* (inductive) solution to the problem posed in Figure 5-7, and governed by equation [5-35].

EMPIRICAL SOLUTION TO THE GOVERNING EQUATIONS OF MOTION

Based on a great deal of experimental data -- obtained by the methods mentioned above, and discussed in some detail by Winter (1979), Davis (1986), and many others -- it has been determined that the functional form of $\Theta(t)$ that seems to be most appropriate to describe a discrete, sagittal plane forearm flexion, that takes the limb from an initial position $\Theta = 0$, to a final position $\Theta = \Theta_o$, in a time period, τ, and in such a way that $a_h = a_v = 0$, is:

$$\Theta(t) = \frac{\Theta_o}{2\pi} \left\{ \frac{2\pi t}{\tau} - \sin \frac{2\pi t}{\tau} \right\}$$

[5-36]

Thus,

$$\dot{\Theta}(t) = \frac{\Theta_o}{\tau} \left\{ 1 - \cos \frac{2\pi t}{\tau} \right\}$$ [5-37]

and,

$$\ddot{\Theta}(t) = \frac{2\pi \Theta_o}{\tau^2} \sin \frac{2\pi t}{\tau}$$ [5-38]

Substituting equations [5-38] and [5-36] into [5-35] gives the empirical result:

$$M_o(t) = \frac{I_o 2\pi \Theta_o}{\tau^2} \sin \frac{2\pi t}{\tau} + m_f \bar{r} g_o \cos \left[\frac{\Theta_o}{2\pi} \left\{ \frac{2\pi t}{\tau} - \sin \frac{2\pi t}{\tau} \right\} \right]$$ [5-39]

All that remains, then, in order to be able to plot $M_o(t)$ vs. t is to provide *anthropometric* information (I_o, m_f, $\bar{r}$, k_r) for our "typical" 75-kg individual, and to specify Θ_o and τ. The anthropometric data is available in many literature references, such as that by Dempster (1961). He lists the distance to the center of gravity of the forearm (*not* including the hand) as being located 43% of the way from the elbow joint to the wrist joint, and that of the clenched hand as being located 50.6% of the way from the wrist joint to the second set of knuckles in the fist, i.e., $\bar{r}_f = 0.43 \, L_f$, and $\bar{r}_w = 0.506 \, L_w$ (see Figure 5-7). Furthermore, the mass, m^*_f, of the forearm *not* including the hand is given as $1.58 \pm 0.13\%$ of the mass of the total body, and the mass, m_w, of the fist is given as $0.63 \pm 0.05\%$ of the mass of the total body. Thus, $m^*_f = 0.0158(75) = 1.185$ kg, $m_w = 0.0063(75) = 0.4725$ kg, and $m_f = m^*_f + m_w = 1.6575$ kg. Finally, with $L_f = 0.24$ meters, and $L_w = 0.13$ meters, we can write,

$$m^*_f \bar{r}_f + m_w (L_f + \bar{r}_w) = m_f \bar{r}$$ [5-40]

from which we calculate $\bar{r} = 0.161$ meters for the forearm-hand combination. Also, k_r for this combination is given as $0.3(L_f + L_w) = 0.111$ meters, so that, $I_o = m_f(\bar{r}^2 + k_r^2) = 0.0634$ kg-m^2.

Assuming that a 140-degree ($\Theta_o = 2.443$ radians) forearm flexion is accomplished in 0.32 seconds (τ), equation [5-39] becomes:

$$M_o(t) = 9.5 \sin \frac{2\pi t}{\tau} + 2.618 \cos\left[0.389\left\{ \frac{2\pi t}{\tau} - \sin \frac{2\pi t}{\tau} \right\} \right] \qquad \text{[5-41]}$$

In Figure 5-8, $M_o(t)$ is plotted as a function of nondimensionalized time, $\frac{t}{\tau}$, for $0 \le \frac{t}{\tau} \le 1$. Note, in particular, that:

1) $M_o(t)$ is not equal to zero at the extremes of the motion because a moment is still required at the elbow to counterbalance (or support) the weight of the forearm even when there is no motion (a situation, incidentally, that we have to worry about less and less as we approach a sub-gravity environment such as that encountered in outer space);

2) Again, because the weight of the forearm *resists* counterclockwise angular accelerations (i.e., interferes with the motion) during the initial phase of the motion, $0 \le \Theta \le 0.41\pi$, but *assists* counterclockwise angular decelerations (i.e., helps slow the arm down) during the early portions of the final phase of the motion, $0.41\pi \le \Theta \le 0.78\pi$, $|M_o(t)|$ has higher values for $0 \le \frac{t}{\tau} \le 0.512$ than it does for $0.512 \le \frac{t}{\tau} \le 1.000$, and,

3) The maximum accelerating moment is on the order of 12 Newton-meters, and the maximum decelerating moment is on the order of 11 Newton-meters.

WORK DONE DURING A DISCRETE FOREARM FLEXION

Let us see how all of the above analysis translates into an equivalent energy-expenditure for the prescribed motion. The work done by a moment in generating an angular displacement is defined to be: $W = \int M_o \delta\Theta = \int M_o \frac{\delta\Theta}{\delta t} \delta t = \int M_o \dot{\Theta} \, \delta t$. From equation [5-37], we have,

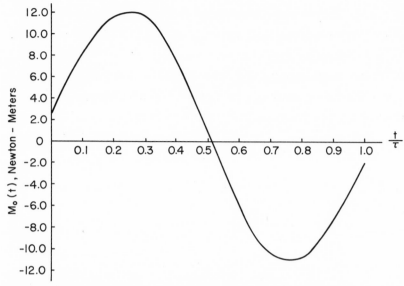

Figure 5-8 Elbow joint torque vs. non-dimensionalized time for a discrete forearm flexion taking place in the sagittal plane.

$$\dot{\Theta}(t) = \frac{\Theta_o}{\tau} \left\{ 1 - \cos \frac{2\pi t}{\tau} \right\} = 2 \frac{\Theta_o}{\tau} \sin^2 \left(\frac{\pi t}{\tau} \right)$$

Using this result, together with equations [5-35] and [5-38], and *approximating* the total flexion as taking place in roughly two equal time segments, we can estimate the work done as follows:

$$\int_0^\tau I_o \ddot{\Theta}\dot{\Theta}\, \delta t = \int_0^\tau I_o \frac{2\pi\Theta_o}{\tau^2} 2 \sin \frac{\pi t}{\tau} \cos \frac{\pi t}{\tau} 2 \frac{\Theta_o}{\tau} \sin^2 \frac{\pi t}{\tau}\, \delta t$$

$$\simeq \frac{16 I_o \Theta_o^2}{\tau^2} \int_0^{\tau/2} \sin^3 \frac{\pi t}{\tau} \left[\cos \frac{\pi t}{\tau} \left(\frac{\pi}{\tau} \delta t \right) \right] = \frac{4 I_o \Theta_o^2}{\tau^2}$$

and, $\int_0^{\Theta_o} m_f \bar{r} g_o \cos \Theta\, \delta\Theta = m_f \bar{r} g_o \sin \Theta_o$, so that:

$$W = \frac{4 I_o \Theta_o^2}{\tau^2} + m_f \bar{r} g_o \sin \Theta_o \qquad\qquad [5\text{-}42]$$

Putting in the appropriate values as defined in the previous section for the right hand side of equation [5-42], we get: $W = 16.46$ N-m, and converting this to power, we get $\frac{16.46}{0.32} = 51.45 \frac{\text{N-m}}{\text{sec}} = 0.069$ Hp.

Thus, we see that to accomplish a 140-degree, sagittal-plane forearm flexion in about one-third of a second requires about as much energy as it takes to power a 50-watt light bulb. If we assume further that cleavage of the terminal phosphate bond in one mole of ATP releases some 7800 calories of usable free energy, and that the efficiency of the conversion of such energy into work by muscles is on the order of 40% (Schneck, D. J., 1984b), then it would take 16.46 Joules $\times$ 0.239 $\frac{\text{cal}}{\text{Joule}}$ $\times$ $\frac{1 \text{ mole ATP}}{7800 \text{ cal}} \times \frac{1}{0.40} = 0.00126$ moles of ATP to provide enough energy for this forearm flexion. And finally, since one mole of glucose can be oxidized aerobically to yield 34 moles of ATP, the prescribed forearm flexion can be accomplished with the aid of 3.7×10^{-5} moles of glucose, or, about $6\frac{2}{3}$ mg of this simple sugar.

If we wish to accomplish the movement in *less* time, it will take more work, and vice versa. That is to say, we see from equation [5-42] that W

goes like $\dfrac{1}{\tau^2}$, so that decreasing τ increases W, and vice versa. At the other extreme, however, as $\tau \to \infty$, W approaches $m_f \bar{r} g_o \sin \Theta_o = 1.6827$ Joules, and so we see that W drops off rather rapidly with increasing τ; several orders of magnitude between $\tau = 0.03$ and $\tau = 0.30$, corresponding to $W = 1{,}683.35$ and $W = 18.5$, respectively, and only one additional order of magnitude, from $W = 18.5$ down to 1.6827 as τ goes all the way out to infinity. One might thus surmise that $\tau = 0.32$ seconds represents, in some sense, a compromise between the desire to minimize the energy expenditure associated with the movement, and the need to accomplish it in a reasonable period of time. This idea of some "optimization scheme" associated with musculoskeletal mechanics is addressed further later in this chapter.

Observe from equation [5-37] that $\dot{\Theta}_{max} = 2\dfrac{\Theta_o}{\tau}$, when $\dfrac{t}{\tau} = \dfrac{1}{2}$. Let this value of $\dot{\Theta}(t)$ be designated as $\dot{\Theta}^*$. Then the first term on the right side of equation [5-42] is recognized to be $I_o(\dot{\Theta}^*)^2 = 2\left[\dfrac{1}{2} I_o(\dot{\Theta}^*)^2\right]$. But, $\dfrac{1}{2} I_o(\dot{\Theta}^*)^2$ is the maximum rotational kinetic energy attained by the forearm during the flexion maneuver; and, since the arm *starts* with zero rotational kinetic energy, and *ends* with zero rotational kinetic energy, $2\left[\dfrac{1}{2} I_o(\dot{\Theta}^*)^2\right]$ represents the *sum* of the muscular effort required to *accelerate* the forearm from $\dot{\Theta} = 0$ to $\dot{\Theta} = \dot{\Theta}^*$, and that required to *decelerate* the forearm from $\dot{\Theta} = \dot{\Theta}^*$ back to $\dot{\Theta} = 0$ again at the end of the motion. In real life, since muscles can only *pull* (i.e., contract) and not *push,* this sum actually represents two pulls: one accomplished by the biceps musculature to accelerate the forearm, and the second accomplished by the triceps musculature to decelerate the forearm. The triceps muscles insert on the other side of elbow joint O (i.e., to the left of point O on Figure 5-7) to generate a clockwise moment around the joint when they contract. These extensors act in concert with the flexors to generate a well-coordinated body movement which involves only "pulls" on the "levers", or bones. In this forearm flexion maneuver, then, the *biceps* do work $\dfrac{1}{2} I_o(\dot{\Theta}^*)^2$, and the *triceps* do work $\dfrac{1}{2} I_o(\dot{\Theta}^*)^2$, so the *total* work associated with the *kinetic* energy changes taking place during the motion is $I_o(\dot{\Theta}^*)^2$, and is accounted for by the first term on the right side of equation [5-42].

The second term on the right side of this equation, $m_f \bar{r} g_o \sin \Theta_o$ is recognized to be the change in *potential* energy of the forearm as a result of the execution of the flexion specified. That is, $\bar{r} \sin \Theta_o$ is the final height of CG relative to its starting point at $y = 0$, and $m_f g_o$ is the weight of the forearm, so $m_f g_o \bar{r} \sin \Theta_o$ = weight $\times$ height = change in potential energy.

Summarizing, then, we see that the work equation [5-42] is simply an energy balance stipulating that the total work done during the movement is the sum of the changes that take place in both kinetic (rotational and translational) and potential energies of the forearm. Moreover, since, from the parallel axis theorem for mass moments of inertia, $I_o = m_f(k_r^2 + \bar{r}^2)$, if we divide both sides of equation [5-42] by m_f, we can define for this motion an intensive scalar potential function,

$$\phi = \frac{4(k_r^2 + \bar{r}^2)}{\tau^2} \Theta_o^2 + \bar{r} g_o \sin \Theta_o \qquad [5\text{-}43]$$

The potential function defined by equation [5-43] is sometimes called the Hamiltonian Function, or, more simply, the Hamiltonian of the motion. It is just the *sum* of potential and kinetic energies of a system at any given point in time and space. This is as opposed to the Lagrangian Function (see later) which is the *difference* between potential and kinetic energies (sometimes called the kinetic potential).

THE HUMAN BODY IS NOT DESIGNED TO DO WORK

The anthropometric data utilized in the preceding forearm flexion example illustrate another basic principle of physiologic function: As a machine, the human body is generally *not* designed to do "heavy work". This is because most muscles concerned with musculoskeletal function insert relatively close to the joint around which they must generate moments. To understand this, we must say a few words about levers.

A lever is a device that consists of a bar or rod (in the case of the body, a *bone*) that is pivoted at some point along its length called the fulcrum (in the case of the body, a *joint*). A force, called the effort -- generated perhaps by a contracting muscle ($\vec{F}$ in Figure 5-7) -- acts on the lever at some point $\bar{\ell}$ measured relative to the fulcrum. This generates a moment that acts against a resistance, or load located at a point $\bar{r}$ (for example) relative to the fulcrum, O. The three possible configurations that relate $\vec{F}$, the resistance (for example, $m_f g_o$), $\bar{\ell}$, and $\bar{r}$ are: (i) the fulcrum can lie between $\vec{F}$ and $m_f g_o$, as is the case for the triceps muscular configuration acting around the elbow joint (i.e., a "seesaw" arrangement) -- in which case we have a lever of the *first class;* (ii) $m_f g_o$, or the resistance, can lie between the fulcrum and $\vec{F}$, as is typically the case for the musculature associated with chewing or mastication (i.e., a "wheelbarrow" or "nutcracker" arrangement) -- in which case we have a lever of the *second class;* and (iii) $\vec{F}$ can lie between the fulcrum and the resistance (i.e., a "tweezer" or "sugar-tong" arrangement), as is illustrated in Figure 5-7 for the biceps musculature of the forearm -- in which case we have a lever of the *third class.*

The "law of the lever" is that the effort and the resistance are inversely proportional to their lever arms, measured relative to the fulcrum, i.e., the mechanical advantage of a lever is defined as:

$$\frac{|\vec{F}|}{m_f g_o} = \frac{\bar{r}}{\bar{\ell}} \qquad \text{[5-44]}$$

If $\bar{\ell} > \bar{r}$ -- as is *always* the case (by definition) for a second class lever, and *may* be the case for first class levers (e.g., crow-bars), but is *never* the case (by definition) for third class levers -- then $|\vec{F}| < m_f g_o$ and the lever is said to "favor" *force*, in the sense that a small force can balance a large resistance because the force has a long "lever arm". If $\bar{\ell} < \bar{r}$ -- as is *always* the case (by definition) for a third class lever, and *may* be the case for first class levers (e.g., diving boards), but is *never* the case (by definition) for second class levers -- then $|\vec{F}| > m_f g_o$ and the lever is said to "favor" *velocity*, in the sense that when a relatively large force succeeds in generating an angular velocity of the lever, the resistance will gain a large speed because the latter is directly proportional to $\bar{r}$ (i.e., $v_\Theta = \bar{r}\dot{\Theta}$). Thus, second-class levers always favor force, (mechanical advantage greater than unity), third-class levers always favor velocity (mechanical advantage less than unity), and first class levers may favor either force or velocity, or, *neither* (in which case the mechanical advantage *is* unity).

With very few exceptions (such as in the jaw musculature), the levers of the body are of the third class (always favoring speed), or of the first class with $\bar{\ell}$ small compared with $\bar{r}$ (also favoring speed). For example, typically $\bar{\ell}$ will be but a small fraction of the distance from a joint to the center of gravity of the corresponding limb, or the distance from the joint to some resistance against which a torque at the joint is being generated. From an engineering point of view, then, the human body is a machine built for speed, agility, and range of motion, at the expense of force. It has a mechanical advantage that allows it to perform easily those tasks which involve fast movements with light objects. But when heavy work is demanded, the organism must use some type of external device, such as a crowbar, to gain a force advantage. We are mobile creatures by design, not powerful torque-generators.

Having said all of that, let us get back to the issue of the indeterminateness that the analyst of human body dynamics inevitably comes face to face with. We got around this dilemma in the previous example by resorting to experimental findings in order to define one of the unknown quantities, $\Theta(t)$. We also mentioned other techniques that are frequently employed (such as dimensional analysis, imposition of additional constraints and assumptions, and so on) to reduce the order of the indeterminateness in mathematical formulations of musculoskeletal function. All of these schemes, however, categorize the problem-solving technique as being an *inductive* one, i.e., concluding what happens in a *general* case from an arsenal of results obtained for *specific* cases.

In physiological research, inductive reasoning has traditionally been the rule, rather than the exception, since purely deductive approaches to the study of human body function invariably lead to more unknown quantities than there are equations to describe them. As we saw above, if there are basically X governing laws of nature that must be satisfied by j^* "unknowns" in any physical system, subject to n well-defined constraints, then a purely analytical formulation leads to a $(j^* - X - n)$'th degree-indeterminate problem. In the case of problems dealing with the kinetics of human movement, $(j^* - X) = J^*$ turns out to be equal exactly to the total number of joints involved in the movement, and, to the corresponding number of unknown moments generated at these joints. That is, it is possible to ascribe values to J^* of the primary unknowns, and still satisfy the laws of mechanics in general. The quantity J^* is thus designated as the number of "degrees-of-freedom" of the system. These degrees of freedom are reduced by the n constraints imposed on the motion, leaving the problem $(J^* - n)$'th-degree-indeterminate.

That this should be the case for musculoskeletal function has not surprised researchers in this field. "After all," they have argued, "the indeterminacy of such problems can be directly attributed to the fact that, if no specific constraints are imposed, the unknown moments at the joints can be generated at 'will' by the normal individual through conscious control of his or her muscles. Since 'will' is at the unpredictable discretion of the individual involved, it thus renders a problem indeterminate in the sense that it is an undefinable variable. Moreover, the degree of indeterminateness (or 'freedom') is exactly equal to the number of joints involved -- not by accident -- but by virtue of the fact that this is precisely where conscious control of movement is manifest."

From an engineering point of view, the above concepts are actually quite novel and fascinating. They add a totally new dimension to what engineers customarily think of when they attempt to analyze a physical system. That is, no "traditional" analyst is normally confronted with the idea that something moves because it "wants to," or that behavior is motivated by pain, fatigue, or discomfort, as well as the natural laws of mechanics. So logical has this argument been over the years, that it seemed unreasonable to question it, and researchers in the field of musculoskeletal mechanics have resolved themselves to always resorting to inductive methods for problem solution. Only recently have techniques been employed that actually allow one to generate *additional* governing equations, rather than attempting to reduce the number of unknown variables by experimentation. This has opened up an entirely new avenue for exploring human body dynamics using purely deductive analysis.

Deductive Analysis of the Kinetics of Human Movement

Moving when something "wants to" -- ("will") can be described as that mysterious, sometimes fickle, often elusive, very personal and magnificent (but abstract) capability that most of us have to exercise conscious control over our posture, or the locomotion of some or all of our body parts. So important to us is this capability, that humankind has sought throughout the ages to expand our powers in order that we might gain control over *all* body functions, not just the musculoskeletal system. Indeed, the reported success of biofeedback, meditation, and other similar "mind-over-matter" techniques attest to the fact that even "involuntary" physiologic activity may be influenced at "will". This seems only to emphasize the futility of attempting a purely deductive physiologic analysis in the presence of "will" as an undefined variable -- or does it?

"WILL" AND THE CONCEPT OF LEAST ENERGY

In 1961, Nubar and Contini (1961) postulated the following: "A mentally normal individual will, in all likelihood, move (or adjust his posture) in such a way as to reduce his total muscular effort to a minimum, consistent with the constraints." In other words, it appears likely, and feasible that, whether conscious or subconscious, the patterns of motion of an individual seem concerned with the reduction of exertion to a minimum at all times, consistent with the task assigned and other restrictions.

When the individual takes the shortest path between two points, when he or she settles into a preferred position while sitting or drifting off to sleep, when s(he) adjusts his or her stance according to some particular rhythmic pattern while swinging a hammer, or jogging, or whatever -- in all of these cases one might surmise that the individual is instinctively obeying some governing law which gives him or her a need to economize physical activity -- or, suffer the consequences of overexertion, which may include discomfort, fatigue, pain, injury, and so on.

Of course, this is not necessarily true in an absolute sense, but in a *relative* sense consistent with certain constraints (Schneck, D. J., 1981, 1984b). For example, there may be a broken bone or other disability or anatomical restriction which constrains movement; there may be a desire to maintain some prescribed velocity, cadence or step length while walking; there may be patterns of motion derived from culture, habit, and experience that dictate how an individual will behave in a given situation, and many more. The point is, however, that these constraints can frequently be well defined both physically and mathematically, and that within and subject to these constraints, the minimum-energy principle will guide a person's desire or "will" to perform a given function in a prescribed manner. Some experimental evidence of this, at least for the case of human gait, is presented in the work of Beckett and Chang (1968), and Beckett and Pan (1971), while Chao and An (1978) explore optimization techniques as applied to the analytic modelling of hand motions, and Milsum (1966) uses optimization as a means for defining the behavior of biological servomechanisms and homeostatic processes. All of these authors, and others (see later), have found reasonable agreement between theory and experiment, which certainly has to be encouraging and must mean something.

Consider further that various forms of "meditation" have been shown to possess the capability for conditioning the human brain to be more highly creative by allowing it to tend naturally to a tranquil state of minimum energy expenditure (Benson, 1975). This state can be characterized by the following attributes:

(1) A spontaneous spread of alpha waves from the back to the front of the brain. These 10 ± 2.5 Hz rhythmical oscillations appearing in an electroencephalogram are known to represent a pleasant, relaxed state of mind; one that is free of anxiety and has an increased awareness of feelings (see further in Chapter 6).

(2) An increased synchronization among the various brain wave patterns measured along different sites of this organ. Such synchronization is known to cause less random thought processes, and is associated physiologically with an increased "orderliness" of the brain.

(3) A more efficient and rapid release of metabolic waste products. Since the latter are known to interfere with normal biochemical pathways, as already discussed, the efficient and effective handling of wastes reduces fatigue and thereby increases mental alertness through a state of deep rest. And,

(4) An integration of the activities of the left and right cerebral hemispheres of the brain (see discussion on pages 70-73 in Chapter 1). This is known to reduce the tendency to distraction, and thus stimulates the imagination through a clearer, less cluttered state of mind.

Although much experimentation is still underway, and there are some conflicting opinions and alternative points of view, at least one concept is receiving favorable reactions worthy of further consideration, viz.: left unto itself, relatively free of internal stimuli and external distractions, the human brain most likely *will* drift naturally towards a minimum energy level. Furthermore, the passiveness of this minimum energy state allows imagery, feelings, and hidden thoughts to drift into one's awareness, where they might otherwise have been masked, subdued, or lost completely. Therein lies the capacity for increasing creativity by exploiting the basic potential of the human mind. Going one step further, the increased awareness of feelings that is associated with mental tranquility provides a basis for allowing the *subconscious* control of body function to come under *conscious* control. This is the essence of biofeedback (c.f., Chapter 6).

In biofeedback, internal physiologic control processes are transducted from a subconscious level at which the individual is unaware of their occurrence, to a conscious level wherein the transduced signals produce an input to the body's various senses. A subject learns to interpret the meaning of this information -- which may be coded into an electroencephalogram, an electrocardiogram, an electromyogram, and so on (see Chapter 6) -- in terms of a corresponding physiologic state of affairs (a "feeling," if you will). He or she is then trained to respond *consciously* in a way that produces a desired alteration of this milieu, and, furthermore, to *recognize* that the corresponding change has occurred by "memorizing" the feeling associated with this change, as reflected in the transduced electromagnetic sensory signals. Once the subject has learned what it "feels like" to have achieved the desired physiologic response, and, furthermore, what (s)he consciously did to *produce* that feeling or response, the "sensory monitoring" of the response may be eliminated. At this point in the training, the individual has developed the conditioned ability to "willfully regulate" his or her internal environment.

Enormous success has been reported (see, for example, Chapter 5 of Bronzino, 1977) in using biofeedback to relieve pain, to regulate muscle spindle behavior, to control the circulation, to relieve migraine headache syndromes, to maintain body temperature, to control metabolism, to regulate blood pressure, and even to provide effective therapy in a wide variety of mental disorders. One cannot deny its potential for increasing our "will" power, but viewed in terms of the minimum energy hypothesis it adds still further credence to the idea that "will" can, indeed, be quantified.

That we should function in a way that minimizes energy expenditure should not come as a complete surprise. In fact, cliches such as, "following the path of least resistance," stem from just such an underlying principle. It is a well established fact that chemical, structural, thermodynamic, mechanical, electrical, hydraulic, and other physical systems that presumably "do not know any better" all seek equilibrium states that are characterized by an attempt to minimize some well-defined energy function or intensive scalar energy potential (Sommerfeld, 1956). In our every-day example of energy barriers for chemical reactions (c.f., Chapter 4), for instance, we had occasion to speak about metastable and stable states of equilibrium for objects such as boulders resting along the sides of hills. We observed there -- and, indeed, most of us have observed in general, that objects acquire a state of stable equilibrium relative to the gravitational field of the earth when they are closest to the center of the earth, i.e., when their potential energy function, $\phi = g_o z$ is at a low point.

Similarly, under the action of a system of conservative forces, Hamilton's principle for determining the equations of motion in classical dynamics asserts that, starting from any reasonable initial conditions, motion ensues such that the time average of the difference between kinetic and potential energies (the so-called kinetic potential or Lagrange function, $\phi =$ Kinetic Energy, KE $-$ Potential Energy, PE) will be a minimum (Morse and Feshbach, 1953). We saw also in Chapter 4 that the condition of equilibrium for an isothermal thermodynamic system of fixed temperature and pressure (or volume, for that matter) requires that there exist a minimum of free energy ($\Delta g^* = 0$, c.f., equations [4-23] to [4-25] for p and T constant) beyond which no spontaneous changes of state are possible.

Indeed, the examples of this minimum energy principle are numerous and well documented in all of the basic sciences. But the idea that our "will" dictates us to perform in accordance with such a minimum energy principle *is* surprising in that it removes the mysterious, seemingly undefinable nature of this single variable that has hitherto made deductive analysis of physiologic function impossible. With the realization that "will" may not, in fact, be a totally random, purely arbitrary quantity that we may designate according to our personal discretion comes the corresponding realization that all of us may be reasonably predictable and thus, deductively analyzed. This is one of the unique contributions that mathematical modelling has made to our continuing efforts to improve the health, comfort, and understanding of humankind, and while it has a historic conceptual basis in the general sciences, its application to humanity has only recently been explored. Let us see how in some more detail.

MATHEMATICAL QUANTIFICATION OF "WILL"

Based on the principle of least energy, one may proceed in several ways to quantify the variable "will" for analytic purposes. One method that has been particularly useful utilizes variational or optimization techniques as a means for seeking additional information for a closed-form analytic solution. The calculus of variations is concerned with maximizing or minimizing functions of points in function spaces (so-called "functionals"), and its physical application relates to the existence of systems capable of achieving an equilibrium state wherein some prescribed energy function has its smallest value. Typically, this energy function will depend on several variables, where the variables themselves are related by one or more equations which represent constraints on the problem. Thus, in principle, variational calculus can be used to generate extra equations which constrain the system to minimize a given energy function. Given such systems, one can employ, for example, the method of Lagrange multipliers (Kaplan, W., 1952, Bertsekas, D. P., 1982) to generate additional equations.

As an illustration, consider the problem of finding the extremal values of an *extensive* energy function, $\phi^*(x, y, z)$, having dimensions $\frac{ML^2}{t^2}$, where $x, y,$ and z are related by the constraint equation $F^*(x, y, z) = h^*$. Lagrange's method directs us to proceed as follows: First form the generalized function, $u^* = \phi^*(x, y, z) + vF^*(x, y, z)$, where v is a constant, as yet undetermined in value. Then $u^* = u^*(x, y, z, v)$ so that the extreme values of u^* can be obtained by setting:

$$\delta u^* = 0 = \frac{\partial u^*}{\partial x} \delta x + \frac{\partial u^*}{\partial y} \delta y + \frac{\partial u^*}{\partial z} \delta z + \frac{\partial u^*}{\partial v} \delta v \qquad \text{[5-45]}$$

Now, treating x, y, and z as independent variables, and realizing that $\delta v = 0$, since v is constant, equation [5-45] can only be satisfied if we impose the following conditions:

$$\frac{\partial u^*}{\partial x} = 0; \quad \frac{\partial u^*}{\partial y} = 0; \quad \frac{\partial u^*}{\partial z} = 0. \qquad [5\text{-}46]$$

The three equations [5-46], along with the equation of constraint, $F^*(x, y, z) = h^*$ can be solved simultaneously to find the appropriate values of the four quantities, x, y, z, and v for which values the energy function $\phi^*(x, y, z)$ will have reached a *relative* minimum (i.e., relative to constraint F^*). In the more general case involving the problem of extremal values of ϕ^* subject to *several* constraints, $F^*_j(x, y, z) = h^*_j$, $j = 1, 2, 3 \ldots n$, form the function:

$$u^*(x, y, z, v_j) = \phi^*(x, y, z) + \sum_{j=1}^{j=n} v_j F^*_j(x, y, z) \qquad [5\text{-}47]$$

and solve for x, y, z and n values of v_j from the n constraint equations and the three partial differential equations on u^* as in [5-46] above.

The calculus of variations goes back to the time of Euler, and variational principles have been applied to physical problems since the time of Huygens. Hamilton, Jacobi, and others developed the Hamiltonian formulation of the variational problem early in the nineteenth century, but perhaps the most significant contribution in recent times was made by L. S. Pontryagin (see Sage and White, 1977). Pontryagin's maximum principle starts with a Hamiltonian function of the form:

$$H^* = f^*_o \psi^*_o + f^*_1 \psi^*_1 + \cdots \qquad [5\text{-}48]$$

and sets $\psi^*_o = -1$. If ψ^*_o were chosen to be positive, then there would result a "minimum principle" which could be applied to the optimization of a given physical process (Pontryagin, et al., 1962).

In the optimization scheme proposed by Nubar and Contini (1961) an energy function, $\phi^*(M_1, M_2, M_3, \ldots, M_{J^*}, t)$ is defined, where t is time and the $M_j's$, $j = 1, 2, 3, \ldots, J^*$ are the moments generated at the J^* joints which are involved in a prescribed movement. The n constraints on the movement are generally given in terms of kinematic quantities or space variables, such as equation [5-34] in the previous forearm-flexion example, where $x = \bar{r} \cos \Theta$ and $y = \bar{r} \sin \Theta$ (c.f., Figure 5-7). However, these can be related back to the moments at the joints through Newton's laws of motion, such as equation [5-35], which can be inverted or solved for $\Theta(M_o)$. Thus, the reconstructed constraint equations become: $F^*_j(M_1, M_2, M_3, \ldots, M_{J^*}, t) = h^*_j$, $j = 1, 2, 3, \ldots, n$. Proceeding according to the method of Lagrange multipliers, one minimizes the effort function, ϕ^*, relative to the constraints, F^*_j, by forming the function:

$$u^*(M_i, v_j, t) = \phi^*(M_i, t) + \sum_{j=1}^{n} v_j F^*_j(M_i, t) \qquad [5\text{-}49]$$

$i = 1, 2, 3, \ldots, J^*$, and setting $\delta u^* = 0$. Treating the moments at the joints as independent variables (like x, y, z in [5-45]) and the Lagrange multipliers as constants, with t acting (like τ) in a parametric sense, one then collects coefficients of $\delta M_1, \delta M_2, \delta M_3, \ldots, \delta M_i$ and sets them individually equal to zero, as in equations [5-46], yielding a total of J^* equations. Elimination of the n Lagrange multipliers from these J^* equations results in a set of $(J^* - n)$ independent relations in the unknowns $M_1, M_2, M_3, \ldots, M_i$ and the parameter t. Recall (page 402) that this is precisely the indeterminateness which arose when only the basic laws of continuum mechanics and the constraint relations were applied to the original problem. Therefore, the additional $(J^* - n)$ equations generated by this variational procedure (where now the constraint equations are *embedded* into equation [5-49]) have reduced the degree of indeterminateness to zero, which allows for the possibility of a purely deductive solution.

Utilizing an effort function, ϕ^*, of the general form, $\beta^*_i M_i{}^2 \Delta t$, where β^*_i ($i = 1, 2, 3, \ldots, J^*$) are "sensitivity coefficients" for the J^* joints involved in the motion (Nubar and Contini, 1961, Schneck, 1984b), Δt is the time duration, τ, of muscular exertion, and the corresponding moments M_i are squared to insure that the energy function will always be positive, Nubar and Contini developed this optimization approach to predict the postural configuration most likely to be assumed by an individual asked to stand on one leg, with the other leg held off the ground at a prescribed distance (constraint, F^*) from the stationary leg. Their results, taking into consideration the simplicity of both their human body model and the sample posture they examined, were quite reasonable in defining an equilibrium position of minimum effort. In fact, so reasonable that others have pursued the approach with great success, using somewhat different methods (see, for example, Milsum, 1966).

For example, Beckett and Chang (1968) and Beckett and Pan (1971) utilized the Hamiltonian formulation to analyze human gait (walking) patterns by minimum energy considerations. In this method, they defined an energy function, ϕ^*, in terms of the kinetic (T^*) and potential (E^*) energies of the musculoskeletal system -- similar to equation [5-42] derived earlier for the simulated forearm flexion problem. Both T^* and E^* were expressed in terms of angular coordinates, Θ_i and $\dot{\Theta}_i$ which specify the spatial orientation and dynamic configuration of the system during each instant of time. Minimization of the energy function relative to constraint functions h^*_j (also written in terms of coordinates Θ_i and $\dot{\Theta}_i$) was accomplished again by the use of Lagrange multipliers. In this case, a new potential energy function, $\overline{E}^* = E^* - vh^*$ is introduced to include the condition of constraint(s), and the joint moments, M_i, are calculated from the Lagrangian formulation relationship:

$$M_i = \frac{\delta}{\delta t} \left\{ \frac{\partial T^*}{\partial \dot{\Theta}_i} \right\} - \frac{\partial T^*}{\partial \Theta_i} + \frac{\partial \overline{E}^*}{\partial \Theta_i} \qquad [5\text{-}50]$$

In equation [5-50], the subscript i identifies the joint under consideration $(i = 1, 2, 3, ..., J^*)$, the coordinate Θ_i is the corresponding angular position of the skeletal part associated with that joint, and the dot over the Θ indicates differentiation with respect to time.

Note that the functional form [5-50] could have been used as an alternate approach to deriving equation [5-35]. That is, since, for pure rotation around a pivot O, $T^* = \frac{1}{2} I_o(\dot{\Theta})^2$, and $E^* = m_f g_o \bar{r} \sin \Theta$, we have, $\frac{\partial T^*}{\partial \dot{\Theta}_i} = I_o \dot{\Theta}$, $\frac{\delta}{\delta t} \left\{ \frac{\partial T^*}{\partial \dot{\Theta}_i} \right\} = I_o \ddot{\Theta}$, $\frac{\partial T^*}{\partial \Theta_i} = 0$ (Θ_i and $\dot{\Theta}_i$ in this formulation are assumed to be independent "generalized" coordinates), and $\frac{\partial \bar{E}^*}{\partial \Theta_i} = m_f g_o \bar{r} \cos \Theta$ ($\bar{E}^* = E^*$ since the constraint has already been embedded into the functional form for both T^* and E^*). Thus, from [5-50],

$$M_o = I_o \ddot{\Theta} + m_f g_o \bar{r} \cos \Theta,$$

which is identical to equation [5-35].

Among the interesting conclusions that Beckett, Chang, and Pan came to regarding the minimum energy principle as it applies to human gait is that, for a given individual, there is a natural gait at which he or she can travel a given distance with a minimum of effort. Given the anthropometric build of the body, one can determine this gait rhythm by pure deductive analysis and it turns out to be somewhere near 80 steps per minute for the "average" person. Above this cadence, the energy consumed to travel the same distance increases, as it does also for cadences below 80 steps per minute. The authors find their results to be in good agreement with the reported findings of several experimental studies available in the literature.

Still a third approach which makes use of minimum energy principles is the optimization scheme proposed by Chao and An (1978) to study the mechanics of the human hand. Instead of defining an energy function, these authors present a generalized theory of muscle force distribution as optimized with respect to individual muscle strength limitations during the performance of a given task. The principle, however, is the same, viz., to minimize effort. Furthermore, their conclusions are consistent with the underlying theme of all such analyses, i.e., that viewed in terms of minimum effort, the seemingly indeterminate problems associated with musculoskeletal dynamics *do* offer unique solutions that take into account the mysterious variable called "will". In this respect, applied mathematicians have made unique contributions to the study of human posture and locomotion -- indeed, to the very essence of understanding physiologic function.

Bearing all of this in mind, if one assumes that the results obtained empirically, as given by equations [5-36] to [5-38] for a forearm flexion, correspond to an individual's instinctive tendency to perform the motion with the least expenditure of energy -- such that the experimental results obtained correspond *exactly* to those that could have been obtained purely by the deductive methods described in this section (specific details for the proof of this are left up to the reader) -- then $\Theta(t)$ given by equation [5-36] and $\tau = 0.32$ must represent, in some sense, an ultimate optimization of the effort defined by equation [5-42]. In other words, the calculated 0.069 Hp should correspond to the optimum means for this particular individual to accomplish this particular task.

Now, earlier in this chapter, we mentioned that the aerobic dephosphorylation of ATP can ultimately provide up to 0.075 Hp per kilogram of tissue involved. Thus, for the forearm, with $m_f = 1.6575$ kg, we should hypothetically be able to generate as much as $(1.6575)(0.075)$ $= 0.1243$ Hp from the forearm-flexion musculature. That we *actually* utilize only 0.069 Hp suggests that the flexion is being accomplished at about $\frac{0.069}{0.124} = 0.556$, or 55.6% of peak capacity -- where, by peak capacity we mean the absolute smallest value of τ at which the motion could have been achieved "at all costs" in terms of energy. Letting horsepower be equal to 0.124, this value of τ is calculated from equation [5-42] to be on the order of 0.26 seconds, which is about 25% faster than the time interval within which the motion is "routinely" accomplished in an everyday sense. In other words, from this rather crude "back-of-the-envelope" type of estimate, one might surmise that, in terms of speed, the body functioning in a minimum energy state may be operating at about two-thirds to three-quarters of peak capacity, which corresponds to utilizing energy at a rate just over half of that potentially available to do work.

Again, this is, at best, a very rough "guesstimate", but within the context of the above discussions, it is not unreasonable to suspect that economy of effort, as a means for efficient survival, carries over into our activities of daily living, i.e., into the "choices" we make as to how certain tasks are to be performed. In fact, one might go still further to suggest that this may be yet another corollary in our growing list of "survival-of-the-fittest attributes", which so far has addressed both mass transport processes and enzyme-controlled biochemical reactions. Given this supposition, human performance can then be analyzed in a purely deductive sense, with the variable "will" being handled through a variety of optimization or variational methods believed to have governed our ultimate evolution. This not only seems reasonable, but, in fact, probable -- especially when viewed in terms of inherent "control" mechanisms such as pain, fatigue, and discomfort. While somewhat abstract in its present level of development, the minimum-energy concept as it applies to physiologic function is certainly deserving of additional input and discussion. Thus it is left to be pondered further.

In summary, then, regardless of which of the approaches suggested above (and possibly others) an investigator chooses to generate a well-posed mathematical problem, the resulting analysis is deductive in the sense that one may draw conclusions about what happens in specific cases from a *generalized* theory valid for *all* cases. A problem is considered to be "well-posed" when the number of unknown quantities exactly equals the number of independent equations available to solve for them, and when such solutions can be shown to be physically meaningful and actually realizable. One of the concerns of the mathematical modeller is to obtain just such a well-posed, deductive, generalized theory for human body dynamics. It is not always possible -- or desirable -- to do a great deal of experimentation to reduce the number of unknown quantities in physiological research, and so the previously discussed inductive approach falls short of providing needed answers in important situations. Design to increase the capability of humans in the aerospace environment is one such example, as is the desire to avoid the needless slaughter of thousands of animals in the laboratory environment.

With this in mind, let us see how the formulation of an analytic model for the study of human body dynamics can be developed still further from the concepts thus far presented.

Development of a Computerized Articulated Total Body Model

Let us return to the forearm-flexion problem considered earlier and relax the assumption that $a_v = a_h = 0$, so that we are back to considering a solution to equation [5-33]. Then the motion of the forearm is coupled through this elbow joint acceleration to the dynamics of the upper arm and shoulder joint, and we have to deal now with a two-link configuration such as is shown schematically in Figure 5-9. If the shoulder joint is designated as O', then we may again write a relative acceleration equation, this time for $\vec{a}_o$ (c.f., equation [5-26]) as:

$$\vec{a}_o = \vec{a}_{o/o'} + \vec{a}_{o'} \qquad [5\text{-}51]$$

where, with the angular displacement, velocity and acceleration of the upper arm being designated, respectively, by $\Theta_1, \dot{\Theta}_1$, and $\ddot{\Theta}_1$, and the length of this limb being represented by L_u, we have (see Figure 5-9):

$$\vec{a}_o = a_h \hat{i} + a_v \hat{j} \qquad [5\text{-}52]$$

and,

$$\vec{a}_{o/o'} = -L_u(\dot{\Theta}_1)^2 \hat{i}'_n + L_u(\ddot{\Theta}_1)\hat{i}'_t \qquad [5\text{-}53]$$

The primed unit vectors indicate directions along (n) and perpendicular to (t) the centerline of the upper arm, just as the unprimed unit vectors

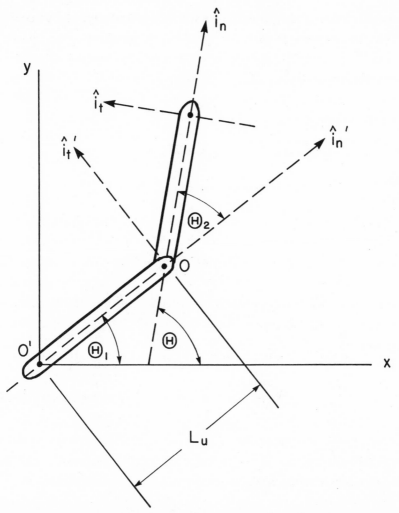

Figure 5-9 Schematic representation of a two-link manipulator system consisting of a fore-arm attached at elbow joint O to an upper arm, which is attached at shoulder joint O' to a torso.

indicate corresponding directions relative to the centerline of the forearm. Thus, we may write further,

$$\vec{a}_{o/o'} = -L_u(\dot{\Theta}_1)^2 \cos \Theta_1 \hat{i} - L_u(\dot{\Theta}_1)^2 \sin \Theta_1 \hat{j}$$

$$- L_u(\ddot{\Theta}_1) \sin \Theta_1 \hat{i} + L_u(\ddot{\Theta}_1) \cos \Theta_1 \hat{j}$$ [5-54]

so that, if we momentarily assume that $\vec{a}_{o'} = 0$,

$$a_h = -L_u(\dot{\Theta}_1)^2 \cos \Theta_1 - L_u(\ddot{\Theta}_1) \sin \Theta_1$$ [5-55]

and,

$$a_v = -L_u(\dot{\Theta}_1)^2 \sin \Theta_1 + L_u(\ddot{\Theta}_1) \cos \Theta_1$$ [5-56]

Substituting equations [5-55] and [5-56] into equation [5-33], then, we obtain,

$$M_o = I_o(\ddot{\Theta}) - L_u(\dot{\Theta}_1)^2 m_f \bar{r} \cos \Theta \sin \Theta_1 + L_u(\ddot{\Theta}_1) m_f \bar{r} \cos \Theta \cos \Theta_1$$

$$+ L_u(\dot{\Theta}_1)^2 m_f \bar{r} \sin \Theta \cos \Theta_1 + L_u(\ddot{\Theta}_1) m_f \bar{r} \sin \Theta \sin \Theta_1 \quad [5-57]$$

$$+ m_f \bar{r} g_o \cos \Theta$$

If we use the trigonometric identities,

$$\sin \Theta \cos \Theta_1 - \cos \Theta \sin \Theta_1 = \sin(\Theta - \Theta_1)$$

and,

$$\cos \Theta \cos \Theta_1 + \sin \Theta \sin \Theta_1 = \cos(\Theta - \Theta_1)$$

together with the observation from Figure 5-9 that $\Theta - \Theta_1 = \Theta_2$, where Θ_2 is the angle that the centerline of the forearm makes relative to the centerline of the upper arm, then equation [5-57] reduces to:

$$M_o = I_o \ddot{\Theta} + m_f \bar{r} [g_o \cos \Theta + L_u(\dot{\Theta}_1^2 \sin \Theta_2 + \ddot{\Theta}_1 \cos \Theta_2)] \qquad [5\text{-}58]$$

Equation [5-58] for the forearm is coupled, then, through L_u, $\dot{\Theta}_1{}^2$, $\ddot{\Theta}_1$ and Θ_2 to the motion of the upper arm. One may then draw a corresponding free-body diagram of the latter, similar to Figure 5-7, and go through an analysis similar to that used to derive equation [5-58], in order to get a corresponding dynamic equation of motion for the upper arm (M_o'). Details of this analysis may be found elsewhere (see, for example, Brady, et al., 1982), and are left up to the reader to pursue.

The point here, is that if we now further relax the assumption that $\vec{a}_{o'} = 0$, the motion of the upper arm may be coupled to the dynamics of the torso to which it is attached. This motion, in turn, may be coupled to the dynamics of the other limbs of the body, and, proceeding in like fashion, one may work one's way through *all* of the limbs, ultimately having generated a complete set of coupled equations that collectively describe the dynamics of a total body model.

Just such a model was originally developed by the Calspan Corporation (Fleck, et al., 1982) in the 1970s for the National Highway Traffic Safety Administration (NHTSA). That initial model, called CVS for "Crash Vehicle Simulator" involved computerized algorithms for describing the dynamics of a 15-linkage system, composed of spherical, circular-cylindrical, truncated-conical, and flat-plate segments, attached through a series of arbitrary "joints" intended to simulate the entire human organism. The model looked something like that illustrated in Figure 5-10, and it was developed primarily for the purpose of simulating via computer graphics the three-dimensional physiologic dynamics of humans involved in automobile crashes.

The model segments included: one head (sphere), one neck (circular cylinder), two circular cylinders representing the upper arms, two truncated circular cones representing the forearms, one hand simulating a fist (sphere) gripping something (like a seat), one hand, fingers extended (flat plate), a torso (circular cylinder or elliptic cylinder), two truncated cylin-

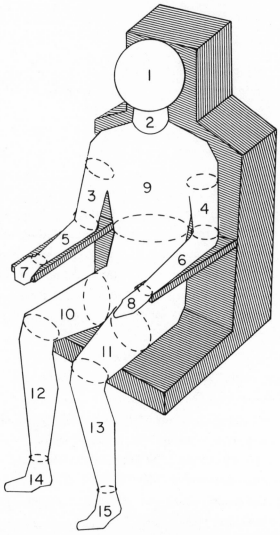

Figure 5-10 Fifteen-linkage articulated total body model used for the computerized analysis of the biodynamic response characteristics of the human organism.

drical cones representing the thighs, two truncated cylindrical cones representing the shanks, and two flat plates representing the feet -- totalling 15 "limbs".

The original CVS was modified in 1975 to include aerodynamic force application and a harness belt capability (Fleck and Butler, 1975, Schneck, 1978, 1980), thus creating the first version of the United States Air Force's Articulated Total Body (ATB) model for the computerized simulation of human body responses to generalized dynamic environments. In 1980, further improvements were made in the area of restraining system modelling, along with the addition of elements from the then three-dimensional CVS program to form ATB-II (Butler and Fleck, 1980). With the incorporation of effects from windblast, ATB-III was generated in 1983 (Butler, et al., 1983). Four additional modifications led to ATB-IV, which was documented in 1988 (Obergefell, et al., 1988a and b). These included:

1) A new wind force option allowing segment contact ellipsoids to block wind;

2) Corrections to prevent angular drift in joints;

3) Improvements allowing the prescription of multi-axis angular displacements; and,

4) A hyperellipsoid option for modelling surfaces in simulations such as corners and edges.

In its current form, the ATB model program contains over 130 subroutines totalling more than 16,000 lines of Fortran source code. It uses as a physiologic simulation, rigid ellipsoids or hyperellipsoids that are intended to correspond to individual body segments coupled by joints at which moments are applied as functions of joint orientations and rates of change of orientations (Obergefell, et al., 1988a and b). The program assumes the presence of only one joint between any two segments of the model. Data that the program requires in order to describe each segment include inertial and material properties, environmental conditions, and various joint parameters. These are provided by an input data file that defines moments of inertia, segment mass and geometry, material properties, the initial orientations of all segments in the system, and the characteristics of each joint and moment actuator (musculature surrounding the joint). Segment resiliency during contact is described through the use of force-deflection constitutive characteristics, which provide for energy losses, permanent offset, and impulsive forces which are experienced during the simulation.

Modeling of material properties is accomplished by the use of tension elements (passive longitudinal muscle), flexible elements (neck, torso, trunk), and actuator elements (contractile muscle). The objective of these elements is to deliver a gross description of body motion, and therefore they have little physiological discretion.

The tension elements are designed to behave statically in a manner similar to linear springs, without the corresponding stiffness, when subjected to a compression force. The ATB model represents the tension elements as a discrete system of $J^* + 1$ particles connected by J^* springs. The tension element is subjected to constraints which insure that all particles lie on a straight line, and that the strains and relative motions within the element are uniquely determined by the positions and motions of the two end elements (Fleck, et al., 1982). As illustrated by the simple forearm flexion example we considered earlier, the equations of motion for these elements are dependent on the positions, velocities, and accelerations of these same two end-points.

The flexible elements are composed of a chain of $J^* + 1$ (e.g., 15 as in Figure 5-10) rigid ellipsoids or hyperellipsoids connected at J^* joints, each having three degrees of freedom with three corresponding stiffness constants. In addition, each of the $J^* - 1$ interior segments of the flexible element is constrained so that its orientation is uniquely determined by the orientations of the end (or outer) segments. These constraints have been introduced to approximate the effects of body muscles which are connected, so that rather than acting on individual joints, they determine the overall flexural characteristics of the represented body member (Fleck, et al., 1982).

In order to determine angular motion, the actuator elements sub-routine requires inputs of position, velocity, and acceleration at the end points of the involved segments. Unlike the tension elements, which offer only *passive* resistance, the actuators can produce an *active* force across a joint. Similar to human muscle, one of the segments attached across the joint is considered to be stationary (the origin), while the other segment (the insertion) moves the bone (lever) to which it is attached through some angle Θ. The moment generated at this junction is determined and controlled by a feedback control equation that can be defined by the program user (Barineau and Schneck, 1988).

The equations of motion used to describe the model are similar to those derived earlier for the simple forearm flexion problem. They are based on the rigid body dynamics of coupled systems with Lagrange type constraint equations. All of the segmental equations of motion are for-mulated in an inertial frame of reference with respect to the center of gravity of the limb involved. The joints between the segments define the "connectivity" of the model through a joint vector. Where that vector is zero, i.e., for a so-called "null joint", the limb segments articulating at that joint are considered to be disjointed segments. That is to say, the artic-ulating segments involved are considered to be positioned as if they were connected, but no force exists to *hold* them in that position during the simulation. The solution of the system also depends on various con-straint equations which are affected by joint type. Some of the possible constraints that the ATB model can accommodate include linear position (inseparable, or "fused" joints), angular position (1-2-or-3-axis type of joint), zero distance (common points shared by more than one segment), fixed distance, and rolling/sliding motion. See Chapter 1, pages 43-51, for a more detailed description of degrees of freedom as they apply to human joints.

As a basis for performing calculations in the ATB program, matrix and vector mathematics are used relative to right-handed orthonormal reference systems (Obergefell, et al., 1988a and b). In this regard, a 3×3 directional cosine matrix is used to define the orientation of each of the $J^* + 1$ segments in the model. As the program proceeds, the directional cosine matrices are updated by the use of quaternions, $\bar{q}$, which are defined (Fleck, et al., 1982, Taylor, 1979, Beeler, et al., 1972, and Hamilton, 1969) to consist of both a vector part, $\vec{m}$, and a scalar part, s^*:

$$\bar{q} = s^* + \vec{m} \qquad [5\text{-}59]$$

Then, rotation of an arbitrary limb vector $\hat{i}_n$, an angular amount Θ about an arbitrary axis in space, defined by the unit vector $\hat{i}_r$, to a new orientation, $\hat{i}_n{}^*$ can be expressed by:

$$\hat{i}_n{}^* = \bar{q}\hat{i}_n\bar{q}^* \qquad [5\text{-}60]$$

where:

$$\bar{q} = \cos\frac{\Theta}{2} + \left(\sin\frac{\Theta}{2}\right)\hat{i}_r \qquad [5\text{-}61]$$

and,

$$\bar{q}^* = \cos\frac{\Theta}{2} - \left(\sin\frac{\Theta}{2}\right)\hat{i}_r \qquad [5\text{-}62]$$

For a null rotation, the scalar parts of $\bar{q}$ and $\bar{q}^*$ would be equal to unity ($\cos 0 = 1$), while the vector terms would be equal to zero ($\sin 0 = 0$), so that $\bar{q} = \bar{q}^*$ and $\hat{i}_n{}^* = \hat{i}_n$ and the relative limb orientation in space would not change (i.e., $\Theta = 0$). In the ATB program, these quaternions are used to update the cosine matrices relating the changes in segment orientations in space. For input and output purposes, the terms yaw (rotation around limb centerline $\hat{i}_n$), pitch (rotation in the sagittal plane -- flexion and/or extension -- around an axis perpendicular to that plane), and roll (rotation in the frontal plane -- abduction and/or adduction around an axis perpendicular to that plane) were chosen over the more technically accurate spin, nutation and precession definitions for signifying the directions of the cosine matrices.

The integrator being used by the ATB program to integrate the equations of motion is called the Vector Exponential Integrator, and is based on a fourth order Runge-Kutta technique with an exponential term for each variable. The integration of a first order differential equation proceeds as follows (Fleck, et al., 1982):

$$\dot{x} = \dot{x}(x, t) \qquad \qquad [5\text{-}63]$$

for which a solution of the following form can be written:

$$x(t) = x(0) + \int_0^t e^{b'(t-t')} \{\dot{x}[x(t'), t'] - b'[x(t') - x(0)]\} \, \delta t' \qquad [5\text{-}64]$$

where b' is a coefficient to be determined. Letting $\dot{x}$ be approximated by:

$$\dot{x}[x(t), t] \simeq b'x(t) + b_o' + b_1't + b_2't^2 = \dot{x}(t), \qquad [5\text{-}65]$$

where b', b_o', b_1', and b_2' are parameters yet to be determined, the equation [5-64] for $x(t)$ can be solved to yield:

$$x(t) = x(0) + [b'x(0) + b_o']te_o(t) + b_1't^2e_1(t) + b_2't^3e_2(t) \qquad [5\text{-}66]$$

where:

$$e_o(t) = \frac{(e^{b't} - 1)}{(b't)} \to 1 \text{ as } b't \to 0 \qquad [5\text{-}67]$$

$$e_1(t) = \frac{(e_o^{b't} - 1)}{(b't)} \to \frac{1}{2} \text{ as } b't \to 0, \text{ and} \qquad [5\text{-}68]$$

$$e_2(t) = \frac{(2e_1^{b't} - 1)}{(b't)} \to \frac{1}{3} \text{ as } b't \to 0. \qquad [5\text{-}69]$$

These exponential terms are responsible for the integrator's name.

The four parameters, b', b_o', b_1' and b_2' determine the behavior of the integrator. They are always chosen so that the equation fits the computed derivatives at the beginning of each integration interval. For a successfully integrated time interval, t', the value of $(t + t')$ is substituted for t into equation [5-65], rewritten as follows:

$$\dot{x}(t) = b'[x(t) - x(0)] + \dot{x}(0) + b_1't + b_2't^2,\qquad [5\text{-}70]$$

which yields:

$$\dot{x}(t + t') = b'[x(t + t') - x(t')] + (b_1' + 2b_2't')t$$

$$+ b_2't^2 + b'[x(t') - x(0)] + \dot{x}(0)\qquad [5\text{-}71]$$

$$+ b_1't' + b_2't'^2$$

The functions are then redefined to preserve the form of the above equations, i.e.:

$$b_1' = b_1' + 2b_2't'$$
$$b_2' = b_2'$$
$$\dot{x}(0) = \dot{x}(t'), \text{ and}$$
$$x(t) = x(t + t')$$

These terms are then used to estimate the value of $x(t)$ at the first half step of the next interval, i.e., when $t = \dfrac{t'}{2}$.

What we have described here is just a very brief outline of the methods by which an investigator might develop a computerized model to simulate human body responses in dynamic environments. The reader is referred to the literature -- specifically, to the work of Fleck, et al., (1982) -- for more complete details of the ATB program. This program provides for a wide variety of outputs with respect to motion, force generation, and damage due to interaction with the environment. Time histories for the entire simulation can be put together and projected as three-dimensional images using a graphics display subroutine, or, they can be presented in the form of tabulated data. Every model segment and joint can be monitored for values of displacement, velocity and acceleration as well as applied and developed torques and forces at the joints. Furthermore, force-deflection characteristics between two or more segments can be analyzed to determine whether or not any damage has occurred during the simulation run. Damage can result from both subject-subject and subject-environment collisions.

Although highly sophisticated in its current form, the ATB model needs significant further modifications to include at least three very important aspects of biodynamic response characteristics: (i) active (i.e., due to muscle contraction) torque generation at the joints; (ii) optimization (based on minimum-energy principles) of the predicted model behavior; and (iii) PID (proportional-integral-derivative) control of musculoskeletal function. Some principles related to the first two have been addressed so far in this text, while the third constitutes much of Chapter 6, which follows after some concluding remarks.

Concluding Remarks

In Chapters 4 and 5, we have addressed some principles related to the second major aspect of the engineering analysis of physiologic function -- *utilization* of mass, energy and momentum by the human organism. Recall that the first major aspect of such an analysis involves the *transport* of mass, energy and momentum, which we examined in Chapters 2 and 3. The third major aspect -- *control* -- shall be discussed in Chapter 6.

While the guiding philosophy governing physiologic transport could be embedded into a generic transport equation (c.f., equation [2-4], for example), with associated "design principles" (LaBarbera and Vogel, 1982), the guiding philosophy governing utilization of mass, energy and momentum seems to be embedded in the concept of optimization, with associated "variational principles". The latter concern themselves with *economy* -- economy in the utilization of mass (c.f., Figure 5-1 and associated discussion); and economy in the utilization of energy and momentum (c.f., earlier discussion related to the deductive analysis of the kinetics of human movement). As we learn more and more about physiologic function, the idea that it is governed by "constrained optimization principles" is gaining more and more credibility. Studies such as those of Nubar and Contini (1961), Beckett and Chang (1968), Beckett and Pan (1971), and Chao and An (1978) have since been backed up and confirmed by Yao (1976), Crowninshield (1978), Crowninshield and Brand (1981), An, et al., (1984), and, most recently -- for the case of muscle forces exerted in the mandible during bite motion -- by Kang, Updike, and Salathe (1990). Each of these investigators is able to show an excellent correlation between the deductive results predicted using optimization theory, and the empirical results based on laboratory experimentation. As the data base of such studies increases, and as they become increasingly verifiable from real-life measurements, researchers are becoming increasingly convinced that some economization-of-effort-scheme is at work in the mechanisms that govern the principles related to human posture and locomotion. Aside from the significance of this as a means for formulating purely deductive methods for the analysis of physiologic function, there is the added potential that such a revelation has to increase our basic understanding of the very nature of life, itself.

Equally important in this respect is the role that homeostatic control mechanisms play in governing physiologic function. Both the transport and the utilization of mass, energy, and momentum by the physiologic system take place within a carefully controlled internal environment. In fact, the concept of optimization, itself, may be thought of as a type of "control" of physiologic function -- in this case "forcing" the function to be carried out within the constraints imposed by certain parameters that need to be "minimized". In the next chapter, we shall formalize somewhat more rigorously a few of the basic principles related to physiologic control, the transport of information, and the concept of homeostasis.

Chapter 6

Basic Principles of Information Transport, Feedback Control, and the Concept of Homeostasis in Physiologic Systems

Introduction

Emphasized throughout this text has been the point that all physiologic function is deliberately and carefully controlled by a variety of very sophisticated mechanisms. In Chapter 1, for example, when we defined homeostasis to be the steady-state internal environment that is required for the maintenance of life, we also stressed that this concept includes as well those processes by which physiologic variables are maintained within well-defined and specifically prescribed limits -- and that inherent to such processes is the corresponding concept of feedback control.

In Chapters 2 and 3, we noted that the mechanisms responsible for the transport of mass, energy, and momentum in physiologic systems depend on a long list of variables, or attributes, through which the organism can intervene to control the transport, itself. Some specific ways by which this is accomplished were addressed near the end of Chapter 3. In Chapters 4 and 5, we saw further how control of the utilization of mass, energy, and momentum by the human body is manifest through cascading sequences of coupled biochemical reactions, governed by complex enzyme systems and optimization schemes (both intrinsic and extrinsic). We have discussed the body's immune response as an example of one type of control; we have spoken about control as it relates to membrane permeability, fluid and electrolyte balance, the oxygen-reserve mechanism, thermoregulation, human posture and locomotion, the generation and propagation of action potentials, and so on. And, indeed, the list *does* go on, and on, and on, to include the organism's need to control as well other physiologic processes, such as cardiac output, heart rate, blood pressure, the anatomical distribution of blood, blood chemistry (calcium ion concentration, glucose concentration, the total number and types of blood cells, and so on), respiration, bone cross-sectional area (functional adaptation), audition, the oculomotor system (tracking and stabilization), the pupillary reflex system's response to light intensity, and appetite, to name but a *few*.

Obviously, if we were to concern ourselves at this point with describing the specifics of all of the control mechanisms associated with even *some* of the physiologic processes listed above, we would be embarking on the preparation of a rather voluminous textbook dedicated to this subject in and of itself. For good literature sources that present the *engineering* aspects of many of these specifics, the reader is directed to the works of Milhorn (1966), and Milsum (1966); while Guyton (1981) provides a fine resource volume that addresses the *physiologic* aspects of specific biological control mechanisms. In this chapter, we concern ourselves less with *specifics*, than with examining some general *principles* of control as they are manifest in the physiologic system. That is, we seek to formalize, in some sense, a generalized *approach* to understanding and mathematically analyzing physiologic control as a *concept*, rather than as a separate and distinct set of actions that are unique to any given physiologic process. In particular, the approach casts basic aspects of life processes and homeostatic mechanisms into an overall scheme of feedback control, with the ultimate intent of providing a unified means for formulating the mathematics of such processes in a systematic, rational manner. For this reason, we begin by first presenting some general principles of feedback control theory. This is then followed by a discussion of how these principles are manifest in the biological processes associated with physiologic function.

General Principles of Feedback Control Theory

Preliminary Concepts and Definitions

OPEN-LOOP SYSTEMS

Refer back to Figure 5-1, which represents schematically the mechanism whereby substrates are converted via enzyme-catalyzed biochemical reactions into products. For now, let us consider only the exergonic reaction, and assume further that it is proceeding only in the direction from raw materials to enzymes, and from substrates to products. Finally, let us conceptualize what is occurring in a "black box" sense: the substrates being designated as the *input* to our "box", the products representing the *output* from our "box", and the "box", itself symbolizing all of the enzyme-catalyzed events that transform the input into the output. In this formulation, the input (substrates) is considered to act essentially as an independent variable -- at least independent of the output -- in the sense that the output does not influence how much substrate comes into the box. The output (products), on the other hand, is a dependent variable, but depends on the input *only* in terms of what happens to the latter as it passes through the box.

The above conceptualization defines an *open-loop* system, and, "what happens to the input as it passes through the box," is called, in the language of control theory, the *Transfer Function* of the system. Mathematically, transfer functions are often defined to be the ratio of the LaPlace Transform of the output of the box to the corresponding LaPlace Transform of the input to the box, and they are generally complex functions. The absolute value of the *Modulus* of these complex transfer functions is called the *Gain* of the system, and basically addresses *how much* substrate becomes *how much* product. It is a type of amplification factor. The associated *Argument* of these complex transfer functions is called the *Phase* of the system, and basically addresses how much of a delay there is between the time something goes into the box, and the time something comes out. It defines the extent to which the output of the system leads, lags, or is in phase with the input to the system.

In an open-loop system, the output and input (or change in output as a function of change in input) are related to one another *only* by the transfer function, and nothing else. Thus, in this simple type of system, the input can be arbitrarily controlled -- which is to say, we can hypothetically eat and drink as much as we care to -- and, at least theoretically, the "box" will continue to manufacture products ad infinitum, at a rate determined by how fast the enzymes can catalyze all of the associated biochemical conversions that take place in the "box". Again, in the language of control theory, you, as the *eater* (i.e., *supplier* of substrates, raw materials, or input) are called the *controlling element* of the open-loop system. To the extent that the box depends on this input in order to carry out its mission i.e., the box, as the *user* of substrates, raw materials, or input, is called the *controlled element* of the open-loop system. Furthermore, since you can independently control how much you eat, the *amount* of substrate presented as input for processing is called the *controlling signal* of the open-loop system. And, finally, the *amount* of product produced by the box is called the *controlled signal* of the open-loop system.

In an open-loop system, the controlling signal, u_c, is independent of the controlled signal, u_o, and u_o depends on u_c through the complex transfer function, K_1 of the controlled element of the system. The latter is defined to be the LaPlace-Transformed ratio u_o/u_c, and has an associated gain, or amplification factor and phase angle.

CLOSED-LOOP SYSTEMS

Now, suppose u_c *does depend* on u_o, to the extent that the latter is somehow monitored and used to "control" (as in through appetite and various other satiety parameters) the input. Then what we have is a configuration such as is illustrated more accurately in Figure 5-1, and which is appropriately called a *closed-loop* system. In the situation depicted in the figure, the "amount" of product is "monitored" by "feeding it back into the black box" to catalyze the breakdown of the enzyme required to metabolize it. Thus, as more product is manufactured, more enzyme is broken down back to raw materials. This reduces the amount of enzyme available to metabolize the substrates, which, in turn, allows less product to be manufactured. In other words, the rate of manufacture of product now depends not only on u_c through K_1, but, also on u_o, itself, i.e., the actual amount of product in existence. Furthermore, as u_c now begins to accumulate because the rate of conversion to u_o declines, the system "back-up" correspondingly acts to suppress the individual's appetite, so that input, u_c, as well, becomes a function of u_o and is no longer completely independent.

To the extent that the output of the above-described system is "fed back" to control the input, this type of closed-loop system is called a *feedback control system,* sometimes referred to as a servomechanism when the controlled variable is mechanical position (such as human posture and locomotion). Figure 5-1 also illustrates the concept of *feedforward control,* in which the *substrates* control their own ultimate destiny by catalyzing the manufacture of the enzymes needed to metabolize them. Thus, we have in the philosophy of physiologic metabolic processes, closed-loop systems illustrating substrate feedforward control of enzyme synthesis, counterbalanced by product feedback control of enzyme catabolism; and all of these systems are coupled through the exergonic-endergonic scheme discussed in Chapter 5.

Further illustrated in the above scenario is the concept of *intrinsic* feedback and feedforward control. That is, the substrate, itself, by its very presence, acts as the feedforward stimulus for enzyme synthesis and the product, likewise, by its very presence, acts as the feedback stimulus for enzyme catabolism. The control is intrinsic in the sense that there are no identifiable "sensors" that measure and respond to substrate or product concentration, per se, and, based on such monitoring, cause something to happen to either of these concentrations by a mobilization of mechanisms external to the system. Rather, the control is inherent to the system, being based on direct substrate and/or product influences on the DNA code in the cell nucleus.

Contrast such *intrinsic* control to that by which substrate accumulation suppresses appetite. In the latter case, biological sensors, called *feedback* or *monitoring elements* in the language of control theory, actually measure and respond to the quantity of substrate present. Biological sensors fall into the general category of devices called "transducers". These are elements of a system that transmit energy, but in so doing, convert that energy from one form into another (c.f., Chapter 1, pgs. 51-64). For the case in point, the transducers are strain-gauge-type stretch-receptors that are particularly prevalent in the walls of the stomach and duodenum. As u_c builds up, it acts to stretch the receptors, which, a la the action of a strain-gauge, convert this "stretch" into a corresponding electromagnetic signal according to the transfer function that characterizes the transducer, itself. The transduced signal, called the *primary feedback signal, u_p*, is a direct measure of the controlled signal (which, in this case is actually u_c), since it is related to the latter through the transfer function (or related gain) that is characteristic of the feedback-element transducer.

The primary feedback signal is now transmitted through the Vagus nerves to the lateral hypothalamus, which normally causes an animal to have an appetite for food. *Vagus* stimulation of this center, however, is *inhibitory*, thus *suppressing* this desire for food while stimulating the corresponding satiety center located in the ventromedial nuclei of the hypothalamus (Guyton, 1981). The result is a decrease in u_c that results from hypothalamic suppression of appetite. In that sense, the control is *extrinsic*, meaning that it becomes manifest not by the inherent properties of the system, itself, but by the actions of the hypothalamus acting as an external mediator. Other examples of extrinsic control involving enzyme systems include endocrine (hormone), vitamin, and trace element-controlled *activation* of *holoenzymes* (c.f., Chapter 4, pgs. 339-340).

In cases where the hypothalamus might, in turn, be acting through another effector organ -- such as the smooth muscle musculature in the case of thermoregulatory mechanisms involving the arterioles of the vascular system (c.f., Chapter 4) -- the effector organs (e.g., smooth muscles) are called the *feedback actuating elements* of the closed-loop system.

REFERENCE SIGNALS IN CLOSED-LOOP SYSTEMS

In most practical situations, servomechanisms are not designed to be simple on/off devices, as implied by the discussion above. Rather, the controlled signal is fed back for the purpose of comparing it with some *reference signal,* or "desired output", u_r. The idea here is for the feedback loop of the system to include a *comparator,* comparing the *actual output,* u_o (or, equivalently, u_p where a sensor is involved), with a *desired output,* u_r. If $u_o = u_r$, the comparator "knows" that the controlled element is performing as desired and it does nothing to alter this situation. If, however, u_o is greater than or less than u_r by some amount (that amount depending on the *sensitivity* of the comparator), then the controlling element sends a control signal, u_c, to the controlled element in an effort to rectify the situation. Normally, the intensity of u_c will be at least *directly proportional* to the *difference* $(u_o - u_r)$, so that the further the actual signal gets away from the desired signal, the more substantial will be the corrective therapy applied. This is called *Proportional Control,* or *Proportional Compensation* (see later). The quantity $(u_o - u_r)$ is called the *error signal* of the feedback control system, and the ratio of the control signal, u_c, to the error signal, $u_o - u_r$, constitutes the so-called proportional control transfer function, K_2', of the comparator element in the feedback loop:

$$\text{Proportional Control, } K_2' = \frac{u_c}{u_o - u_r} \qquad \text{[6-1]}$$

Proportional control or compensation, by responding to the difference between u_o and u_r, gives the controller the ability to affect a *smooth* transition to u_r, thus minimizing inertial overshoots in the system and preventing wild undulations around u_r. By making the intensity of the response directly proportional to the deviation of u_o from u_r, it provides smooth and uniform accelerations and decelerations of system components to desired target values. We shall have more to say about this later.

Ideally, in an equilibrium situation, once u_o reaches u_r -- as, for example, body temperature reaching 37°C, no further compensation should be necessary and the system should have reached a steady-state performance level. In real life, however, this is rarely the case. More likely, some *disturbance*, or *disturbing signal*, u_d, will tend to shift the system operating point away from u_r and a correction will become necessary. In Chapter 4, we discussed how some of these corrections are handled by the body in the case where u_d might represent thermal disturbances (hot or cold stress). Later in this chapter we shall address how corrections are handled in a more generalized sense.

Stability of Feedback Control Systems

The concept of stability has two aspects: First, one must address the issue of whether or not the output of a feedback control system is always finite, or limited, for any finite input (i.e., does the system "blow up" in any sense) -- and, if it is, does the controlled signal actually approach the desired signal in some sense when the system is upset by some disturbance. An unstable system would be one for which a disturbance causes the controlled signal to drift without limit *away* from the reference signal when a disturbing signal upsets the equilibrium of the system. Of course, to "drift away without limit" (i.e., approach infinity) is strictly a mathematical concept. What happens in real life is that the controlled signal increases (or decreases) until some other component in the system breaks down (e.g., the individual dies at given extremes of body core temperature), or, until some nonlinearity intervenes to constrain the output, perhaps allowing the organism to "adapt" in some sense to a new value of u_r (see later).

Second, in discussing stability of feedback control systems, one must address the issue of how many oscillations are required to bring the system to an equilibrated state, and to what extent each fluctuation "overshoots" that state before sufficient damping occurs to stabilize matters. These considerations are called the *Transient Response* characteristics of the feedback control system, and they are represented basically by two parameters: the *overshoot* and the *decay time*. Recall that, in general, we want the intensity of u_c to be at least proportional to the difference $(u_o - u_r)$ so that corrective therapy is applied in direct proportion to the amount of disturbance created by u_d. Ideally, our system should respond to a disturbance by quickly adjusting u_o (the controlled signal) so that it approaches u_r (the reference signal) as smoothly and as expeditiously as possible. This can be described mathematically by letting:

$$u_o = u_r(1 - e^{-\frac{t}{\tau_d}}) \qquad \qquad [6\text{-}2]$$

where τ_d is called the *decay time-response-constant* of the system. Furthermore, since we also want the intensity of the controlling signal to be proportional to the deviation of u_o from u_r, we may write, from equation [6-1]:

$$u_c = K_2{}'(u_o - u_r) \qquad \qquad [6\text{-}3]$$

Eliminating u_r from equations [6-2] and [6-3] we get, as the ideal response characteristic for our system,

$$u_o = \frac{u_c}{K_2{}'}(1 - e^{\frac{t}{\tau_d}}) \qquad \qquad [6\text{-}4]$$

Equation [6-4], or equation [6-2] both assume that the system starts from $u_o = 0$ at time $t = 0$, at which point it is commanded to take on the reference value u_r. The system does so exponentially, i.e., $u_o \to u_r$ as $t \to \infty$, with some characteristic time response constant, τ_d, and with no "overshoot". That is, ideally, the system does not go beyond its assigned value u_r before equilibrating to this value, but, rather, it approaches this value smoothly and asymptotically. Furthermore, the smaller is the value of the decay time constant, the faster will u_o approach u_r, i.e., $u_o = 0.99u_r$ by the time $t = 5\tau_d$, as determined from equation [6-2]. In other words, the decay time response constant determines how fast u_o will come however close we desire to u_r. Note in equation [6-4] that the limit of u_o as t approaches infinity is still finite because u_o approaches u_r at this point (c.f., equation [6-2]), causing u_c also to approach zero (c.f., equation [6-3]), so that the limit of [6-4] as $t \to \infty$ is of the form $\frac{0}{0}$, which has a finite limit.

OVERDAMPED, UNDERDAMPED AND CRITICALLY DAMPED SYSTEMS

The idealizations of the previous paragraphs and equations [6-2] to [6-4] define an *Overdamped* Feedback Control System, where the transition from $u_o = 0$ to $u_o = u_r$ takes place nice and smoothly, over an infinite period of time, with no system overshoots. Unfortunately, such idealizations are not very realistic, because the physical systems we deal with in real life have, among other things, inertial properties that tend to cause *overshoots*. That is, the desire to have a system respond *quickly* (we normally cannot wait an infinite period of time for u_o to reach u_r) without overshooting is almost self-contradictory. To get a *fast* response -- say, for example, attempting to get a baseball player from home plate to first base in the absolute minimum amount of time -- generally requires imparting to a physical system (the baseball player) a great deal of inertia. This usually produces overshoots -- the player runs well past the base to make sure he or she gets there before the ball does -- and in some cases the amount of overshoot is of no great concern. On the other hand, to get an *accurate* response -- i.e., if the player tried to avoid over-running the base -- generally requires decelerating a physical system such that it approaches a limit slowly, sacrificing speed for accuracy ($t \to \infty$ in equations [6-2] to [6-4]). An example of this might be elevator systems which carry people. They will accelerate and travel relatively rapidly between floors, but as they approach a particular level, they will be decelerated to come to a stop ever so slowly on the desired landing.

Furthermore, in real systems, set-point disturbances u_d might not be clear-cut, discrete events that require simple, finite corrections. More likely, disturbances to an equilibrium situation will be *continuous*, and the feedback control networks will have to maintain a given set point in the face of constant steady-state errors. A good example of this is the constant tendency that gravity has to make us fall down, and the role that the physiologic myotatic reflex arc (monosynaptic reflex arc, Schneck, 1984b) plays in allowing us to maintain a balanced erect posture in the face of this ever-present disturbing force.

The net response of a feedback control system therefore represents the designer's attempt to compromise response time with accuracy, by allowing the system to experience overshoots that can be damped out quickly, and that have small relative amplitudes. The existence of such overshoots leads to system oscillations that allow u_o to approach u_r in some cyclic fashion. This defines an *Underdamped* Feedback Control System, where the adverbs "quickly" and "small" can be further quantified if we let the error signal, $u_o - u_r$, be of the form:

$$u_o - u_r = u_a e^{-\frac{t}{\tau_d}} \left[\sin(2\pi \frac{t}{\tau_t} + \theta_o) + C_o \right] \qquad [6\text{-}5]$$

The quantity τ_t represents the "period" of the damped response of an underdamped feedback control system. The quotes are inserted around the word "period" because the response of the system is not actually a repeating, cyclic sine wave with an identifiable period in the strict mathematical sense. Rather, the response defined by equation [6-5] is oscillating with a diminishing amplitude, and the "period" is taken to correspond to the time interval within which the slope of the curve represented by equation [6-5] passes through zero three times -- the first time at $t = 0$, the second time at the point of *maximum overshoot* (see later), and the third time at $t = \tau_t$. Forcing the curve to pass through $t = 0$ with zero *slope* is more physically meaningful and insures, along with the requirement that u_o also equals zero at this time, that the system does, indeed, begin its response from an equilibrium situation at rest. These constraints, i.e., $\dfrac{\delta u_o}{\delta t} = 0$ at $t = 0$ and $u_o = 0$ at $t = 0$ also fix the value of the phase angle, θ_o, in terms of the other parameters in equation [6-5]. Performing the necessary calculus, one arrives at the conclusion that:

$$\sin \theta_o = \pm \sqrt{1 - \left[\frac{u_r}{u_a}\right]^2 \left[\frac{\tau_t}{2\pi\tau_d}\right]^2} \qquad [6\text{-}6]$$

where, in most cases, the negative sign will be required to give proper angles, in terms of mathematically meaningful quantities and self-consistency with the rest of the analysis. The constant, C_o, may also be determined from the conditions at $t = 0$, yielding:

$$C_o = -\left[\sin \theta_o + \frac{u_r}{u_a}\right] \qquad [6\text{-}7]$$

Equation [6-5] describes a damped, oscillating wave that begins at $u_o = 0$ initially and approaches $u_o = u_r$ asymptotically, all the while taking on values that vacillate around this limit. The value of t for which u_o passes through u_r for the *first* time during these excursions, i.e., the time,

$$t = \left[\text{Arc Sin}\,(-C_o) - \theta_o\right] \frac{\tau_t}{2\pi} \qquad [6\text{-}8]$$

is called the *buildup*, or *rise time*, τ_b, of the underdamped feedback control system. The *decay time*, τ^*_d, of the system (not to be confused with the decay time-response *constant*, τ_d) is taken to be the interval of time from the point where $u_o = u_r$ for the first time (i.e., $t = \tau_b$) to the point $t = 5\tau_d$ where the exponential function in equation [6-5] is down to less than 1% of its initial value of unity at $t = 0$, from which point on, u_o is virtually equal to and oscillating with just the slightest perturbation around u_r for all subsequent time.

A servo system with good relative stability characteristics may be defined to be one having a comparatively short decay time, equal to about four times the buildup time. Furthermore, the sum of the decay time plus the buildup time (which equals five time-response constants as per the discussion above) should not exceed twice the period of the damped motion, so that the amplitude of the oscillatory response will have been reduced or damped to less than 1% of its initial value within two complete cycles of the function defined in equation [6-5]. Thus, for good stability, we should have: $\tau^*_d = 5\tau_d - \tau_b = 4\tau_b$, which suggests that the corresponding decay time response constant for the system should be equal to, or at least on the same order of magnitude as the buildup or rise time of the system, i.e.,

$$\tau_d = \tau_b = \frac{1}{4}\tau^*_d \qquad [6\text{-}9]$$

Moreover, $\tau^*_d + \tau_b = 5\tau_d \le 2\tau_t$, or,

$$\frac{\tau_d}{\tau_t} \le 0.40, \text{ or, } \frac{2\tau_d}{\tau_t} \le 0.80 \qquad [6\text{-}10]$$

As an illustration, suppose $(\tau_d/\tau_t) = 0.325$. Then, equations [6-5], applied at $t = \tau_b = \tau_d$ (i.e., when $u_o = u_r$ for the first time), together with the initial ($t = 0$) conditions [6-6] and [6-7], can be solved simultaneously to yield:

$$\theta_o = 243.63°(1.3535\pi \text{ radians}),$$
$$C_o = -0.011, \text{ and,}$$
$$\frac{u_r}{u_a} = 0.907,$$

where u_a is the maximum value of the exponentially decreasing amplitude of the oscillations of u_o around u_r (c.f., equation [6-5]). Substitution of these quantities into equation [6-5], together with equations [6-9] and [6-10], yields the equation:

$$\frac{u_o}{u_r} = 1 + (1.1026)e^{-\frac{t}{\tau_d}}[\sin(0.65\frac{t}{\tau_d} + 1.3535)\pi - 0.011] \qquad [6\text{-}11]$$

Equation [6-11] is plotted in Figure 6-1 as nondimensionalized response, u_o/u_r vs. nondimensionalized time, t/τ_d. Observe, again, that the buildup time, τ_b is equal to the decay time response constant, τ_d (by design), so that $u_o/u_r = 1$ when $(t = \tau_b)/\tau_d = 1$; and that the decay time, τ^*_d is about four times as large as the rise time, τ_b. Furthermore, as long as the decay time-constant is *positive,* events such as those described by equations [6-2], [6-4] and [6-5] *will* die out with time, thus insuring that the controlled signal will approach the desired signal in due time. "In due time" is quantified by specifying a *steady state error,* which is essentially a measure of the amplitude of the oscillation of u_o around u_r after either five decay time constants have elapsed, or, after two complete cycles have transpired, or, if appropriate, in the presence of a continuous disturbance.

One can estimate the stability of the system, and thus determine the steady state error by determining the decay in overshoot amplitude per cycle of oscillation. That is, note in Figure 6-1 that the overshoot amplitude decays exponentially according to the equation:

$$\frac{u_o}{u_r} = 1 + (1.1026)e^{-\frac{t}{\tau_d}} \qquad [6\text{-}12]$$

which forms an envelope for the attenuating wave (compare this equation, for example, with equation [6-2]). Note further that the first "overshoot" reaches its maximum value (1.2095) for a nondimensionalized time value of 1.5432, and that the second "overshoot" (really an *undershoot* on the graph) reaches its maximum value (0.9539) for a nondimensionalized time value of 3.0769 -- the difference between these being $3.0769 - 1.5432 = 1.5337$ time constants, during which the absolute value of the amplitude changed by $1.2095 - 0.9539 = 0.2556$ times the reference quantity u_r. Thus, the amplitude decayed from an initial overshoot of some 21% (i.e., $1.2095 - 1, \times 100$), to an overshoot of only 4.6% (i.e., $1 - 0.9539, \times 100$) in essentially half of one complete cycle. The latter, i.e., the nondimensionalized damped period, τ_t/τ_d, of the response is calculated based on the previous definition to be 3.0769, and

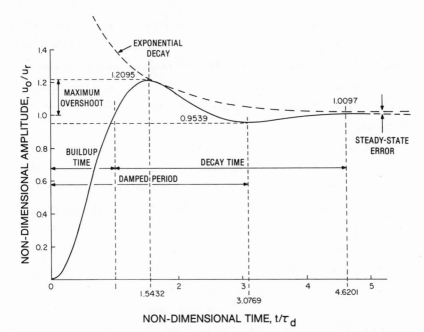

Figure 6-1 Stable response characteristics of an underdamped feedback control system.

half of that would be around 1.5385 (compare this value with 1.5337). Thus, this decay amounts to about a 78% decrease in overshoot percentage $[(20.95 - 4.61) \times 100/20.95]$ per half-cycle, which is excellent, because an accepted *minimum* value of this quantity to insure a suitably stable servo is around 50%. One hundred percent would, of course, be very nice, indeed.

A further measure of the exponential decay in overshoot amplitude is the so-called *Logarithmic Decrement,* which is defined to be the natural logarithm of the ratio of two successive *overshoots* (not undershoots). In the current example, the first overshoot has a nondimensionalized amplitude of 1.2095, for $t/\tau_d = 1.5432$, and the second overshoot has a nondimensionalized amplitude of 1.0097 for $t/\tau_d = 4.6201$, i.e., one period (3.0769) later. Thus, the logarithmic decrement is $\ell n \dfrac{1.2095}{1.0097} = 0.18 = \delta_\ell$. The significance of the logarithmic decrement is that it is also related to the *damping factor,* or *damping ratio* of the feedback system. For a second-order physical system, this ratio, ζ^*, is given by:

$$\zeta^* = \frac{\delta_\ell}{\sqrt{(2\pi)^2 + \delta_\ell^2}} \qquad [6\text{-}13]$$

which, for the sample problem being considered, turns out to be equal to 0.0287. The damping ratio, in turn, can be used to get an estimate of the *natural period,* τ_n, of the feedback control system, from the relationship:

$$\frac{\tau_d}{\tau_t} = \frac{\tau_d}{\tau_n} \sqrt{1 - \zeta^{*2}} = 0.325, \qquad [6\text{-}14]$$

from which, $\dfrac{\tau_d}{\tau_n} = 0.3251$ -- very close to $\dfrac{\tau_d}{\tau_t}$ for this illustration.

The importance of knowing τ_n, the reciprocal of which gives the *natural frequency,* f_n, of a physical system, is related to system instabilities that arise if u_d excites the system at a frequency f_n. Such excitation could lead to *resonance* phenomena which cause the amplitude of the response of the system to the given disturbance to grow large enough to damage the system. At resonance, such *amplitude amplification,* known as the *amplitude magnification factor,* is given by $\frac{1}{2\zeta*}$, which, again for the example we are considering, has the value 17.4. This means that, if we disturb the controlled element of our feedback system at its natural frequency, such disturbances will be magnified over 17 times before they eventually get damped out (if, indeed, they get damped out at all). This type of phenomenon is what causes glass to break when we "sing" (excite the glass at sonic frequencies) or sustain tones of specific pitches that happen to correspond to the resonant frequencies of the glass material. It may also lead to body tissue damage that results from "vibrating" various organs at their natural frequency. Various measurements suggest that the natural frequency of the whole human body is in the 4-8 Hz range, so that excitation disturbances involving these, and the primary harmonics of this range could be potentially damaging due to the effects of resonance. Fortunately, the amplification factor drops off dramatically as we move away from f_n in either direction, and as we increase the amount of damping in a physical system.

And finally, a servo system is considered to have relatively good stability characteristics if it has a maximum overshoot that does not exceed 30% of the value of the reference signal. From previous calculations and Figure 6-1 we see that the maximum overshoot for this sample system is 1.2095 (i.e., 20.95%). Thus, we conclude that this is, in fact, a stable and well-damped system because, in summary:

(1) it has an *overshoot coefficient,* u_a that keeps $\left(\frac{u_o}{u_r} \right) < 1.3$ for all time;

(2) it has a buildup time approximately equal to the decay time-constant of the exponential function that envelopes the decaying amplitude of the response signal;

(3) it has a total decay time equal to about four times the buildup time;

(4) the decay in overshoot amplitude percentage is better than 50% per half-cycle of oscillation;

(5) the amplitude of the oscillatory response (steady-state error) will have reached less than 1% of the target value u_r in less than two complete cycles of the total unsteady motion; and, in general,

(6) it rapidly approaches the reference value to which it has been commanded, with a very small steady-state error.

This is precisely what the manufacturer is after in designing good shock absorbers -- and it is what the designer of feedback control systems dreams of when he or she speaks of good response characteristics. Mother nature does pretty well, herself, in this respect, since virtually *all* of the feedback control systems in the human body hover around being overdamped or just slightly underdamped with the above stability specifications. The overdamped criterion requires $\zeta^* > 1$; the underdamped criterion requires $\zeta^* < 1$; and the cross-over point, i.e., $\zeta^* = 1$, defines a *critically damped* system. In general, a critically damped system, when disturbed, will approach equilibrium faster than will an overdamped system, approaching u_r for large time and without periodicities. Bearing this in mind, let us say a few words about the functional form of equation [6-3].

Proportional, Integral, and Derivative (PID) Control

DIFFERENTIAL COMPENSATION

In the previous section, we saw that, like the controlled-element transfer function, K_1, defined earlier, the proportional control transfer function, K_2' is *also* generally a complex function -- as is evident when one substitutes an amplitude and phase angle equation such as [6-5] (or [6-11]) into [6-1], for example. This means that K_2' has a corresponding gain and associated phase, the existence of which signifies that there is a lag time between a change in u_o away from u_r (i.e., the generation of an error signal, $u_o - u_r$, due to a disturbance, u_d) and the system's *response* to that change (i.e., the generation of a control signal, u_c).

Ideally, we want to minimize (or eliminate completely, if possible) such a lag between the generation of $u_o - u_r$ and the generation of u_c. That is, we would like to speed up the *Response Time* of the system to disturbances u_d. This could be accomplished if our system could somehow *anticipate* the error signal and, in a sense, start correcting for it almost before it actually happens. Indeed, this type of behavior is characteristic of control elements that respond to the *rate-of-change* of $u_o - u_r$, as well as to its magnitude. In the same sense that an acceleration (derivative of velocity) precedes the velocity, itself, or that a velocity (derivative of displacement) precedes the displacement, per se, the derivative of any function precedes the function, itself. Thus, if the response characteristics of the control element are proportional to the derivative of $u_o - u_r$, they can respond to the actual value of $u_o - u_r$ almost exactly while it is occurring. This is what makes a shock absorber, for example, effective. It responds not to displacement, but to velocity (rate-of-change-of-displacement), and so can generate a force to resist displacement *before* the latter actually occurs -- providing a "smooth ride" for the occupants of a vehicle.

We thus introduce into our feedback control system, *differential compensation*, characterized by an associated transfer function of the form,

$$\text{Derivative Control, } K_2{}'' = \frac{u_c}{\dfrac{\delta}{\delta t}(u_o - u_r)} \qquad \textbf{[6-15]}$$

Equation [6-15] defines what is called *reset* control, *lead compensation*, or derivative control, in the same sense that equation [6-1] defines proportional control. When u_c is proportional to the derivative of $u_o - u_r$, it will lead the latter, allowing the system to anticipate the magnitude or "strength" of the error signal. This not only prevents system response delays, but also damps the response characteristics of the system to prevent it from *over-reacting* if the rate-of-change of the error signal is too large. In other words, it represents dissipation of any tendency for the system to "run away".

INTEGRAL COMPENSATION

In the section on Stability of Feedback Control Systems, we also noted that following the stable transient response of the system, there always results some *steady-state error*, u_e. This is essentially due to the fact that we cannot normally wait an "infinite period of time" for u_o to eventually stabilize at *exactly* u_r -- especially when there is a continuous disturbance u_d upsetting the equilibrium of the system constantly. Well, if we could somehow record the *history* of u_d, or, more importantly, the history of the response of our system *to* the family of u_d's most likely to be encountered, it would be possible for us either to reduce considerably, or totally eliminate steady-state errors that result from long-term, continuous disturbances.

Mathematically speaking, the recording of history implies an integral over time -- just as "anticipation" implies a time-derivative. Thus, to prevent long-term offsets between u_o and u_r, we seek, in some sense, a transfer function of the form,

$$\text{Integral Control,}\ \ K_2{'''} = \frac{u_c}{\int (u_o - u_r)\delta t} \qquad \qquad \textbf{[6-16]}$$

Equation [6-16] defines what is called *lag compensation*, or integral control, which is responsible for driving the system back to its original set point if a constant disturbance or continuous error is present. When u_c is proportional to the *integral* of the error signal, it will lag the latter, just as displacement (integral of velocity) *lags* velocity, which, in turn, lags acceleration (of which velocity is the integral). That is, one must have a velocity *before* one can have a displacement, and one must have an acceleration *before* one can have a velocity. Lag compensation allows a feedback control system to maintain a certain inertia, which keeps it on a steady course despite constant disturbances. It represents a resistance to *change,* just as mass (e.g., a flywheel) does by storing kinetic energy of motion. In other words, the system "builds-up" a resistance to u_d.

A feedback control system that has good stability characteristics should thus also possess proportional, integral, and derivative control elements -- known as "PID-control", and the controlling signal, u_c will then satisfy a more general form of equation [6-3], i.e.,

$$u_c = K_2\,'(u_o - u_r) + K_2\,''\,\frac{\delta}{\delta t}\,(u_o - u_r) + K_2\,'''\int (u_o - u_r)\delta t \qquad [6\text{-}17]$$

where such a control signal is said to possess proportional characteristics with lag-lead compensation. The form of equation [6-17] is remarkably consistent with discussions in previous chapters which spoke of both electrical circuits or thermodynamic networks that contained resistance, (proportional control), capacitance, (derivative control), and/or inductance, (integral control) elements; and of the concepts of dissipation (resistance), charge (capacitance), and flux (inductance), as they relate to fundamental dimensions of state. Again, there appears to be a set of common denominators, or fundamental guiding philosophies that under-lie our physical world, in general, and our physiologic system, in partic-ular. Not unrelated to this is the generic form of any feedback control system, which may be put into a canonical pattern as summarized below.

The Canonical Form of Any Feedback Control System

From the brief discussions of this chapter, one may conclude that a spe-cific distinction of feedback control systems is that they all may be made to fit some canonical pattern, such as that shown in Figure 6-2. This scheme is "canonical" in the sense that it provides the simplest possible general description of the functional behavior of the system. Each path-way, each "signal", each box, and each loop may, in fact, contain several pathways (in series or in parallel), several signals, several subsets of

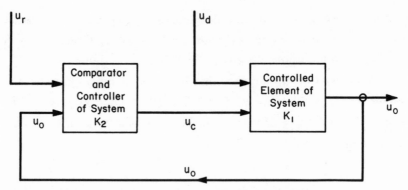

Figure 6-2 Canonical representation of a typical feedback control system.

interacting boxes, and several embedded loops, but the typical schematic configuration shown illustrates the ultimate manifestation of all of these into a basic input/output design.

The reference signal, u_r, is simply a standard to which the system attempts to conform -- or, a target output to which the system is commanded. The feedback signal, u_o, represents the output variable that is monitored in order to assess the performance of the system relative to u_r (and, perhaps, to *set* the standard, as described later). The signals u_o and u_r both enter a comparator, which is part of the controlling section of the feedback control system. Having an overall transfer function, K_2, this part of the system is where decisions are made, and where appropriate action is taken, depending on whether u_o is less than, bigger than, or equal to the reference signal, u_r, and how fast and how consistently u_o is changing relative to u_r.

If $u_o \neq u_r$ (perhaps due to some disturbance, u_d), then the comparator and controller of the servomechanism issues forth a control signal, u_c, in an attempt to alleviate this situation. The control signal will generally be related to the error signal, $u_o - u_r$, through an equation of the form [6-17], representing stable PID control. The basic assumption of feedback control theory is that the control signal can bring the controlled signal to within an acceptable range of the reference signal, depending on the value of the disturbing signal. That is, there is presumed to be some *envelope* of disturbances u_d within which the controller can regulate effectively and efficiently the controlled elements of the system. Should u_d fall outside of this envelope, then one or more of the controlled systems may become unstable, leading to a breakdown in one or more components of the systems, or to the intervention of some nonlinearity in the systems, or to a combination of both, as discussed previously.

In a sense, one may think of the control signal as having a *functional* dependence on the error signal, as defined, for example, by equation [6-17], and a *parametric* dependence on the disturbing signal (c.f., equation [6-21] later), to the extent that u_d determines the range of values of u_o within which the feedback control system will be in stable equilibrium. If u_o falls outside of this range, then u_c will not be effective in bringing u_o to within some acceptable neighborhood of u_r. The steady-state error of the system will then grow without limit, and it will eventually break down. Alternatively, some nonlinearity may intervene, or, the system may be reset to some new value of u_r to perhaps accommodate a new "window" for u_o in terms of u_d. More on this later.

The schematic representation of any feedback control system includes, of course, the controlled element, which is under the influence of both u_c and u_d, and which provides the monitored output signal, u_o. The latter is related to some linear or nonlinear combination of u_c and u_d through the overall transfer function, K_1, of the controlled element, i.e., $u_o = u_o(u_c, u_d)$, from which:

$$\delta u_o = \frac{\partial u_o}{\partial u_c}\, \delta u_c + \frac{\partial u_o}{\partial u_d}\, \delta u_d = K_1{}'\, \delta u_c + K_1{}''\, \delta u_d \qquad \text{[6-18]}$$

With these preliminary concepts in mind, the text which follows demonstrates how certain fundamental aspects of physiologic function may be made to fit this scheme for the purpose of mathematical formulation, modelling, and analysis.

Reference Signals in Physiologic Systems

"Floating" Set Points

The human organism establishes values of u_r for itself based first and foremost on its *metabolic* and *thermoregulatory* requirements (see Chapter 4). These are further fine-tuned to economize on the expenditure of energy that is necessary to meet those requirements (see Chapter 5). Within this framework, the body sets for itself reference quantities that constrain its various subsystems to operate within limits that insure stability (see General Principles of Feedback Control Theory in this chapter), competence, and safety, while achieving some desired behavioral pattern, some well-defined metabolic goal, or some specific task -- these being limited only by the envelope of physiologic performance capabilities. The levels of desired performance are based on millions of years worth of evolutionary experiences that have taught the organism what it needs to know, and *do* to survive.

Indeed, one might think of physiologic "reference quantities" or "set points" as being the actual manifestation of evolution, which teaches us that physiologic systems which operate in the most *fitting* way are those that will ultimately survive. "Fitting" translates into "desired performance" -- whether that performance be related to mass, energy and momentum *transport*, or mass, energy and momentum *utilization*, or *control*, or whatever -- and, in turn, "desired performance" translates into "reference quantities," u_r. Going one step further, the human organism is capable of *coding* these reference quantities into genetic material that provides repositories for the information that characterizes a certain species. This allows succeeding generations to inherit all of the standards of performance established through years and years of experience, so that the wheel is not being constantly reinvented in the physiologic sense.

Biological servomechanisms, or homeostatic feedback control systems do differ, however, in one important way from manmade, engineering-type servomechanisms, in that the desired performance (or reference quantities) to which the system attempts to conform do *not* necessarily remain absolutely constant for all time. Rather, one may think of physiological feedback control systems as having "floating" set-points (like a thermostat) that "float" in accordance with *changes* (on a moment-to-moment basis) in the metabolic or thermoregulatory requirements of the organism -- or, with a re-prioritization (on an organ-to-organ basis) of the metabolic requirements of certain specific tissues; or, on the state of evolution of the body -- or, on any of a number of other factors that may justify a change in u_r. These may include, for example, set-point changes called for by: (i) the persistence of a disturbance u_d, which may originate in the environment within which the body finds itself (see later), (ii) the value of the output, u_o, itself (as may occur in the case of fever), (iii) the state of health of the organism, (iv) an individual's desire to achieve some particular pattern of behavior, (v) the envelope of capability of the body in the performance of a specific task, (vi) the associated cost of achieving desired levels of performance, (vii) certain limiting safety and stability considerations, or (viii) a long list of other variables, many of which have perhaps not even been identified as yet.

The significant difference, then, between inanimate and animated systems is that output, u_o, in physiologic systems tends to be monitored *twice*: once by "higher" control centers (such as the brain) to *establish* the reference signal, u_r; and a second time by biological sensors, comparators and controlling systems to get u_o to *equal* u_r as closely and as stably as possible. In other words, looking at Figure 6-2, biological servomechanisms sense u_o in accordance with their needs, and then put in the proper reference quantity, u_r, *automatically* to satisfy those needs, which may change as often as on a moment-to-moment basis.

Some of the centers of the brain in humans that are responsible for establishing values for u_r are those that receive sensory information from the *special senses* (sight, hearing, smell, taste and touch), from *exteroceptors* that respond to the external environment (cutaneous pain sensors, temperature sensors, and sensors that monitor balance and equilibrium), and from *interoceptors* that respond to the internal environment (visceral sensors, baroreceptors, chemo-and-pH-receptors, proprioceptors, muscle spindles, Golgi tendon organs, kinesthetic receptors, strain-gauge-type ECF detectors, piezoelectric-type detectors such as bone, and so on -- see Chapter 1). Note how many of these are actually concerned with dimensions of perception, i.e., they relate to the body's ability to sense inertia (dimension, M; senses of balance and equilibrium); three-dimensional space (dimension, L; senses of sight and touch); thermodynamic events (dimension, T; thermal sensors); electromagnetic events (dimension, electric charge, q, or, for light, lumens, or for sound, pitch [frequency] and loudness [decibels]; senses of touch, sight, sound, pain); and time.

The brain centers involved include the subconscious or lower levels (reticular substance of the medulla, pons and mesencephalon; the cerebellum and basal ganglia; and the thalamus and hypothalamus), and the conscious or higher levels that have to do with memory, wakefulness and abstract processes of thought, i.e., the somesthetic areas of the cerebral cortex (Jacob, Francone, and Lossow, 1982). The sensory information involved originates from a variety of biological sensors, the transfer functions of which act to transduce the input into a corresponding action potential output. The latter, in turn, is transported via a neurological network of excitable tissue, which transfers the coded information as described more fully below. Thus, physiological information is digitally coded into and transported in the form of action potentials, the generation of which was addressed in Chapter 5, and the transmission of which is examined in great detail by Plonsey (1969), Deutsch and Micheli-Tzanakou (1987), and, to a much more abbreviated extent in the sections that follow.

Sensory Transduction of Information

SELECTIVE STIMULATION OF PHYSIOLOGIC RECEPTORS

In Chapter 5, we mentioned that the resting equilibrium situation established by the mechanisms discussed in Chapter 3 (pgs. 231-241) could be altered in several ways, each leading to the generation of an action potential. Among these ways are included a variety of different forms of energy, such as light, sound, heat, cold, pressure, and so on. A *receptor* is a cell, or part of a cell, which is specialized to generate electrical activity in response to being stimulated by very small changes in certain types of energy. In short, they are *transducers* which simply convert the particular energy to which they are sensitive into a corresponding action potential. Receptors are generally organized together with other types of cells to form a *sense organ,* such as are described in Chapter 1. The receptors in the eye, for instance, are the rods and cones of the retina, but the eye, itself a sense organ, also includes a lens, cornea, iris, and other structural units which are *not* excitable.

That form of energy to which a receptor is especially sensitive is called the *adequate stimulus* for the receptor. Some receptors may respond to more than one type of stimulus, but when this is the case they are usually expected to sense *primarily* only one of the stimuli, and to be excited only secondarily to the others. For instance, you see "stars" when you get hit by a blow to the eye because the increase in intraocular *pressure* caused by the impact induces the neurons in the retina to "fire" -- but pressure is not normally considered to be an *adequate* stimulus for the eye because a lot of pressure is necessary to fire these delicate retinal receptor cells, compared with their response to *light*. They are thus considered to be adequately or primarily light-sensitive, and only secondarily or incidentally pressure-sensitive.

Frequently, the receptor forms the end of a *neuron* or *nerve cell*. Some stimulus, such as a stretch or deformation that causes minute pores in the receptor membrane to widen; or a sound pressure pulse that accomplishes the same thing; or light which produces a chemical photosensitive response -- causes the permeability of the receptor membrane to sodium to increase, which depolarizes the cell ending. This depolarization is called the *receptor potential*. The receptor potential increases in strength in direct proportion to the intensity and rate of change of the disturbance (e.g., increasing stretch). If it gets large enough, action potentials are generated and transmitted through the afferent sensory nerve fibers attached to the receptors. Thus, the *rate* of firing of action potentials increases with corresponding increases in receptor potentials. The receptor potential -- unlike the all-or-none response characteristics of an action potential, however, is continuously *graded* to be directly proportional to both the intensity *and* the rate-of-change of intensity of the stimulus.

For example, located in the right and left internal carotid arteries are sinus stretch receptors that send, via the sinus Nerve of Hering feeding into the IX'th cranial (Glossopharyngeal) Nerve, pressure information to cardiovascular regulatory centers in the medulla of the brain. The firing frequency, f_B, of these so-called "baroreceptors" (baro = pressure) is proportional not only to the magnitude, p, of arterial pressure, but also to the rate-of-change with respect to time, $\frac{\delta p}{\delta t}$, of such pressure according to the relationship:

$$f_B = G_1 \frac{\delta p^+}{\delta t} + G_2 \frac{\delta p^-}{\delta t} + G_3 [p(t) - p_t], \qquad \text{[6-19]}$$

where p_t is the threshold pressure required to fire the receptor (around 40 mm Hg), G_1, G_2, and G_3 are sensitivity coefficients, and the superscripts $+$ and $-$ are used to designate rates of increase ($+$) or decrease ($-$) in pressure. Since these have associated with them different sensitivity coefficients, G_1 and G_2, respectively, the reader will note that the sinus receptors respond differently to rates of pressure *increase* than they do to pressure *decrease* (a hysteresis effect). Furthermore, G_1, G_2, and G_3, as well as p_t can vary with the level of mean pressure in the artery, and f_B tends to be higher under cyclic loading than it is under steady-state loading. Baroreceptors are also located in the aortic arch, at the bifurcation of the brachiocephalic artery into the right-subclavian and right common carotid arteries, and at the entrance to the left common carotid artery. These are connected to the medulla by the Xth Cranial Nerve (the Vagus), which receives its information from the Aortic Depressor Nerve.

Both the Aortic and Carotid Baroreceptor Reflex Systems work primarily through the arteriolar system to regulate peripheral resistance, (see Chapter 2, equation [2-10]) and hence, blood pressure. They appear to respond to blood pressure up to a maximum of about 200 mm Hg, beyond which the extreme pressure damages these sensitive transducers and thereby makes them nonfunctional.

FACILITATION AND ADAPTATION

Each time a particular sensory signal passes through a sequence of pathways, these pathways become more capable of transmitting the *same* signal the next time it comes through -- a process called *facilitation*, or, "memory of sensation". One may think of this as "programmed learning",

akin to programming a computer to do a specific task. Facilitation ultimately allows the system to *adapt*, in both a long-term and short-term
sense, to stay alive, as described later.

If a *constant* stimulus is applied and maintained, the receptor potential (and consequently the rate of firing of action potentials) will
gradually decrease. This decrease in response with prolonged application
of the stimulus is called *adaptation*. Some receptors adapt very rapidly,
and are consequently sensitive almost entirely to *changes* in stimulus (i.e.,
they are predominantly rate-sensitive, or so-called *phasic-types* of
transducers). Others adapt hardly at all, and all gradations in-between
are seen (i.e., these are *amplitude-sensitive*, or so-called *tonic-types* of
transducers). In phasic-type transducers, G_1 and G_2 tend to dominate
over G_3. Examples include both cold and warm cutaneous temperature
receptors (Milsum, 1966). In tonic-type transducers, G_3 tends to dominate over G_1 and G_2. Examples include some receptors concerned with the
maintenance of balance and equilibrium (Schneck, D. J., 1984b).

Adaptation needs to be examined from two points of view. On the
one hand, it is the process whereby, if a *constant* (or perpetual) stimulus
is applied to a receptor, the receptor potential will gradually decrease, so
that its response will require a stimulus of gradually increasing intensity.
In other words, *sensory inputs* to the physiologic system develop response
thresholds that depend on how persistent the disturbance is, such that if
it becomes constant, the system essentially "ignores" it, and a greater
stimulus is required to elicit a response. One may view this as the "boy
who cried wolf" scenario.

On the other hand, facilitation and adaptation in many instances
may gradually develop into conditioned reflexes, and may even reach the
point wherein *no* sensory input is required anymore to elicit a particular
response. Taking this even one step further, the conditioned reflex soon
may evolve into a newly defined set point for u_o. In this case, the process
of biological adaptation can therefore be modelled mathematically by
noting that it is nothing more than the process of changing the set-points
of a living system as necessary (based on experience, "programming" or

any of the other parameters listed earlier), to allow it to continue functioning as economically as possible. These set-points, for various subsystems of the body, may also be functions of *each other*, in an integrative sense, to insure that the *whole* organism functions in a synchronized, directed fashion. One may view this aspect of adaptation as the "Pavlov's Dogs" scenario, which illustrates the idea that *motor outputs* in the physiologic system develop response thresholds that also depend on how persistent the disturbance is. Perhaps the two points of view presented above can best be reconciled by example.

PHYSIOLOGIC CONTROL OF ERYTHROPOIESIS

Consider the physiologic control of red blood cell production (known as "erythropoiesis"). Recall from Chapter 2 that the primary function of red blood cells (erythrocytes) is to transport oxygen to all the cells of the body; a function accomplished by the erythrocyte's carrier molecule, hemoglobin. Red blood cells are produced in bone marrow, their rate of production being controlled by a plasma stimulant called erythropoietin (E), which, in turn, is activated by an "erythropoietic factor" (R), secreted from the kidneys in response to anoxia (low oxygen concentration in the blood).

The effect of erythropoietin on bone marrow tissue further illustrates the concept introduced in Chapter 1 of *selective stimulation* of organs and tissues as a result of their having specific receptors, or "loading docks", if you will, that give the stimulant in question access to the target organ or tissue involved. In Chapter 1 we mentioned this as it relates to alpha-and-beta-sites on target organs stimulated by the secretions of the sympathetic nervous system. In the case in point here, the "target organ" involved is bone marrow tissue; specifically, *stem cells* and *hemocytoblasts* (committed stem cells), which are precursors of erythrocytes and which have receptor sites that are specifically compatible with the alpha-globulin hormone *E* (Guyton, 1981). The significance of this concept of organ receptor sites within the present context is that it, too, participates in the process of adaptation, as we shall see later.

At sea level, where the partial pressure of oxygen in moist alveolar air is about 100-103 mm Hg (c.f., Chapter 2, page 157), red blood cell hemoglobin molecules are nearly (97.5%) completely saturated with oxygen as they pass through the lungs, and so the organism's need for oxygen (some 20 mℓ O_2/dℓ blood) can be met satisfactorily by maintaining the number of cells/mm^3 of whole blood at about 5 million (ibid.). Since the cells have an average volume (mean corpuscular volume, MCV) of 90 cubic microns, 5 million cells/mm^3 of blood occupy some 45% by volume (the so-called hematocrit, Hct) of the fluid. Furthermore, since each blood cell contains some 30×10^{-12} grams of hemoglobin (mean corpuscular hemoglobin, MCH), and since the average individual contains a total blood volume (V_B *) of approximately 5 liters, at *sea level* the total need for hemoglobin is around 750 grams. This can be calculated from the equation:

$$\text{total hemoglobin, } TH = \frac{\text{MCH}}{\text{MCV}} \times \frac{\text{Hct}}{100} \times V_B * \qquad \textbf{[6-20]}$$

In terms of Figure 6-2, one may define, for erythropoiesis: $u_o = $ *actual* blood hematocrit, at any instant of time; $u_r = $ *desired* blood hematocrit (at *sea level,* some 45%); $u_c = $ circulating level of plasma erythropoietin, E, an alpha-globulin hormone whose origin can be traced to the kidney and whose target tissue is bone marrow; $K_2 = $ the kidney; and $K_1 = $ the bone marrow. Disturbing signals, u_d in this case would be normal daily red blood cell destruction rate of about 1% (2 - 10 million cells per second), or hemorrhaging (blood loss) due to some accident, or perhaps some anemic pathologic condition, or anything else that would cause u_o to deviate from u_r, giving rise to an error signal, $u_o - u_r$. The adequate stimulus, then, is anything that results in *anoxia,* which is sensed by the kidney. It is thus a chemical stimulus, making the kidney, in a sense, a *chemo-receptor.* The key point, however, is that the *reference signal,* $u_r = 45\%$ does *not* change with u_d at sea level because the disturbance lies *not* in the hematocrit *required* to meet the metabolic needs of the organism, but in the *maintenance* of that hematocrit in the face of upsetting influences. In other words, nothing acts to change any of the variables given in equation [6-20] because hemorrhaging is usually an *acute* (as opposed to a persistent, or *chronic*) phenomenon; daily destruction of red blood cells and removal from the circulation by the Spleen is already taken into account in the establishment of u_r; anemia and other pathologic conditions fall into the category of placing u_d beyond the range where u_c can effectively correct for the disturbance; and the oxygen concentration at sea level remains relatively constant. Likewise, K_2 and K_1 do not change in response to u_d, and u_c stays generally at some optimum operating level.

Consider now what happens if the kidney senses *prolonged anoxia* (i.e., $u_o = 45\%$ is not now sufficient to satisfy the needs of the organism), generated, for example, by *continuous* exposure to an environment associated with high altitudes, where the oxygen concentration remains consistently low. Acting on the kidney and brain, as well as the bone marrow, this environmental disturbance now creates a situation wherein the *reference quantity, u_r* = Hct = 45% is no longer adequate to satisfy the metabolic needs of the body, i.e., u_r now *depends parametrically* on u_d, which is sensed as a condition of persistent anoxia. Responding to this sustained lack of oxygen, the kidney gain, K_2 drives u_c (renal erythropoietic factor, R, and consequently, erythropoietin, E) higher, increasing red cell production. All the while that this is happening, u_r begins to drift slowly and asymptotically (depending on the individual, over a period from several weeks to years) towards a new value that will eventually provide a higher hematocrit to carry sufficient oxygen to all of the cells of the organism.

Furthermore, the hypothalamus of the brain, acting through the pituitary gland (anti-diuretic hormone, ADH) and, ultimately, the kidney, again (vasopressin hormone), drives the reference value for total blood volume, V_B*, higher, as well. Lastly, MCH is increased slightly -- all three of these, i.e., MCH, Hct, and V_B*, acting to increase *TH*, in accordance with equation [6-20]. In fact, at prolonged exposure to an altitude of 22,500 feet, where arterial oxygen saturation drops to only 50% as air pressure drops from 760 mm Hg at sea level to 349 mm Hg, MCH increases $1\frac{2}{3}$% to about 30.5×10^{-12} grams, Hct increases $44\frac{1}{2}$% to 65% and V_B* increases 25% to 6.25 liters, leading to an increase in TH of nearly 84%. All of this can be modelled mathematically (see, for example, Mylrea and Abbrecht, 1973, Goldstick, 1973, and Guyton, 1981) because: (a) the functional relationship between alveolar oxygen partial pressure and altitude is known; (b) the functional relationship between alveolar oxygen partial pressure and arterial hemoglobin-oxygen saturation is known; (c) the functional relationship between hemoglobin-oxygen saturation and kidney release of R is known; (d) the functional relationship between plasma erythropoietin concentration and R is known; (e) the functional relationship between plasma erythropoietin concentration and bone marrow production of red blood cells is known; and, (f) so, too, are known functional relationships between total blood volume and vasopressin concentration, vasopressin concentration and concentration of ADH, and concentration of ADH vs. hypothalamic activity.

The important point to be emphasized is that these changes in u_r are long-term adaptations to sustained failures of u_o to meet the *needs* of the organism, as perceived by a chronic anoxia. In other words, u_r is *set* by sensing u_o (adaptation), which is then sensed again to keep it at u_r through u_c in the face of u_d. The adaptation is gradual. In the situation depicted above, for example, virtually no effect is observed for some 2-3 weeks following continuous exposure to a high-altitude environment. About half the effects begin to be manifest after about a month, and it takes several months before the effects become fully developed. Then, interestingly, over a further period of time, the E levels in plasma *drop* back to their normal sea-level values, but the bone marrow *continues* to manufacture red blood cells at the *new* (higher-altitude) Hct value. That is to say, the kidney gain, K_2, drops back to its sea-level value, while the bone marrow gain, K_1, goes up, and stays at its high-altitude value -- so the stimulus, E, is no longer necessary to elicit the response (an increase in u_o). One might say that the *change in* K_1 reduces (or eliminates entirely) the organism's *need to respond* (via K_2) to the sustained anoxia. This is the "Pavlov's Dogs" scenario mentioned earlier, which can be explained as follows.

Under prolonged anoxia, such as relocation to a high-altitude geographic environment, or in the case of residents *native* to that environment, one finds characteristically a greater concentration (number per unit area) of Erythropoietin *receptor sites* ("docks") on the bone-marrow tissue of these individuals, than would normally be the case for individuals residing at sea level. In other words, the adaptation process provides for this tissue to be even more receptive to E (u_c) than it would be otherwise, so that the *same amount* of sea-level hormone produces a correspondingly *greater* response (K_1 increases) because it has more ways to gain access to its target organ. By this same mechanism, these individuals can survive equally well with *lower* E-levels in their blood than you or I could at considerably *higher* levels because, again, their target organs are more *receptive* to the hormone. In effect, one might think of controlling the number of receptor sites on a target organ as a means of *biasing* the sensitivity of that organ to the respective hormone or stimulus involved; and *increasing the bias* as the process of adaptation and selective tissue stimulation as it applies to the target tissues of the erythrocyte-control system, i.e.:

> "If a persistent stimulus is applied to organ receptors, that organ eventually develops more receptor sites, so that it becomes more responsive to hormone stimulation, requiring less transmitter substance to elicit a correspondingly greater effect. This is a conditioned reflex."

Now, the other side of the coin is that there is a limit as to how many receptor sites the tissue can accommodate, just as there is also a limit to how much E the organism can manufacture. When either one of these reaches the point of saturation, *no* amount of stimulation, regardless of how intense, can elicit a further response. As saturation levels are approached, it takes progressively stronger and stronger stimuli to generate an adaptation response. Moreover, if persistent disturbances are slow and deliberate, neurological accommodation phenomena (see Chapter 5) can interfere with the response of the tissue, as can refractory effects (ibid.) when the continuous disturbance occurs at a very rapid frequency. All of these events -- receptor site and hormone manufacture limitations, accommodation, refractory states, and saturation -- lead to a situation wherein it takes *more and more* of a stimulus to *elicit* an adaptation *response* (the "boy who cried wolf" scenario), but virtually *no* stimulus once the response is in place (the "Pavlov's dogs" scenario). One might say in this case that there is a progressive reduction in the *response to the need,* i.e., in the ability of the organism to change u_r as easily anymore. The two points of view related to adaptation can thus be summarized by the following statement:

"Adaptation acts both to reduce the organism's *need* to respond to a stimulus, and, to reduce the organism's *response* to that need."

Taking this analysis to its logical limit, the *reason* one finds more receptor sites in the bone marrow tissue of individuals native to high-altitude environments is that their genes are *coded* for it. That is, over an extended period of time, all of the adaptive changes discussed above eventually generate a new coding in genetic material, such that subsequent generations of high-altitude natives automatically possess greater quantities of hemoglobin, more efficient hemoglobin-oxygen dissociation curve characteristics, greatly expanded pulmonary capillary systems, larger hearts (especially the right side that feeds the pulmonary circulation) and a higher ratio of ventilatory capacity to body mass (increased chest size coupled with a somewhat decreased body size) to increase metabolic efficiency.

The hierarchy of events, then, which lead to the establishment of reference set-points, u_r, may be viewed within the context of evolutionary theory as proceeding from the idea of "floating" target values for homeostatic regulators, to facilitation, to adaptation, and, ultimately, to

genetic coding. Physiologic examples of floating set-points and adaptation are numerous, and the literature on adaptive control systems is vast
(see, for example, Guyton, 1981, Schneck, 1984b, Apter, 1970, Milsum,
1970, Milhorn, Jr., 1966, Brown and Gann, 1973, and Huffaker, 1974).

Is the Establishment of u_r Reversible?

The concepts of floating set-points, facilitation, programmed learning,
adaptation, conditioned reflexes, and genetic coding may also be factors
in developing one's self-image and attitudes towards life. Thus, for example, programmed (conditioned) negativism, which begins very early in
life with such words as, "no", or "do not", or "don't touch", or "you can't",
or "you must not", or "that's stupid", and so on, may be viewed as a
conditioned response to persistent *negative* disturbances to the psyche.
These very quickly establish self-image, self-belief or self-esteem set-
points that *expect* defeat, and, therefore, "will" it to happen whether an
individual realizes it, or not. There is an underlying truth in the concept
that our bodies can be programmed, by constant reinforcement of certain
attitudes and perceptions, to make these self-fulfilling. Brainwashing is a
perfect example of this. But in a less-dramatic, though more-everyday
sense, we are constantly being brainwashed to think the worst, expect the
worst, hope for the best (almost *against* hope), but plan for disasters!
Those disasters are almost sure to come because we *virtually plan* (i.e., *set*
u_r) it that way. Expect the worst (u_r), and you will *get* the worst (u_o). This
is programming, conditioning and establishing negative emotional set-
points (feelings) that are sure to generate corresponding behavioral patterns and actions.

But all is not lost. If one accepts the idea of programmed learning
-- even to the extent that it may be genetically coded -- then one can also
accept the corollary of "deprogramming" which can *change* the set-points
with *positive* reinforcement as a persistent input, rather than negative as
a persistent "disturbance". At any stage in life, one can effectively "erase"
the old disk, and program in a new one using the *same* principles of re-

petition, conditioning and adaptation.   The relatively new science of biofeedback, as introduced briefly in Chapter 5 may offer a means for doing this, as may some other new approaches termed, "self-talk", or "transcendental meditation", or "the relaxation response", or "dyanetics", or "autohypnosis" and so on. Indeed, with a fundamental understanding of physiologic function as it relates to the establishment of u_r, humankind could embark on a campaign to ultimately genetically code a "kinder, gentler, more positive" generation of compassionate, caring individuals. The only prerequisite is really the *desire* to do so, and a further comprehension of physiologic information as it is coded and transported via the "body language" of action potentials. These are discussed further below.

Feedback Signals in Physiologic Systems

Thinking in terms of information transport in feedback control systems, one may view reference signals as being coded temporarily into conditioned or adaptive responses, and permanently into genetic repositories. Similarly, one may view feedback signals, u_o or u_p, as being coded into electromagnetic sensory information that may be termed, "body language" (not to be confused with the more common physical implications of such terminology). Sensory information is generated as described previously by specialized transducers that monitor the controlled elements of physiologic systems, or the internal and external environments of the body, or both. The "body language" generated is simply based on biological signals that are transmitted as action potentials, such signals providing a communications network that continuously tells the organism what its current condition is, so that this condition may be compared to the desired goal and appropriate action taken accordingly.   The physiologic functional unit of information transport is the *neuron*, or nerve cell, the anatomy and physiology of which we now examine briefly. A more comprehensive treatise on the subject from the physiologic point of view may be found in Guyton (1981), and from the engineering point of view, in Deutsch and Micheli-Tzanakou (1987).

Anatomy and Physiology of a Neuron

All neurons have four basic components, illustrated in Figure 6-3. These are:

1) One or more *input elements,* called *dendrites.* In the generation of an action potential, these branched, protoplasmic processes of nerve tissue are depolarized first, and conduct impulses from the "receptor zone" of the neuron to the cell body, or soma. Thus, action potentials originate at the dendrites, which may receive information from a sense organ or from another nerve, and which may be thought of as the *data acquisition* elements of the information transport system.

2) An integrative cell body, or *data processing* and analysis element called a *soma.* The soma includes the nucleus, or CPU of the nerve cell. Here, all inputs are received, processed, stored (if necessary), and evaluated before being transmitted further to the attached axon. There are some 20×10^9 nerve-cell bodies in an average individual, representing 0.02% of all the cells in the organism.

3) An active transmission line, or *Axon,* sometimes called a *fiber.* Axons are long slender extensions of the neuron. They may be as short as a few thousandths of an inch, to as long as six feet or more, and they frequently branch extensively. In vertebrates, all axons larger than about 10^{-3} mm (1 micron) in diameter have an insulation covering in the form of a myelin sheath surrounded by Schwann Cells. They are called *myelinated nerve fibers.* Myelin is actually many layers of fused cell membranes which cover the membrane of an axon, decreasing the effective membrane capacity, C^*_m, and increasing its resistance, R^*_m. The latter effect, however, is more than offset by the fact that axon resistance is inversely proportional to axon diameter, so that the large diameters of myelinated nerve fibers causes them to have a smaller *net* resistance *and* capacitance than their much smaller unmyelinated counterparts. At regular intervals (about every millimeter) the myelin sheath is interrupted, and tiny gaps called *Nodes of Ranvier* expose portions of bare membrane of the axon to interstitial fluid. Transmission of an action potential thus

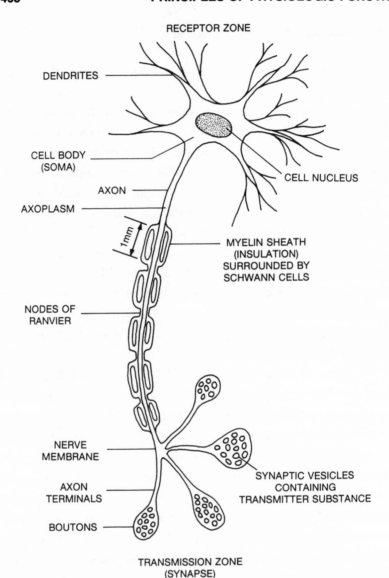

Figure 6-3 Schematic illustration showing the essential features of a "typical" myelinated nerve cell.

"jumps" from node to node along the axon, in a process called *Saltatory Conduction.*

The effect of myelination is to increase the conduction velocity through nerve fibers. Since the membrane capacity is much less, it takes less time to charge it up; and since the net resistance of the larger diameter axon is much smaller, it presents less of an obstacle to the propagation of action potentials through it, i.e., the larger the fiber diameter, the greater its conduction velocity. An action potential is conducted down an axon at a constant velocity, so long as the axon maintains a constant fiber diameter (which it usually does). Furthermore, neither the shape, nor the size of the action potential change as it is conducted down a single axon.

4) And finally, the typical "generic" neuron contains one or more output elements called *Terminals,* the ends of which flare out into bulb-like expansions called *boutons* (or boutons terminaux). These are bulbous endings that contain synaptic vesicles, within which are stored chemical transmitter substances, such as *Acetylcholine* (Ach). More about that later.

A group of axons wrapped in a connective-tissue cover called the *Endoneurium* constitutes a *nerve bundle.* Several of these, encapsulated by a laminated cover called the *Perineurium* comprise a *nerve trunk;* and nerve trunks are bunched together and wrapped with still another loose connective tissue sheath called the *Epineurium* to form the basic Nerve of the peripheral nervous system. Note that any given nerve contains a *mixture* of axon types, each having a different conduction velocity, or Fourier component. Thus, whereas neither the shape, nor the size of the action potential change as it propagates down a *single* axon, the *compound action potential* wave shape of a *nerve does* change with distance from the stimulation site. The compound action potential, which is what one generally measures in the laboratory, thus exhibits a phenomenon known as *wave dispersion,* which is caused by Fourier components of the composite wave travelling with different velocities. Since they will be at different locations along the nerve at different times, the composite wave shape will correspondingly look different at different points along the nerve.

In the earliest voltage amplitude vs. time recordings of a compound action potential (measured at a particular point along the nerve) a convention was established wherein successive amplitude peak components (wave elevations) were labelled with the Greek letters, $\alpha, \beta, \gamma, \delta$, and, sometimes, ε. Since these elevations correspond to activity in individual fibers conducting at different velocities, which, in turn, is directly proportional to fiber diameter, the graphical Greek-letter designation has also been adopted as a categorization of fiber diameters: α corresponding to a propagation speed in the range from 70 to 130 meters per second, characteristic of myelinated nerve fibers having diameters ranging from 12 to 22 microns; β representing 6 to 12 micron-diameter myelinated nerves conducting at 30 to 80 m/sec; γ designating 3 to 7 micron-diameter myelinated nerves along which action potentials travel at speeds ranging from 10 to 50 m/sec; δ being used to define 1 to 5 micron-diameter myelinated or unmyelinated nerve fibers which conduct impulses at the rate of 0.7 to 30 m/sec; and ε being reserved to categorize any unmyelinated nerve fiber having a diameter less than about 1 micron, along which action potentials propagate at speeds less than around 6 meters per second.

More intense systematic examination of the compound action potentials of various nerves of different composition reveals that axons can be further classified into four distinctive types, known as A, B, s.C and d.r.C fibers. A-fibers are myelinated, somatic, afferent and efferent neurons; B-fibers are thinly myelinated, efferent, preganglionic axons found in autonomic nerves; C-fibers are unmyelinated -- the s.C group being the efferent postganglionic sympathetic axons (hence the prefix s before the C), and the d.r.C group being the small unmyelinated afferent axons found in peripheral nerves and dorsal roots (hence the prefix d.r. before the C). The properties of these fiber types are summarized in Table 6-I. Note that the A-fibers are further subdivided based on diameter and conduction velocity into type I (70 - 130 m/sec propagation speed,

12 - 22 micron-diameter axon, corresponding approximately to the A-α-type); type II (30 - 80 m/sec propagation speed, 6 - 12 micron-diameter axon, corresponding approximately to the A-β-type); type III (5 - 50 m/sec propagation speed, 1 - 6 micron-diameter axon, corresponding approximately to the A-γ-and-δ-myelinated-type); and type IV (usually including all of the C-fibers, both d.r.C and s.C, unmyelinated subgroupings). This unfortunate hierarchy of interchangeable Greek-letter and Roman-numeral designations can sometimes be quite confusing! Such is life, however, when one tries to maintain custom and tradition while progressing into the future.

Note also in Table 6-I the *inverse* relationship between conduction velocity and action potential threshold, the more rapidly conducting axons being more easily excited than the slower ones. In general, the faster axons are associated with *motor* functions (emanating *out* from the central nervous system), and the slower ones with *sensory* functions (generally propagating *in* towards the central nervous system) -- although this is not a hard and fast rule. Moreover, as a compound nerve trunk is stimulated progressively from its lowest threshold axon to its highest threshold axon (maximal intensities), the successive components, α, β, γ, ... , appear in the compound action potential recording *in order,* thus allowing one to identify *which* fiber is responding to *which* stimulation intensity. This type of coding allows the organism to discriminate the *intensity* of the stimulus on the basis of *which* neuron is firing at any given time (c.f., Chapter 5). Finally, although each individual nerve cell, or neuron, follows the "all-or-none" response law for the generation of an action potential, the fact that not *all* the cells of any given nerve are necessarily firing at the same *time,* or responding to the same *stimulus,* allows one to elicit a *graded intensity* response from that compound nerve. Thus, the information coded into action potentials is based on *which* neuron is responding to *what stimulus* (firing threshold), at *what frequency,* and conducted at *what speed.* Moreover, such sensory body language may actually include several "dialects", or even consist of many *different* languages, as is illustrated in Table 6-II, and discussed further below.

Table 6-I

Axon Types

Fiber Type And Diameter (Microns)	A (Myelinated) 1 - 22				B (Thinly Myelinated)	C or IV (Unmyelinated) 0.3 - 1.3			
	I (Muscles) 12 - 22 (α)	II (Somatic) 6 - 12 (β)	III (Motor) 1 - 6		≤ 3.0	s.C (Sympathetic) 0.3 - 1.3		d.r.C (Dorsal Roots) 0.4 - 1.2	
			3 - 7 (γ)	1 - 5 (δ)		1 - 1.3 (δ)	0.3 - 1.0 (ϵ)	1 - 1.2 (δ)	0.4 - 1.0 (ϵ)
Conduction v (m/sec)	70 - 130	30 - 80	10 - 50	5 - 30	3 - 15	0.7 - 2.3	0.7 - 2.3	0.6 - 2	0.6 - 2.0
Action Potential Spike (msec)	0.4 - 0.5				1.2	2.0			
R^*_m	50,000 to 50 Megaohms per cm.				30,000	180,000 to 9,000 Megaohms per cm.			
C^*_m	8.9 Picofarads per cm.				$\simeq 200$	80 to 400 Picofarads per cm.			
Absolute Refractory τ (m sec)	0.4 - 1.0				1.2	2.0			
Activation to Threshold (mv)	20 - 30				40	50			
τ^*_m	445 to 0.445 msec per cm^2				6	14.4 to 3.6 seconds per cm^2			

Sensory Information Dialects (Syntax)

Body language -- action potentials -- feedback control signals -- may indeed contain the key to understanding the very basis of physiologic function and control. We know, for example, that heart muscle tissue "obeys" commands issued from its "pacing" control center, which is the sino-atrial (SA) node. The SA-node "speaks" in the language of *electrocardiosyntax* (i.e., what we measure as an electrocardiogram) and the heart muscle, "understanding" this language, responds. And the effectiveness of the results obtained by electrocardiosyntax is monitored in its transduced form -- which is blood pressure -- by baroreceptors placed at strategic sites (such as the carotid artery and the aorta, as previously mentioned) in the vascular system. These baroreceptors send feedback signals (electro-baro-syntax, see Table 6-II) to higher control centers in the brain, where comparisons are made between "actual" blood pressure (u_o) and "desired" blood pressure (u_r). This is the language of the baroreceptor reflex arc that ultimately attempts to regulate blood pressure.

Going one step further, we also know that the human eye does not see -- the brain does; and the human ear does not hear -- the brain does; and the human nose does not smell -- the brain does! Indeed, all of our "special" senses or sense organs are nothing more than transducers that merely convert one form of energy (e.g., light or sound) into electrochemical energy (action potentials). The latter is transmitted via respective nerve pathways to appropriate CPU centers in the brain, where the incoming "language" or syntax is decoded and translated into a corresponding image, or sound, or taste, or smell, or feeling, or whatever. In other words, our sense organs and various other extremities may be viewed as "peripheral devices" for the CPU, each designed to provide input (*which* nerve, at *what* frequency, and so on) information for subsequent interpretation.

Imagine, now, that *we* could "speak" electroretinosyntax, or, more specifically, that we could design and build a transducer that has precisely

Table 6-II

Some Examples Of The Different Dialects
Associated With Sensory Body Language

1. Electroencephalosyntax (Brain Waves)	11. Proprioceptive-syntax (Posture & Locomotion)
2. Electroretinosyntax (Sight, Vision)	12. Electrothermal Syntax (Heat, Temperature)
3. Electroaudiosyntax (Sound, Audition)	13. Electrochemosyntax (Biochemical Reactions)
4. Electrosalivosyntax (Taste, Gustation)	14. Electro-kinesthetic-syntax (Joint Position & Orientation
5. Electrotactilesyntax (Touch, Tactual)	15. Muscle Spindle Electrosyntax (Stretch, Strain)
6. Electrobarosyntax (Pressure)	16. Golgi Tendon Organ Syntax (Force, Stress Perception)
7. Electro-olfactosyntax (Smell, Olfaction)	17. Visceral Sensorisyntax (ECF and Stretch Perception)
8. Electrovestibulosyntax (Balance & Equilibrium)	18. Somesthetic Syntax (Consciously Perceived Sensations)
9. Electrocochleosyntax (Hearing and Balance)	19. Extero-and-Intero-Syntax
10. Electro-axono-syntax (Action Potentials)	20. Electronystagmosyntax (Eyeball Movements)

the same input/output characteristics as the human optical transducer, which is the eye. Then, if we could properly interface such a peripheral device with the optic nerve to form a continuous pathway to the brain, our synthetic "eye" would be capable of "speaking" to the visual centers of the brain in a language they are designed to understand. As far as these centers are concerned, then, they would be "seeing" just as if the *natural* eye were transmitting the signals instead of the *artificial* one. The same argument could be presented for the development of an artificial ear, or for the development of an artificial limb that is operated by the same alpha-motoneurons that operated the corresponding natural limb, or for the inclusion of sensory feedback information (i.e., a sense of "touch") in artificial limbs -- or, for the development of *any* prosthetic device that is to be used to replace a part of the human body. If we could only understand the proper body *language* that is native to the part we are replacing, then we could, indeed, replace the exact *function* of the part, as well as its *structure*.

The above reasoning is not as abstract as it may sound at first. Biological signals, from an engineering point of view, are remarkably reproducible (Plonsey, 1969, Bendat and Piersol, 1971, De Coulon, 1986, Deutsch and Micheli-Tzanakou, 1987). This means that electromagnetic recordings of physiological action potentials can be subjected to sophisticated techniques of biological signal analysis (see later in this chapter) in order to "decode" (hence, gain the ability to reproduce) the feedback language of the human feedback control systems. Once understood, it becomes a relatively straightforward matter to incorporate these languages as software integral to the microprocessor hardware of prosthetic devices, thus moving us closer to being able to communicate with the organism. All that is left, then, is to develop techniques for interfacing synthetic devices with physiologic tissue (biocompatibility, biomaterials), and for *powering* them, and the process of organ-replacement is complete. When that day finally arrives -- and it is not all that far away when viewed in terms of modern-day technological accomplishments (see, for example, Sayers, 1970) -- what an exciting time it will be!

One area in which the decoding of biological signals has been very successful is the development of upper-extremity myoelectric prosthetic devices (Childress, et al., 1974). Another relates to the piezoelectric functional properties (constitutive modelling) of bone tissue, which has paved the way for the corresponding treatment of bone ailments through electromagnetic induction (Brighton, 1977). Progress towards being able to attach synthetic nerves to physiologic ones is proceeding along the lines of using metal (Na, K, Ca) monomolecular-layers-coated sutures to bind the living with the artificial (Colmano, et al., 1983). Hardly a day goes by that one does not hear or read about major technological breakthroughs that bridge the gap between science fiction and reality, thanks in great part to the work of biomedical engineers. But in the interest of space, it is best to move on.

Before doing so, however, one more important point should be re-emphasized, and that has to do with the role of *biofeedback* in control systems theory. Recall from Chapter 5 that the intent of biofeedback is to make us aware of what it "feels" like when such control is manifest, and thus, by bringing to the level of consciousness an awareness of such feelings, give us the ability to affect control over bodily processes that heretofore have traditionally been considered to be uncontrollable at the conscious level. This brings us to the subject of comparator and controlling sections of biological feedback control systems or servomechanisms.

Control Signals in Physiologic Systems

Homeostatic Regulators

Comparator and controlling systems of biological servomechanisms are called homeostatic regulators. Recall that homeostasis is the process through which variables critical to life, and others not so critical, are maintained within prescribed limits, where, as reasoned earlier, these limits may be based on anatomical constraints, on the laws of physics, on associated "cost functions," and so on.

Thus, a homeostatic regulator will first *compare* the achieved performance, u_o, of a physiologic subsystem with the desired performance, u_r, of that system and, within certain restrictions, *decide* upon and *execute* the best course of action in order to bring the two into correspondence. Of course, if $u_o = u_r$ (i.e., the error signal, $u_o - u_r$ to which the regulator is responding in accordance with equation [6-17], is zero), then everything is fine and nothing happens. But rarely, if ever, is this the case, for we live in an environment that is constantly challenging the equilibrium of our physiologic systems; and if the environment is not doing it, we do it ourselves by the very life styles that we lead. Therefore, the more likely situation is one in which homeostatic regulators are continually making corrections (homeokinetics is a descriptive term that has been coined to describe this) in order to bring u_o to within an acceptable range of u_r, when the equilibrium of the system is upset by some disturbance, u_d.

When $u_o \neq u_r$, then the homeostatic regulator issues forth a control signal described by equation [6-17] in an attempt to alleviate this situation, as discussed earlier. In the physiologic context, the guiding principle is that there is some normal "envelope" of u_d within which homeostatic regulators can function effectively to keep the various subsystems of the human body in states of stable equilibrium. Should the disturbing signals fall outside of this envelope, then one or more of these systems may become unstable and fail. In real life, instabilities of a feedback control system -- i.e., those situations when there is no associated decay time for u_c, with a corresponding stable overshoot and phase relationship -- manifest themselves in the breakdown of one or more components of the system, or the intervention of some nonlinearity ($K_1, K_2, \ldots$, become variable), or a combination of both. Thus, one might look for such malfunctions in physiologic subsystems that become unstable, and, these malfunctions may become manifest in what we have come to call *disease*, or *pathology*, or *allergic reactions*, or other forms of substandard physiologic performance.

One theory of disease, then, is that it represents an inability of the body to cope with or handle disturbances that fall outside the operating ranges of its homeostatic regulators -- or that insult these regulators too frequently or too persistently, in both space and time -- such that the system cannot adapt quickly enough to prevent damage. Furthermore, these regulator operating ranges may not have fixed limits. Rather, they may depend on an individual's lifestyle, emotional state, age, physical condition, diet, and environmental factors such as climate. Moreover, the operating range of any one homeostatic regulator may be a function of the operating range of other regulators in the body, since they undoubtedly influence one another as they all act in harmony to maintain stable equilibrium.

As also addressed earlier, control signals, while having a *functional* dependence on the error signal, may also have a *parametric* dependence on the disturbing signal. In other words, u_d may determine the range of values of u_o within which the feedback control system will be in stable equilibrium. If u_o falls outside of this range, then the steady-state error of the system will grow essentially without limit, and it will eventually break down (or get "sick"). Alternatively, the system may attempt to *adapt* (evolve) to change u_r in order to accommodate a new "window" for u_o in terms of u_d. In that case, one could write:

$$u_c = u_c(u_r, u_o) = u_c[u_r(u_d), u_o(u_d)]$$ [6-21]

and proceed to treat the control signal as a *functional* whose value (or range of values) depends on the relationship between u_r and u_d, as monitored and regulated through "higher centers." It may even be possible to show that there is some optimization scheme wherein, of all the allowable functions that can relate u_r to u_d, the ones upon which physiologic function is based are chosen such that a minimum amount of control, u_c, is required to maintain a homeostatic state of affairs. This would make considerations related to control self-consistent with those related to both the transport and utilization of mass, energy, and momentum by physiologic systems.

Some current thinking suggests that our body's immune system may function according to the above hypotheses (May, 1974). That is to say, the system does well as long as it receives insults spaced far enough apart in time and in space, but it breaks down otherwise if the insults are superimposed too close together, which one might call the "straw that broke the camel's back" scenario. These insults become cumulative, then, "tipping the scales" in favor of physiologic malfunctions that may range in severity from a simple allergic attack to dreaded diseases such as cancer. Given enough time, however, or helped along by successful immunotherapy, the system may adapt or evolve to change its reference signals (u_r), its transfer functions $(K_1, K_2, \ldots)$, or its control signals (u_c) to accommodate persistent values of disturbances (u_d) that affect system outputs (u_o). It is important to note that such accommodation may ultimately be manifest by changes in system transfer functions, as discussed earlier in the example involving adaptation to a high altitude environment, and the ultimate effects of such adaptation on the kidney and on bone marrow.

Controlling signals may *also* be coded into various types of "body language", which is now classified as motor signals, rather than feedback sensory signals. It is known, for example, that there is a language that seems to govern healing. Send the proper piezoelectric signal (electro-osteosyntax) to the osteoblasts in the region of a bone fracture and they "understand" to start manufacturing new bone tissue (Brighton, 1977). In fact, such signals are responsible for the servomechanisms that allow bone to function at an optimum stress level in a process called *functional adaptation*. Moreover, "injury potentials" have been measured at the site of damaged physiologic tissue, and these electromagnetic signals have been shown to stimulate the growth of *new* tissue in the injured area (e.g., the growth of a new finger from a joint at which an existing finger has been severed). It is known further that various of the body's immune reactions are related to electrochemical potentials that establish the

Table 6-III

Some Examples Of The Different Dialects
Associated With Motor Body Language

1.	Electrocardiosyntax (Heart Muscle)	7.	Electro-Osteo Syntax (Functional Adaptation)
2.	Electromyosyntax (Striated Skeletal Muscle)	8.	Electromagnetic Induction Of Cell Proliferation (Injury Potentials)
3.	Post-Ganglionic Electro-Axonic-Syntax Of The Autonomic Nervous System	9.	Electromagnetic Induction Of Lymphocyte Activity (Immune System)
4.	Electro-Gastrosyntax (Smooth Muscles, including Peristaltic movements of the Gastrointestinal Tract, and the Gastro-Colic Reflex)	10.	Autonomic Electromagnetic Motor Stimulation Of The Endocrine Glands Of Secretion
		11.	Electroarteriolar Syntax (Control Of Smooth Muscle In Arterioles)
5.	Electrotropism Syntax	12.	Electromagnetic Induction Of DNA and RNA Synthesis
6.	Electro-Oculosyntax (Motor Function of the Muscles of the Eyes)	13.	Electromagnetic Speech Synthesis

stereospecificity of lymphocytes. The list goes on and on, to include, for example, some of the dialects illustrated in Table 6-III. The point is that there is a sophisticated communications network in the body, whereby homeostatic regulators can "speak" (motor control) to their constituents -- exciting them (inciting them to action), or inhibiting them (preventing them from "doing their thing"), or otherwise manipulating them.

Again, if we could somehow learn to understand (decode, as in Bendat and Piersol, 1971, de Coulon, 1986, and later in this chapter) and speak (reproduce in synthetic form) this "body language" of metabolic processes that control life, then we would be brought one step closer to being able to *intervene* in these processes when they malfunction. Thus, by. appropriate intervention -- perhaps electromagnetically, perhaps electrochemically, perhaps pharmaceutically, or whatever -- we could control disease, affect cures for many (if not all) pathologic conditions, replace ailing or malfunctioning body parts, and even learn how to improve upon and exceed the present limitations imposed on human performance. But the story is not quite that simple. In the human organism, the propagation of an action potential (or wave of depolarization) along one nerve fiber is only a part of the real-life situation. Nerves are normally arranged in an end-to-end relationship with one another. The connecting region between two (or more) nerves is called a *synapse* -- where terminal branches of one nerve influence the dendrites of the connecting nerve. Likewise, the ends of nerves which conduct impulses to muscles have a special, very close contact with the membrane of the muscle cell. This region is called the *myoneural junction*.

Inhibition and Excitation: Synaptic Transmission of Information

Both synapses and myoneural junctions have special capabilities to allow a depolarization wave to "jump" -- across a 100 to 300 Angstrom gap -- from one cell to the next. These capabilities involve *transmitter substances* carried by the synaptic vesicles (see Figure 6-3); and this is how depolarization waves (propagated action potentials) can move from nerve to nerve or from nerve to muscle.

A *synapse,* from the Greek, "syn", meaning, *with,* and "aptein", meaning, *to touch,* is a specialized functional and anatomical junction between two neurons, where the depolarization activity of one "upstream" (*presynaptic*) cell affects the excitability of another "downstream" (*postsynaptic*) cell. Since the same kinds of mechanisms that occur in synapses can also occur more generally between other types of excitable cells, as well, one can define a *synaptic-like relation* as: "a specialized functional and anatomical junction between any two excitable cells, where the electrical activity in an upstream (prejunctional) cell affects the excitability of a neighboring downstream (postjunctional) cell." For our purposes, we may think in terms of basically four types of physiologic synaptic-like relations: chemical postsynaptic excitatory information transmission, (1) with and (2) without signal modulation; (3) chemical postsynaptic inhibitory transmission of information; and (4) chemical presynaptic inhibition of information transport. Let us describe each of these briefly (see also, Guyton, 1981, and Deutsch and Micheli-Tzanakou, 1987).

CHEMICAL POSTSYNAPTIC EXCITATION WITHOUT SIGNAL MODULATION

Chemical postsynaptic excitation without signal modulation is typical at the neuromuscular junction between motor nerve fibers and muscle cells, and between certain neurons in the nervous system of most animals. It works something like this:

First, an action potential arriving in the prejunctional terminal branches of, for example, an alpha-motor-neuron, causes a small amount of a chemical transmitter substance -- in this case, *Acetylcholine* -- to be released from the vesicles in the terminal boutons of the prejunctional cell. Just exactly *how* this is accomplished is not yet entirely understood, but it is known that *each* action potential corresponds to a prescribed quantity ("unit") of transmitter substance-release. Thus, action potentials can be "summed" at the neuromuscular junction, such that the *total* amount of transmitter substance released becomes a function of the firing *frequency* of the prejunctional neurons.

Second, the transmitter substance, Ach, diffuses across the synaptic cleft (up to 300 angstroms), in about 0.05 milliseconds or less (about 50 microseconds), to reach the dendritic receptor sites on the postjunctional cell. Since it takes on the order of 50×10^{-6} seconds for the signal to traverse a distance of some 300×10^{-10} meters, the *synaptic* transmission speed is around 0.6 mm/sec. Compare this with *nerve* propagation velocities on the order of .6 to as much as 130 *meters* per second (Table 6-I), and it becomes immediately obvious that the synaptic region actually *delays* the propagation of an action potential by at least three orders of magnitude, and up to a factor of 2-or-3 $\times 10^5$! Because of the very small distances involved, however, this delay is short-lived.

Third, acetylcholine causes the postjunctional membrane permeability to both sodium and potassium to increase by a factor of about 1000, locally, on that portion of the muscle cell which is part of the *neuromuscular junction* (which is $\frac{1}{1000}$'th to $\frac{1}{10,000}$'th of the total muscle cell surface). Thus, at this site on muscle, or on the corresponding postsynaptic dendrites of postjunctional nerve membranes, ionic current starts to enter the cell, which starts a wave of depolarization propagating down the cell membrane. The *net* depolarization which occurs in a neuromuscular junction is up to 40 millivolts, and the junctional current flow lasts for about 1-2 milliseconds, although the actual potential lasts longer, decaying passively with the time constant of the membrane. This *excitatory potential* is called an *end plate potential,* abbreviated EPP.

Fourth, as already mentioned, the EPP produces an action potential in the downstream, postjunctional muscle or nerve cell (c.f., Chapter 5). Normally, a single presynaptic action potential in an alpha-motorneuron will cause a postsynaptic EPP large enough to cause one action potential in the muscle fiber to which it is attached.

Fifth, the action potential of the muscle cell causes the cell to *contract.* Each skeletal muscle cell normally has only one nerve fiber coming to it, but each nerve fiber can go to anywhere from 6 to 200 muscle fibers, forming what is known as a *motor unit.* Every time a nerve fiber "fires", *all* of the muscle cells to which it goes fire along with it, i.e., the entire motor unit fires. Therefore, this neuromuscular junction is a method of transferring a chemical excitatory action potential from one cell to the next without modulation, i.e., without regulating the one-to-one correspondence between stimulus and response.

Finally, the synaptic acetylcholine is destroyed by an enzyme, *cholinesterase,* which is located on the outer surface of the postjunctional membrane. Suffice it to say, that anything which has the ability to affect the presence, absence, or functional state of this enzyme, has the corresponding ability to control neuromuscular synaptic transmission of information. A group of drugs called "Parasympathomimetics" (for example, Neostigmine) are one such family of "parasympathetic-like" cholinesterase enzyme inhibitors. By preventing the destruction of Ach and thus allowing it to accumulate in the synaptic region around cholinergic receptors, such drugs promote the increased stimulation of these receptors, and hence prolong or amplify the total response. For this reason, they are also called "cholinergic" drugs.

CHEMICAL POSTSYNAPTIC EXCITATION WITH SIGNAL MODULATION

In the central nervous system, the situation involving chemical excitatory synapses between neurons is quite different. Here, not only does a single presynaptic fiber go to many postsynaptic cells (as is the case in muscle), but, each postsynaptic cell also receives excitatory inputs from *many* presynaptic neurons (which is *not* the case in muscle motor units). Moreover, whereas a single presynaptic action potential in an alpha-motorneuron will normally suffice to produce a single contraction of the affiliated motor unit, a single action potential in one presynaptic neuron of the central nervous system is normally *not* of sufficient strength to fire the postsynaptic cell. Thus, in the latter case, we can effectively *modulate* the transfer of a chemical excitatory action potential from one cell to the next. That is to say, the postsynaptic cell acts as a summation point in the information network, at which point "either/or" types of decisions can be made. If the *sum* of all excitatory ($+$) and inhibitory ($-$) inputs (see later) to the postsynaptic cell exceed its firing threshold, the cell will fire. Otherwise, it will not. The mere existence of any given excitatory input to the postsynaptic cell does not, in and of itself, guarantee that an action potential will be generated. In other words, for motor units, a threshold EPP from just a single α-motoneuron is generally both a necessary *and* a sufficient condition to induce contraction; for information transport in the CNS (Central Nervous System, e.g., inputs *to* an alpha motoneuron) a presynaptic action potential is (or multiple upstream stimulations are) necessary, but not normally *sufficient* to necessarily evoke a downstream response. Thus, once an alpha motoneuron *fires*, its motor unit normally fires with it. But the alpha motor neuron will not *necessarily* fire just because it is excited by any one input.

The *process* of transfer of information in the synaptic region is the same here as it was in the neuromuscular junction described earlier. However, in the central nervous system, the transmitter substance may include not only acetylcholine, but norepinephrine, dopamine, and others, as well, some of which have not even been identified as yet. Nevertheless, the synaptic transmitter substance increases the permeability of a small area of the postsynaptic cell to sodium and potassium, which, in turn, depolarizes the cell.

This depolarization is called an *excitatory postsynaptic potential*, abbreviated EPSP. From a single synapse, an excitatory postsynaptic potential has been measured to be in the range 0.2 to 0.5 millivolts, with synaptic current flow for most chemical excitatory synapses lasting anywhere from 1 to 2 milliseconds. The duration of an EPSP will depend on the duration of synaptic current flow and the time constant of the membrane. This duration can be as short as 5 msec, or as long as 100 msec.

EPSP's can be added, as mentioned above. The addition can be *spatial*, wherein inputs from two or more different groups of presynaptic fibers superimpose in the synaptic region; or, it may be *temporal*, wherein the same presynaptic fibers are stimulated in rapid succession, so that a second EPSP occurs before the first one has had time to completely decay. With the synaptic region acting as a summation point, if a postsynaptic cell is depolarized enough (to threshold), it will fire an action potential. Because of the long duration of some EPSP's, a single action potential in a group of presynaptic cells may cause many downstream action potentials to be generated. Thus, the firing rate of a downstream neuron will increase with the EPSP time (depolarization time).

From the above discussion, it is evident from an information-processing point of view that the human body is a *digital*, rather than an *analog* computer. Excitable cells either "fire" (ON = generation of an action potential), or they don't (OFF = stimulus below threshold). If they do, the information is coded into how *often* they are firing (the action potential frequency), and *which ones* are firing (a function of their respective threshold potential). Synaptic regions, and target organs and tissues act as "summers", simply *counting* impulses per second of an excitatory or inhibitory nature. If excitation dominates inhibition (see below), a response ensues. If inhibition dominates excitation, a status-quo situation is preserved. The compound action potential may thus be viewed in terms of its corresponding frequency spectrum, which is a conglomerate of excitatory and inhibitory impulses travelling through any given nerve.

CHEMICAL POSTSYNAPTIC INHIBITION

A *postsynaptic inhibitory response* is one which: (a) decreases the firing rate of the postsynaptic cell; or (b) increases the amount of excitatory input which is necessary in order to fire the downstream cell; or (c) hyper-polarizes the downstream cell membrane, thus decreasing its resting potential; or (d) alters the downstream cell membrane permeability to raise its threshold potential; or (e) all of the above. In this case, an action potential in the presynaptic cell causes the release of an inhibitory chemical transmitter substance, such as Glycine, Serotonin, Gamma Aminobutyric Acid (GABA), and, sometimes, Norepinephrine. The substance diffuses across the synaptic region to the postsynaptic cell, where it acts to markedly increase membrane permeability to K^+, but *not Na^+*

Recall from Table 3-I that K^+ concentrations in the ICF are higher than they are in the ECF. The increased membrane permeability to K^+ thus allows it to escape from inside the cell, making it even more negative with respect to the outside, i.e., *hyper-polarizing* it up to 10 to 15 mv compared with its normal resting potential. Such hyperpolarization is called an *inhibitory postsynaptic potential,* IPSP. IPSP's are usually no more than a few millivolts, and they also last from about 5 msec to more than 100 msec.

In addition to being hyperpolarized, the postsynaptic membrane, by virtue of its increased permeability to both K^+ and Cl^- also has increased transmembrane conductance to both these ions. This, too, inhibits the transmission of an action potential in that, if one tries to excite the membrane with a rapid influx of sodium, any tendency that this may create for the membrane potential to shift away from its resting value is effectively neutralized by a rapid flux of potassium and chloride ions the other way. That is, for a sodium influx to "offset" the "neutralizing effect" of increased membrane conductance (to K^+ and Cl^-) such influx required to cause excitation may have to be as much as 5 to 20 times normal (Guyton, 1981). This type of inhibition due to increased conductance of the postsynaptic membrane is called resting potential "clamping", and is believed to be more of a major factor in inhibiting the generation of an action potential than is the hyperpolarization of the membrane, itself.

CHEMICAL PRESYNAPTIC INHIBITION

Chemical presynaptic inhibition refers to the decrease, or total elimination of the amount of excitatory chemical transmitter substance released from a presynaptic fiber when the latter is stimulated by an action potential. It is believed that such attenuation of transmitter substance release is accomplished by *presynaptic synapses,* i.e., synaptic sites located on the terminal fibrils and synaptic knobs of the presynaptic cells in regions upstream of where the knobs themselves terminate on the following neuron (Guyton, 1981).

Presynaptic inhibition selectively decreases the efficacy of *certain* excitatory inputs, and not others. In this sense, it differs fundamentally from postsynaptic inhibition, which inhibits the cell to *all* excitatory inputs. In mammals, presynaptic inhibition seems to occur primarily in sensory systems, rather than in motor ones, and its purpose appears to be to control the *sensitivity* of the response to the sensory input. That is, the greater the intensity of the sensory input signal -- the stimulus generating the action potential -- the greater the negative feedback inhibition (presynaptic chemical inhibition). Thus, increased stimulus strength to the dendrites of a sensory nerve reduces the response at the output end of the nerve (i.e., effectively makes the nerve less "sensitive" to the stimulus, see earlier discussion of the adaptation phenomenon). The duration of such inhibition is several hundred milliseconds.

The four types of chemical synapses described above are, so far, the only ones known to occur in mammalian organisms. Much of the work explicating these mechanisms was done by Eccles, Fatt, and Katz, with Eccles having received the Nobel Prize for his research in 1963.

Controlled Elements in Physiologic Systems

In physiologic terms, very little is known about the inter-relationships and functional dependencies that exist between the disturbing signals, u_d, the control signals, u_c, and the feedback signals, u_o, that constitute the input/output characteristics of the controlled elements in physiologic systems. Indeed, if more was known, then the practice of medicine would be more *preventive* (or "proactive") than *symptomatic,* retrospective and *ex post facto* (or "reactive"). That is to say, a greater knowledge of the parametric dependence of u_c on u_d (c.f., equation [6-21]) in terms of ultimate values of u_o for which homeostatic regulators become unstable, would allow us to *anticipate* and intervene where necessary to prevent breakdowns. Anticipation implies functional dependence on derivatives of $u_r(u_d)$, u_d, and u_o (see earlier discussion), as well as on the instantaneous values of these variables. The medical community is not generally trained to be sensitive to *rates-of-change,* as opposed to *amounts-of-change,* and so the health care delivery system tends to be out-of-phase with real-time physiologic events.

To wit: much of the practice of medicine today treats *symptoms* (which usually appear after-the-fact), not *causes* (which are usually inconspicuous or subtle as they act before-the-fact). The health-care-delivery system tends to be reactive, not proactive. In fact, when a treatment does affect a cure, or provide relief to the patient, the chances are very likely that the physician may not even have diagnosed the condition exactly, or at all, for that matter. He or she is intent first, and foremost, on alleviating what it is that is happening to the patient *now,* and the treatment for the symptoms of discomfort is prescribed accordingly.

Fortunately, because the same type of treatment may be equally effective in managing several disorders which present similar symptoms, it is of less immediate concern to the physician to establish *exactly* what is wrong with the individual -- unless the first line of defense fails to produce satisfactory results. Then, and only then, does it become necessary to probe deeper. This is not to put down or otherwise denigrate medical practitioners. They are basically supplying what the public wants. Nor do we mean to imply that health-care providers are not concerned with an accurate diagnosis of the condition of their patients. But, short of charging the patient a small fortune for diagnostic tests (some of which may cause considerable discomfort or be exposing the patient to some risk and danger), and spending a great deal of time interviewing, examining, contemplating and evaluating, the physician first makes an "educated guess" as to what the problem is, based on a more superficial assessment of the patient's symptoms.

Again, fortunately for the physician, statistics are also on his or her side, for they reveal that, in as many as 70 to 80% of all cases, an educated guess is good enough. This is due primarily to the fact that most ailments tend to disappear on their own, given enough time, even if they go untreated, and, in some instances *in spite of* (rather than *because of*) the prescribed treatment. In many ways, the human body is designed and well-equipped to heal itself, so, in a sense, the medical practitioner is "buying time" in his or her first volley against the symptoms of a physical ailment, expecting that the organism will soon take care of itself, anyway. Where the guess, or first approximation in mathematical terms, does not solve the problem, the medical team goes through several more iterations in an effort to help the patient. In about 10 to 20% of all cases, this second iteration works if the condition is treatable. At other times, it does not and subsequent successive approximations or iterations are necessary. Sometimes, even given the proper diagnosis, nothing further can be done for the patient anyway. *Always,* there is a great deal of trial and error involved -- and a little bit of luck! Indeed, so much of medicine is still an art, not a science.

On the other hand, if we had a greater comprehension of cause-and-effect relationships, based, for example, on a more complete understanding of some engineering and control aspects of physiologic function, then we could formulate clinical *warning signals*, u_{warning}, or u_w, which would alert us to impending malfunctions or breakdowns:

$$u_w = u_w(u_d, \frac{\delta u_d}{\delta t}, \frac{\delta^2 u_d}{\delta t^2}, \dots, u_r, \frac{\delta u_r}{\delta t}, \dots, u_c, K_1, K_2, u_o, \dots); \quad [6\text{-}22]$$

where, essentially, the warning signals take into account: (i) the response characteristics of the respective homeostatic regulators, (ii) the ability of the tissues to *respond* to u_c (i.e., receptor sites, sensitivity, constitutive properties, excitability, selectivity, and other parameters already discussed), (iii) how well the organism can *adapt* (including such consider-ations as accommodation, receptor site saturation, facilitation, conditioning, and programmability), (iv) the *envelope* of normal physiologic function, and (v) the actual *state* of the system -- all viewed with respect to the characteristics of the disturbance.

In principle, the above is not necessarily an altogether new concept. Indeed, medicine has tried for many years to define "correlates" and warning symptoms for many pathologic conditions. But only recently have attempts been made to establish such listings based on a systems control-theory approach to the analysis of physiologic function. This is where mathematical modelling and engineering concepts have made such great contributions to the health-care delivery system, and where these disciplines promise to make even more headway as they continue to use the scientific method to define, quantify and inter-relate biological control processes.

Transfer Functions in Physiologic Systems

Among other things, the transfer function (and associated gain and phase) of a feedback control system are measures of the *sensitivity* of that system to a disturbance. For a supersensitive system, for example, very slight fluctuations of u_d may produce wild undulations in u_o -- even though the overall system is fundamentally stable. A good physiologic example of this is the famous "knee-jerk" reflex, wherein a mere tap of the quadriceps musculature tendons where they insert just below the knee cap results in a rather dramatic contraction of this musculature, causing a sudden extension ("jerk") at the knee joint of the lower leg.

On the other hand, for a relatively insensitive system, u_d might have to change by orders of magnitude, and persist for a very long period of time, to affect even a slight change in u_o. A good physiologic example of the latter is the functional adaptation mechanism in the skeletal system, wherein bone tissue can alter its cross-sectional area to reduce overall tissue stress in response to a long-term and persistent increase in external loading. Such an increase needs to be rather extensive before any permanent, readily measurable change in cross-sectional area is manifest, and the change progresses rather slowly in rather small increments. So, too, does that associated with bone marrow responses to hypoxia, as discussed earlier.

Other than by long-term adaptive mechanisms and evolution, such as has been already addressed, the sensitivity of the controlled element of a feedback control system can be adjusted in a variety of more immediate ways. First, there are a number of pre-and-post synaptic excitatory and inhibitory mechanisms by which membranes can be hyper-or-hypo-polarized to change their "excitability thresholds." These were described earlier in this chapter and elsewhere in this text. Suffice it to say here that the threshold potential at which an excitable membrane will fire determines its sensitivity to a stimulus, and one of the ways that this can be controlled is by hyper-or-hypo-polarizing the membrane with respect to this threshold.

Another means for affecting sensitivity is to put a "bias" on the feedback sensor that monitors u_o, effectively changing K_1. To illustrate this type of short-term "pre-loading" of the monitoring transducers, one can cite γ-nerve innervation of the spindles of striated skeletal muscles (Schneck, D. J., 1984b). That is, muscle spindles are basically strain-gage type stretch transducers that convert changes in muscle length into a proportional excitation signal that stimulates α-motoneuron activity -- causing the corresponding muscle to contract (as in the "knee jerk" example used earlier, where the change in muscle length is produced by the "tap" on the muscle tendon below the knee, and the response is the contraction of the quadriceps musculature). Gamma-nerve stimulation of the intrafusal muscle fibers of the spindle receptors effectively puts a "pre-stretch" on these transducers, making them highly sensitive to any additional deformation or disturbance. In fact, that is why the knee jerk reflex is so ultra-sensitive to the slightest stimulus, i.e., the spindle receptors of the quadriceps musculature are so highly biased (pre-stretched) that the *tiniest* additional stretch of the tendons -- such as a light tap below the knee where they insert into the shank of the lower leg -- produces a total contraction of the muscle, and a consequent jerk at the knee joint. In terms of Figure 6-2, this *myotactic reflex arc,* as it is known, can be modelled mathematically by letting:

$u_r = f_t$ = firing frequency of an α-motoneuron necessary to bring a muscle fiber membrane to threshold, so it will fire (i.e., generate a contraction);

$u_o = f_{Ia}$ = actual firing frequency of the muscle spindle type *Ia* (see Table 6-I, a = afferent) sensory fibers, which excite the α-motoneuron of the corresponding muscle fiber;

$u_c = f_\alpha$ = actual firing frequency of the α-motoneuron fibers;

u_d = a combination of f_γ, the actual firing frequency of the biasing γ-motoneurons, which *actively* (i.e., by contracting the intrafusal fibers they innervate) sensitize the muscle spindles to stretch by preloading it; and, ξ^*, a *passive* stretch-disturbance applied to the muscle spindle;

K_2 = the transfer function of the α-motoneuron, which is acting as the comparator and controller of the system, as well as the actuating element of the system; and,

K_1 = the transfer function of the motor unit controlled by the corresponding muscle spindle receptor being stimulated; this is the controlled element of the system, which, you will note, contains within it the muscle spindle sensing elements as well.

Then,

$$u_o = u_o(u_d, u_c) = u_o(f_\gamma, \xi^*, f_\alpha) = f_{Ia} \qquad [6\text{-}23]$$

and,

$$\delta f_{Ia} = \frac{\partial f_{Ia}}{\partial f_\gamma}\, \delta f_\gamma + \frac{\partial f_{Ia}}{\partial \xi^*}\, \delta \xi^* + \frac{\partial f_{Ia}}{\partial f_\alpha}\, \delta f_\alpha\,, \text{ or,}$$

$$\delta f_{Ia} = K_\gamma\, \delta f_\gamma + K_{\xi^*}\, \delta \xi^* - K_\alpha\, \delta f_\alpha \qquad [6\text{-}24]$$

where:

K_γ = the muscle spindle transfer function for active γ-innervation (biasing);

$K_{\xi*}$ = the muscle spindle transfer function for passive stretch; and,

K_α = the muscle spindle transfer function for active contraction (note that this transfer function is negative because the spindle fires at a lower frequency as the muscle contracts, i.e., *decreases* its length).

Also, $K_1 = K_1(K_\gamma, T, \ldots)$. Specific details describing the mathematical treatment of the above schematic representation are given in Schneck (1984b), along with other examples relevant to many of the points made in this chapter.

Alteration of the gains and transfer functions of elements in a feedback control loop is what was meant earlier when we said that "some nonlinearity may intervene in determining the response of the system." Through proper research, one can actually establish functional relationships that define changes in u_o which are produced by changes in K_1, or in K_2, or in the sensitivity of the controlled element. Many of these have already been identified and labelled *response functions* for homeostatic regulators (see, for example, Mylrea and Abbrecht, 1973, Schneck, 1984b, Milsum, 1970, Milhorn, 1966, Brown and Gann, 1973, Huffaker, 1974, and Milsum, 1966). Unfortunately, *most* response functions remain basically unknown for the majority of homeostatic regulators, but it is hoped that works such as this will stimulate research based on this line of reasoning.

Engineering Methods in Biological Signal Analysis

Several times in this chapter we have talked about analyzing and reproducing biological signals. It is appropriate at this time, then, that we say a few words about how this is done from the engineering perspective. For a more comprehensive discussion of this topic, the reader is directed to the fine works of Bendat and Piersol (1971) and F. de Coulon (1986), among others.

The Autocorrelation Function

Biological signal analysis begins by establishing the periodicity, or lack thereof, of the "body language" involved. This is accomplished by the use of *Correlograms.* Suppose Figure 6-4 represents a "typical" $u_c(t)$ vs. t graph obtained by monitoring, for example, the physiologic syntax associated with brain function, i.e., an electroencephalogram or EEG. To obtain a correlogram for this signal, one proceeds as follows:

First, *duplicates* of the EEG recording shown in the figure are fed simultaneously into an electronic time-delay instrument. This unit contains circuitry which is capable of displacing the two identical signals from one another by a predetermined time delay, τ, which usually ranges from 0 to 100 milliseconds or more in increments of 0.10 msec or less, depending on the specifications of the instrument used and the characteristics of the signal being analyzed. The rate at which the time delay increases is controlled by a sweep-drive unit. The data collected is $u_c(t)$ for a given t, and $u_c(t + \tau)$ for a given τ, both t and τ being variable. The sweep drive unit also *plots* τ along the x-axis of an x-y recorder.

Second, the time-displaced signals, $u_c(t)$ and $u_c(t + \tau)$, with τ acting as a parameter and t an independent variable, are processed by an analog correlator to compute an *autocorrelation coefficient*, defined by the equation:

$$R_f(\tau) = \lim_{t_p \to \infty} \frac{1}{t_P} \int_o^{t_p} u_c(t)u_c(t + \tau)\, \delta t \qquad \text{[6-25]}$$

where t_P is the total range in time over which the physiologic signal has been measured -- hypothetically infinite, but realistically, some arbitrary period of, say, at least several seconds. The analog correlator also *plots* $R_f(\tau)$ along the y-axis of the x-y recorder.

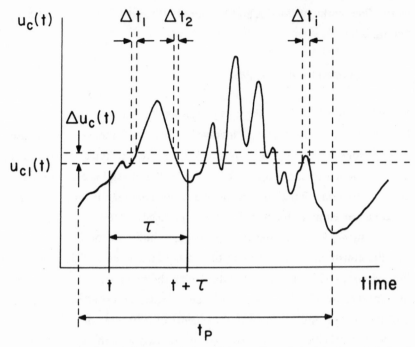

Figure 6-4 Hypothetical electroencephalographic recording of physiologic control signal syntax as a graph of action potential voltage vs. time.

The resulting plot of $R_t(\tau)$ from the analog correlator vs. τ from the sweep drive unit is called a correlogram. This graph contains information that essentially tells an investigator to what extent the value of a function at any given point in time is related to the value of that *same* function at some other point in time (hence the name, *auto*correlation function, the Greek prefix meaning, "self"). Note that $R_t(\tau)$ defines the time-averaged mean value, $\overline{u_c(t)u_c(t+\tau)}$ of the function $u_c(t)u_c(t+\tau)$, and that the autocorrelation coefficient is a maximum at $\tau = 0$ -- i.e., the value of a function at any given time is perfectly correlated with the value of the *same* function at the *same* time. For $\tau = 0$, equation [6-25] defines the *mean-square* value of $u_c(t)$:

$$R_t(0) = \lim_{t_P \to \infty} \frac{1}{t_P} \int_o^{t_P} u_c{}^2(t)\,\delta t = \overline{u_c{}^2(t)} \qquad \text{[6-26]}$$

and the square-root of equation [6-26] defines the *root-mean-square*, or "RMS" value of $u_c(t)$:

$$\text{RMS } u_c(t) = \sqrt{R_t(0)} \qquad \text{[6-27]}$$

Combining equations [6-25] and [6-26], one can also define a *normalized* autocorrelation coefficient,

$$R_t{}^*(\tau) = \frac{\overline{u_c(t)u_c(t+\tau)}}{\overline{u_c{}^2(t)}} = \frac{R_t(\tau)}{R_t(0)} \qquad \text{[6-28]}$$

whose value is unity for $\tau = 0$, i.e., $R^*{}_t(0) = 1$.

At the other extreme, the time-averaged *mean* value of $u_c(t)$ is obtained by taking the square-root of equation [6-25] as the time-displacement, τ, becomes very *large*, hypothetically approaching infinity:

$$\overline{u_c(t)} = \sqrt{R_t(\infty)}\,,$$

so that,

$$R_f(\infty) = \left[\overline{u_c(t)}\right]^2 \qquad \text{[6-29]}$$

The autocorrelogram thus starts at the mean-square-value of $u_c(t)$ for $\tau = 0$, and approaches the mean-value-squared of $u_c(t)$, for $\tau \to \infty$. If the function $u_c(t)$ is further cyclic in time, or has features that are repeating in nature, its corresponding correlogram will likewise be cyclic in time and show a certain degree of periodicity as it approaches its limit. On the other hand, if the function is aperiodic, or noncyclic, then its correlogram will approach a limit in some arbitrary fashion, indicating that values of the function displaced in time τ from some value at a given time, t, will not necessarily continue to be affected by that value. In fact, totally and purely random data with a zero mean-value has a corresponding autocorrelogram which is almost like a Dirac delta function (a sharp single spike) at zero time displacement ($\tau = 0$) that rapidly goes to zero for $\tau > 0$, meaning that the value of the function at any given time has virtually *no* relationship to its value at any other time, i.e., it is completely random.

These features of the autocorrelogram make it a convenient tool for distinguishing between periodic and nonperiodic information. If the autocorrelogram shows a tendency towards periodicity, it suggests that the signal being examined is likewise periodic and may be analyzed by techniques of Fourier Analysis.

Fourier Analysis

If $u_c(t)$ shows a tendency towards periodicity, it may be expanded into a Fourier series, i.e., it may be written as:

$$u_c(t) = \frac{a_o}{2} + \sum_{i=1}^{i=\infty} \left[a_i \cos 2\pi i f_o t + b_i \sin 2\pi i f_o t\right] \qquad \text{[6-30]}$$

where:

f_o = the fundamental frequency of the periodic function, which is the number of regular intervals (cycles) per unit time that the waveform of the time-varying function, $u_c(t)$ exactly repeats itself (this is the inverse of τ_P, the time period required for one full fluctuation or waveform interval to elapse); and,

$$a_i = \frac{2}{\tau_P} \int_o^{\tau_P} u_c(t) \cos 2\pi i f_o t \; \delta t, \quad i = 0, 1, 2, \dots \qquad \text{[6-31]}$$

$$b_i = \frac{2}{\tau_P} \int_o^{\tau_P} u_c(t) \sin 2\pi i f_o t \; \delta t, \quad i = 0, 1, 2, \dots \qquad \text{[6-32]}$$

Let:

$$a_i \cos 2\pi i f_o t + b_i \sin 2\pi i f_o t = X_i \cos \; (2\pi i f_o t - \theta_i)$$
$$= X_i \cos \theta_i \cos 2\pi i f_o t$$
$$+ X_i \sin \theta_i \sin 2\pi i f_o t$$

Then:

$$a_i = X_i \cos \theta_i \quad \text{and,} \quad b_i = X_i \sin \theta_i \qquad \text{[6-33]}$$

From equation [6-33], one can construct a *phase triangle,* which is defined as a right triangle wherein:

θ_i = the phase angle, or, simply, the *phase* of the i'th component of the Fourier series expansion of $u_c(t)$, defined in terms of the Fourier coefficients [6-31] and [6-32] as:

$$\theta_i = \text{Arc Tan} \; \frac{b_i}{a_i} \qquad \text{[6-34]}$$

and X_i = the hypotenuse of the phase triangle of sides a_i and b_i, known as the *amplitude* of the i'th component of the Fourier series expansion of $u_c(t)$, and defined by:

$$X_i{}^2 = a_i{}^2 + b_i{}^2 \qquad \qquad \text{[6-35]}$$

Finally, with $X_o = \dfrac{a_o}{2}$, equation [6-30] becomes:

$$u_c(t) = X_o + \sum_{i=1}^{\infty} X_i \cos (2\pi i f_o t - \theta_i) \qquad \qquad \text{[6-36]}$$

Each value of i in equation [6-36] is called a *harmonic component* of the Fourier series representation of $u_c(t)$, and has associated with it a corresponding amplitude, X_i (c.f., equations [6-31,32,35]), and phase, θ_i (c.f., equations [6-31,32,34]). A graph displaying X_i vs. if_o, the corresponding frequency of the harmonic component, expressed as a multiple of the fundamental frequency, f_o, is called a *Frequency Spectrum* for the signal $u_c(t)$. If one thinks of $u_c(t)$, the resultant waveform as a finished product, like a cake, then the Frequency Spectrum of u_c may be thought of as the ingredients of the cake -- if_o telling you *what* ingredient to use, and X_i telling you *how much* of it to use. The phase angle θ_i, then, would be telling you in what *order* the ingredients are to be mixed together, i.e., by how much each should lag or lead the other in the process of being mixed.

It is evident from the above analogy, that if one has the correct amounts of the proper ingredients, and mixes them in the right order, then one should be able to reproduce the *same* cake every time. In other words, defining the frequency and phase spectra of a periodic signal allows one to *reproduce* that signal exactly in accordance with equation [6-36]. In terms of physiologic syntax, the Fourier spectrum is akin to knowing all of the proper verbs, nouns, adverbs, adjectives, and so on appropriate to "speaking" body language by constructing the right sentences. Thus, the frequency spectrum effectively allows one to "decode" body language.

For example, suppose, again, that Figure 6-4 represents brain activity as recorded in an electroencephalogram (EEG). We find, then, that the frequency spectrum of this signal ranges from 1 Hertz (cps) to 50 Hertz, and includes components having amplitudes lying between 50 and

200 microvolts, and phase angles between 0 and 90°. Furthermore, various studies reveal that there is a correlation between certain of the spectral ranges and the corresponding levels of consciousness of the individual involved, i.e., the lowest frequency, greatest amplitude regions of the EEG representing states of deep sleep, and the highest frequency, lowest amplitude portions of the signal representing complete wakefulness. Inbetween, one can define several graded "levels of consciousness" as follows:

1) 1-3 Hz δ-waves, very synchronized ($\theta \sim 0°$) along all regions of the cerebral cortex, dominant in deep sleep and having the slowest alternating pattern (long period) with the greatest amplitude (around 200 microvolts);

2) 4-7 Hz θ-waves, associated with drowsiness, day-dreaming, and certain "creative problem-solving" states (amplitude $\sim$ 100 microvolts);

3) 8-12 Hz α-waves located in the occipital lobes of the cerebral cortex, with amplitudes in the 50-100 microvolt range, corresponding to pleasant feelings of well-being, tranquility, an increased awareness of feelings, relief from anxiety, and general complacency. It is these waves that one is trained to identify and "generate" through biofeedback techniques; and it is these waves that conspicuously disappear from the EEG when the brain displays any type of attention or alertness, or when it focusses on objects in the visual field; or when it is otherwise distracted by environmental activity (known as the *arousing influence* of external stimuli);

4) 12-15 Hz, low-amplitude "spindle bursts" that are associated with sensori-motor biorhythms (so-called SMR activity); and,

5) 13-25 Hz (usually) to as high as 50 Hz (rarely), 50 microvolt β-waves, highly disorganized ($\theta \rightarrow 90°$) and desynchronized, and seen typically during periods of cortical waking activity, i.e., focussed attention, intense mental activity, visual scanning, anxiety or apprehension, worry, concentration, fear, anger, frustration, excitement, hunger, and surprise. Interestingly, β-wave activity is also manifest during periods of so-called "REM (Rapid-Eye Movement) sleep," during which active dreaming takes place. Although individuals experiencing REM sleep are

actually more difficult to arouse than those in deep, low-frequency, δ-wave sleep, their brain activity is at the very active level normally associated with complete wakefulness.

If we distinguish between β-wave activity associated with complete wakefulness (for which an awareness of the outer world is consciously present), and that associated with REM-sleep (for which an awareness of the outer world is purely hallucinatory), then the above classifications define basically six "states of consciousness" that range from an abstract awareness of *self* to a conscious awareness of the external *environment*. Sometimes, a seventh, somewhat hypothetical level of consciousness is hypothesized, wherein an individual can experience a *simultaneous* sense of self *and* environment. This is a kind of *transcendental* state in which both the subconscious and the conscious mind co-exist at a common level of awareness. This, again, is the state of mind that biofeedback attempts to exploit.

The above example illustrates how effective Fourier analysis can be in decoding "body language" which is periodic, with fundamental frequency f_o. The latter is determined in practice by finding the *highest* common divisor for all frequencies in the spectrum, and periodicity is confirmed if the ratios of all possible pairs of frequencies form *rational* numbers (i.e., if these ratios can be expressed as the quotient of two integers). If the latter condition is true for *most* of the ratios, but not all, then the data involved can be classified as *almost periodic*, and treated by a combination of Fourier analysis and the methods described further below.

Fourier Transform Methods for Transient Signals

As we progress from purely periodic data that can be handled by the techniques of Fourier series to transient signals such as that illustrated in Figure 6-1, and defined by equations [6-11] or [6-12], the corresponding frequency spectra go from being *discrete*, such as is defined by the *summation* sign in equations [6-30] or [6-36], to being *continuous*. In other words, rather than being defined by a *series* whose terms progress in *stepwise* fashion ($i = 1, 2, 3, \ldots$) with *discrete* step sizes ($\Delta i = 1$), transient data needs to be defined by a *function* which progresses *continuously* with *differential* step sizes (because it is essentially composed of all frequencies). Thus, the summation sign becomes an integral sign, the cosine and sine functions coalesce into the more general exponential function, $e^{-i2\pi ft}$ ($i = \sqrt{-1}$), and the Fourier *Series* reduces to a Fourier *Transform* integral defined by:

$$u_c(f) = \int_{-\infty}^{+\infty} u_c(t)e^{-i2\pi ft}\delta t \quad (i = \sqrt{-1}) \qquad [6\text{-}37]$$

Note the absence above of a fundamental frequency, f_o (since this signal is *not* periodic); and that the finite frequency *spectrum*, X_i vs. if_o for $u_c(t)$ as defined by equation [6-36], with $i = 1, 2, 3, \ldots$, has now become a continuous frequency *function*, $u_c(f)$ as defined by equation [6-37]. It is sometimes convenient to replace the limits of integration from minus infinity to zero, and from plus ∞, to: $t = $ some *finite* time (usually the signal sampling time). This defines a *Finite Fourier Transform*, from which a *Fast Fourier Transform* (FFT) algorithm may be developed. Considering this, again, to be beyond the scope of the current work, the reader is referred to Bendat and Piersol (1971) and other literature references for further details.

Descriptive Properties for Random Data

Fourier series and Fourier transform techniques are valid only for *deterministic* data, i.e., biological signals that can be uniquely defined by an explicit mathematical equation that allows one to *predict* (determine) values of the data from the form of the equation. Unfortunately, not all such information is deterministic. In many cases, it is purely random, (i.e., the correlogram approaches a Dirac delta function), and can be described only in terms of statistical averages and probability considerations. When this is the case, one may still proceed to analyze the information as follows.

THE PROBABILITY DENSITY FUNCTION

The probability density function for random data describes the *probability* (rather than the deterministic *predictability*) that the data will assume some given value at any arbitrary instant in time. For example, returning to Figure 6-4, suppose we wanted to know the probability that $u_c(t)$ takes on the specific value $u_{c1}(t)$ at any given time. One way to establish this probability is to define a small "window", $\Delta u_c(t)$, around $u_{c1}(t)$; see how much "time", on the average, $u_c(t)$ spends in this window, and then take the limit of this as the width of the window approaches zero. In other words, a probability density function is constructed from a graph such as Figure 6-4 as follows:

First, the biological signal is fed into an electronic sweep drive instrument, which allows one to preselect values of $u_{c1}(t)$, i.e., "target" values of interest. An amplitude comparator then selects the window width, $\Delta u_c(t)$, which may be arbitrarily varied around the reference value, $u_{c1}(t)$.

Second, the amplitude comparator simultaneously acts as a discriminating circuit to allow only that portion of $u_c(t)$ which has values in the range $\Delta u_c(t)$ to pass through to the signal processing stage of the instrumentation. An on-off timing circuit also records the elapsed intervals of time, Δt_i $(i = 1, 2, \ldots, n)$ during which the discriminating circuit allows $u_c(t)$ to pass through.

Third, the signal processor calculates the probability of finding $u_c(t)$ in the interval $\Delta u_c(t)$ around $u_{c1}(t)$ at any given time as: "probability of

finding $u_c(t)$ in the *neighborhood* of $u_{c1}(t)$" $= \dfrac{\sum\limits_{i=1}^{i=n} \Delta t_i}{t_P}$, where t_P is the total signal sampling time. The probability density function, $p[u_{cj}(t)]$, $j = 1, 2, 3, \ldots$, is then defined to be:

$$p[u_{cj}(t)] = \lim_{\Delta u_c(t) \to 0} \left[\frac{1}{\Delta u_c(t)} \right] \lim_{t_P \to \infty} \left\{ \frac{\sum\limits_{i=1}^{i=n} \Delta t_{ij}}{t_P} \right\} \qquad [6\text{-}38]$$

A graph of $p[u_{cj}(t)]$ vs. $u_{cj}(t)$ as $\lim_{\Delta u_c(t) \to 0} \dfrac{\Delta t_{ij}}{\Delta u_c(t)} \to \dfrac{\delta t}{\delta u_c}$, i.e., the plot of:

$$p(u_c) = \frac{1}{t_P} \frac{\delta t}{\delta u_c} \qquad [6\text{-}39]$$

vs. u_c as u_c varies continuously throughout the sampling interval t_P (hence we can drop the subscript j above) provides important information concerning the temporal distribution of the random data being analyzed. If the graph shows a typical bell-shaped distribution around a mean-value that does not "drift" or change with time, the data is said to be *stationary*. Otherwise, it is *nonstationary*, meaning that (among other things) the autocorrelation function [6-25] becomes a function of t as well as τ. Probability density function plots are also a convenient means for distinguishing between deterministic periodic information and nondeterministic random data, and they provide additional insights into the so-called "central tendencies" of the data as follows.

THE MOMENTS OF A PROBABILITY DENSITY FUNCTION

The m'th moment of a probability density function is defined to be:

$$\overline{u_c^{\,m}} = \int_{-\infty}^{+\infty} u_c^{\,m} \, p(u_c) \, \delta u_c \qquad [6\text{-}40]$$

For $m = 0$, using equation [6-39] we see that:

$$1 = \int_{-\infty}^{+\infty} p(u_c)\,\delta u_c = \int_0^{t_p} \frac{\delta t}{t_p} = \frac{t_p}{t_p} \qquad \text{[6-41]}$$

which simply states that the probability of finding *any* value of u_c somewhere in the time interval t_p is 100%. For $m = 1$,

$$\overline{u_c(t)} = \int_{-\infty}^{+\infty} u_c\,p(u_c)\,\delta u_c = \lim_{t_p \to \infty} \frac{1}{t_p} \int_o^{t_p} u_c(t)\,\delta t \qquad \text{[6-42]}$$

Equation [6-42] defines the time-averaged *mean value* of $u_c(t)$, which is related to the autocorrelation coefficient by equation [6-29]. The *difference* between $u_c(t)$ at any instant of time, t, and the mean value of $u_c(t)$ as defined by equation [6-42] is called the *deviation* of $u_c = u_c(t) - \overline{u_c(t)}$. The time-averaged mean of the *square* of the deviations, i.e.,

$$\overline{[u_c(t) - \overline{u_c(t)}]^2} = \overline{u_c{}^2(t) - 2u_c(t)\,\overline{u_c(t)} + [\overline{u_c(t)}]^2}$$

$$= \overline{u_c{}^2(t)} - 2[\overline{u_c(t)}]^2 + [\overline{u_c(t)}]^2,$$

or,

$$\lim_{t_p \to \infty} \frac{1}{t_p} \int_o^{t_p} [u_c(t) - \overline{u_c(t)}]^2\,\delta t = \overline{u_c{}^2(t)} - [\overline{u_c(t)}]^2 \qquad \text{[6-43]}$$

is called the *variance* of u_c, which, as can be seen from equation [6-43] is equal to the difference between the *mean-square-value* of $u_c(t)$ (i.e., equation [6-26] for $R_t(0)$) and the *mean-value-squared* of $u_c(t)$ (i.e., equation [6-29] for $R_t(\infty)$). The mean-squared-value of $u_c(t)$ is also given by equation [6-40] for $m = 2$, i.e.,

$$\overline{u_c{}^2(t)} = \int_{-\infty}^{+\infty} u_c{}^2\,p(u_c)\,\delta u_c = \lim_{t_p \to \infty} \frac{1}{t_p} \int_o^{t_p} u_c{}^2(t)\,\delta t \qquad \text{[6-44]}$$

Again, the square-root of the mean-square-value is called the *root-mean-square*, or RMS value of $u_c(t)$. The square-root of the variance is called the *standard deviation* of $u_c(t)$ (i.e., the square-root of equation [6-43]).

The *mode* of $u_c(t)$ is that value of u_c that appears most often in the sampling period t_p, and the difference between the *mean* of u_c and its *mode* is called the *skewness* of the data. It is defined by the third moment, $m = 3$ of the probability density function:

$$\overline{u_c^{\,3}(t)} = \int_{-\infty}^{+\infty} u_c^{\,3}\, p(u_c)\, \delta u_c = \lim_{t_p \to \infty} \frac{1}{t_p} \int_o^{t_p} u_c^{\,3}(t)\, \delta t \qquad \text{[6-45]}$$

The skewness is sometimes normalized with respect to the cube of the standard deviation.

The fourth moment ($m = 4$ in equation [6-40]), usually normalized with respect to the mean-square-value, $m = 2$, squared) is used to define the *flatness factor*, or *kurtosis* of the plot of $p(u_c)$ vs. u_c. This is a measure of the degree of taper of the normal distribution (bell-shaped) curve discussed in the previous section, i.e., to what extent the data tends to "hover" around its mean value, just as the variance is a measure of the variability or dispersion of the random variable. A related parameter, the *width factor*, which is the standard deviation divided by the mean, measures the range over which most of the data tend to occur (i.e., the "width" of the bell-shaped curve).

Proceeding in like fashion, one can compute all of the moments of the probability density function, and thus completely characterize the nature of the biological signal being analyzed. Conversely, if one *knew* all of the moments of $p(u_c)$ ahead of time, one could, from these, completely characterize the probability density function, itself. Quite often, one can get at least two of these -- the mean, $\overline{u_c(t)} = \sqrt{R_f(\infty)} = \overline{M}$, and the standard deviation, S.D., from equation [6-43] -- with relatively little effort. If this is the case, then *Tchebycheff's Inequality* may be employed, which asserts that, for any real number, c^*, the probability that a random variable, $u_c(t)$ lies in the interval $\overline{M} + c^*$S.D. and $\overline{M} - c^*$S.D. is larger than $1 - \dfrac{1}{c^{*2}}$. Thus, for example, if $c^* = 3$, there would be nearly a 90% chance that $u_c(t)$ would fall in the interval between $\overline{M}$ and 3 standard deviations on either side of $\overline{M}$. A more accurate prediction of such probabilities when $p(u_c)$ *is* known can be obtained by putting *finite* limits on equation [6-40] with $m = 0$, thus defining the so-called *Probability Distribution Function, $P[u_c(t)]$*. The latter states that the probability of finding $u_c(t)$ somewhere within any given interval, u_{c1} and u_{c2} is given by:

$$P[u_c(t)] = \int_{u_{c1}}^{u_{c2}} p(u_c) \, \delta u_c \qquad \qquad [6\text{-}46]$$

If we let u_{c1} go to minus infinity, equation [6-46] yields the probability that $u_c(t)$ is instantaneously less than u_{c2} at any given time. If we let u_{c2} go to plus infinity, we get from this equation the probability that $u_c(t)$ is instantaneously greater than u_{c1} at any given time. If we let *both* of these limits be approached, then, of course, we recover equation [6-41] identically.

POWER SPECTRAL DENSITY FUNCTIONS

The random data equivalent to a Fourier analysis for periodic data is the power spectral density function, which transforms the analysis from the time domain into the frequency domain. the "PDF" may be obtained from the Fourier Transform of the autocorrelation function as follows:

$$PDF(f) = 4\int_{0}^{\infty} R_f(\tau) \cos 2\pi f \tau \, \delta \tau \qquad \qquad [6\text{-}47]$$

This reduced form of the more general expression [6-37] derives from the fact that the autocorrelation function is an even function of τ (Bendat and Piersol, 1971).

A plot of *PDF(f)* vs. frequency, f, results in an energy spectral diagram which describes the general frequency composition of the data being considered. As such, then, it may be used like a Fourier spectrum for periodic data to characterize the biological signal under investigation.

SUMMARY

The four descriptive properties discussed, i.e., correlation analysis, probability density functions, measures of central tendency (moments of probability density functions), and power spectral density functions, are the major ones involved in the characterization of nondeterministic, stationary random signals (sometimes called *stochastic processes*, from the Greek, "stokhiastikos," meaning, "to guess at."). These functions allow us to "guess" (based on probability theory) what a particular random variable is "most likely" to do when it cannot otherwise be exactly described by a mathematical equation. Our guesses are most often (or, we would *like them* to be) based on *time* averages of the random data, and this is allowable as long as the data is *ergodic*. In order to explain what this means, consider the following:

Suppose you were interested in determining the average temperature of the human body. Strictly speaking, to get a precise value for this thermodynamic variable, you should measure the body temperature of *each and every* human being (theoretically, past and present) inhabiting this planet -- to insure that you will have accessed *all* of the possible manifestations of this particular event. Such global temperature measurements should be recorded continuously over a very long period of time; averaged for each individual; and then the time-averages should be spatially (geographically) averaged across the entire population of the earth. What you will have calculated is called the *ensemble* average, and it is the "proper" way to get the information you seek. However, it is obviously not the most *desirable* way -- especially since, to be further precise, you should have done this since the beginning of humankind ($t = -\infty$), lest you may have *missed* one possible manifestation of this variable, and, you should continue to do it ad infinitum forever more ($t = +\infty$).

Well, you are lucky. Because people all over the world are basically physiologically the *same* (at least as it relates to body temperature), the chances are pretty good that if you simply measure the temperature of *one* (or, to be on the safe side, perhaps a few) individual(s) for a long enough period of time, you will probably get to witness sooner or later the entire spectrum of values that this variable is most likely to assume. In other words, your *time average* for a *portion* of the ensemble (as $t_P \to \infty$) will probably be very close to, if not exactly equal to, the *ensemble average* for the *entire* population (sample size $\to \infty$). Data for which this is true is called *ergodic* data, and, to be ergodic, the data must at least be stationary, i.e., its time-averaged properties must not change.

On the other hand, suppose, now, that what you wished to measure instead, is the average temperature on the surface of the *earth*. If you were to use the same approach as above, i.e., park yourself in one (or perhaps several) location(s) with a thermometer in hand and continuously monitor T, you would obviously *not* eventually encounter *all* of the possible manifestations of environmental temperature. This is because climate all over the world is not the same, and neither are its time-average properties (i.e., it is nonstationary). Thus, the continuous *temporal* average of a random sample size will never be the same as the *ensemble* average -- which is to say, the random data is *Nonergodic*.

Philosophers argue over whether or not there really *is* such a thing as purely random, nonstationary, nonergodic data. The classification of such is presumed by some to be merely our admission of ignorance as to an understanding of the basic physical principles that govern our existence -- or, as to our ability to quantify these for the purpose of determinism and predictability. There is undoubtedly some merit in this point of view in that it keeps us asking questions, searching for answers, and continuing in our quest to comprehend life.

On the other hand, there is *also* a school of thought that proposes unpredictability to be a *fundamental* part of our existence. *Nothing*, they argue, can be entirely certain under any circumstances because there is an inherent "dynamic freedom" that neither will, nor ever *can* be totally and completely described, to the point of making things predictable. Regardless of which point of view suits you best, the fact of the matter is that, in the absence of a unified theory of our universe, the methods of statistical analysis have provided us with the means to get meaningful estimates (such estimates having associated with them a "standard error" of measurement) of life's processes. These shall have to suffice while we continue to travel the road to ultimate knowledge.

One last point to summarize the essential features of information processing as it relates to physiologic function:

Based on the mechanisms by which excitable tissue in physiologic systems transport information -- in the form of digitized, discrete action potentials -- and within the context of Fourier series for periodic signals and power spectral density functions for random data, the human body may be considered from an information processing point of view to be one large frequency analyzer.

The fact that tissues have specific excitation thresholds, that they obey digital all-or-none (on/off) response laws, that they transmit information coded into *which* particular courier (neuron) is firing at *what* particular frequency and propagation speed, and that they can be summed into compound action potentials to be delivered to specific distinations -- all suggest that the system is performing some very sophisticated frequency analysis on the information being transported; and that the previously discussed "body language" is, in fact, coded into the frequency spectra of compound action potentials. In fact, apropos to the previous comments concerning the "determinism" of random data, the human body may know more than we do about how to handle such data; and it would behoove us to study the organism in much greater detail to gain some insights as to how this can be done.

The concept of viewing the human organism as a frequency analyzer goes along with that which, from a thermodynamic perspective, sees the body as an isothermal, isobaric, electrochemical engine designed more for speed and locomotion than for doing work. These two concepts really embody most of the engineering principles related to physiologic function, and that is what the material in this text has addressed. We thus close with the following remarks.

Concluding Remarks

In this chapter, we have put forth the premise that physiologic function is carefully controlled by homeostatic regulators, and that such regulators may have a corresponding syntax, or language that is uniquely their own. The language is coded into electromagnetic information that is transmitted through the medium of compound action potentials. These are digitized bits of data that are sensory (informative) or motor (causative), depending upon whether they are being transmitted to (afferent) or from (efferent) the central nervous system. Homeostatic regulators can adapt to persistent environmental disturbances, to change system reference or "set points" depending on the needs of the organism. Conversely, these regulators, together with their respective target organs or tissues can disappear (atrophy) if they become no longer useful to the physiologic system. This aspect of physiologic adaptation has come to be known as the "use it, or lose it" syndrome. It is most prominent, for example, in the case of muscles that lose their ability to develop proper tone when not exercised regularly (as is the case with paraplegics and other victims of spinal column injuries); and in the case of bones that literally have a tendency to self-destruct if not called upon to function in accordance with their load-bearing physiologic purpose (as is the case with astronauts in a sub-gravity environment, and individuals that are bedridden or otherwise confined to wheel chairs). It is *less* prominent, though perhaps even more devastating, in a society whose educational system stresses left-brain, cognitive-rational modes of information processing (see Chapter 1) to the exclusion of (even the specific *discrimination* against) right-brain, spatial-creative modes of information processing. This issue is addressed elsewhere (Schneck, J. K., and Schneck, D. J., 1989) and we shall not pursue it here.

The systems approach to physiologic function, aspects of which have been discussed in this text, has also received a great deal of attention by other investigators. Most notable among these are Apter (1970), Clynes (1970), Milsum (1966, 1970), Milhorn (1966), Brown and Gann (1973), Huffaker (1974), Kline (1976), Cooney (1976), and Brown (1985). The uniqueness of this approach is that it reduces seemingly complex, unrelated sequences of physiologic processes into more readily identifiable sets of common denominators that illucidate some basic principles of biologic function. Appropriate interpretation of these biologic principles can be expected to move us closer to success in our continuing efforts to improve the health, comfort and understanding of humankind. Indeed, the ideas formulated herein -- which have addressed the spectrum of physiologic issues ranging: (i) from survival of the fittest attributes (selective mass transport processes, enzyme-controlled biochemical reactions, optimization schemes, adaptation and control) to pathologic processes; (ii) from mass, energy and momentum transport to mass, energy and momentum utilization; (iii) from variational principles involving the mechanics of human motion to biological signal analysis; (iv) from genetics to psychological conditioning; (v) from normal metabolism to allergic reactions; and (vi) from purely analytic to numerical, nondimensional, and experimental techniques -- have all been developed with the ultimate intent of stimulating new and creative approaches to the analysis of physiologic function. Such approaches should proceed from the point of view that *all* such function basically fits into an underlying pattern, i.e., that it is deterministic and capable of being deductively analyzed. The details then become simply special cases (or different sets of boundary conditions) within the framework of a unified guiding theory. If, and when, we ever reach that stage, it is the hope of this author that this textbook will have made some reasonable contribution toward that end.

References

Ad Hoc Committee On Polar Biomedical Research, (1982), "Polar Biomedical Research: An Assessment (Part I) and Appendix On Polar Medicine -- A Literature Review (Part II)," National Research Council, Commission On Physical Sciences, Mathematics, and Resources, Polar Research Board Report, Washington, D. C., The National Academy Press."

Adamson, A. W., (1973), *A Textbook Of Physical Chemistry*, New York, Academic Press.

Albus, J. S., (1975), "Data Storage In The Cerebellar Model Articulation Controller (CMAC)," *J. Dynamic Systems, Measurement, Control*, Vol. 97, pp. 228-233.

Altman, P. L., and Dittmer, D. S., (1961), *Blood And Other Body Fluids*, Bethesda, Maryland, The Federation Of American Societies For Experimental Biology.

Altman, P. L., and Dittmer, D. S., eds., (1971), *Respiration And Circulation*, Bethesda, Maryland, Federation Of American Societies For Experimental Biology.

Amdur, I., and Hammes, G., (1966), *Chemical Kinetics: Principles And Selected Topics*, New York, McGraw-Hill Book Company.

An, K. N., Kwak, B. M., Chao, E. Y., and Morrey, B. F., (1984), "Determination of Muscle and Joint Forces: A New Technique to Solve the Indeterminate Problem," *J. Biomech. Engrg., Trans. ASME*, Vol. 106, pp. 364-367.

Apter, J., (1970), "Biosystems Modeling," in: *Biomedical Engineering Systems*, (Edited by M. Clynes and J. H. Milsum), New York, McGraw-Hill Book Company.

Astarita, G., and Marrucci, G., (1974), *Principles Of Non-Newtonian Fluid Mechanics*, New York, McGraw-Hill Book Company.

Atkins, A. R., (1962), "A Method Of Simulating Heat Flow And Control In A Nude Man With An Analogue Computer," M.S. Thesis, University Of The Witwatersrand, Johannesburg.

Barineau, D., and Schneck, D. J., (1988), "Improvements In The Control of Robotic Motion Simulations Using The ATB Model," Virginia Polytechnic Institute and State University Technical Report Number VPI-E-88-38; Blacksburg, Virginia.

Barrow, G. M., (1974), *Physical Chemistry For The Life Sciences*, New York, McGraw-Hill Book Company.

Becker, H. A., (1976), *Dimensionless Parameters: Theory And Methodology*, London, Applied Science Publishers, Ltd.

Beckett, R., and Chang, K., (1968), "An Evaluation Of The Kinemamatics Of Gait By Minimum Energy," *J. Biomech.*, Vol. 1, pp. 147 - 159.

Beckett, R., and Pan, K. C., (1971), "Analysis Of Gait Using A Minimum Energy Approach," Aerospace Medical Research Laboratory Technical Report Number AMRL-TR-71-29, Paper No. 33, pp. 823-841, Ohio, Wright-Patterson Air Force Base.

Beeler, M., Gosper, R. W., and Schroeppel, R., (1972), "Hakmem," MIT Artificial Intelligence Laboratory, AI Memo 239, Cambridge, Massachusetts.

Bejan, A., (1988), "A Second Look At The Second Law," *Mech. Eng.*, Vol. 110, No. 5, pp. 58-65.

Benati, M., Gaglio, S., Morasso, P., Tagliasco, V., and Zaccaria, R., (1980), "Anthropomorphic Robotics," *Biological Cybernetics*, Vol. 38, pp. 141-150.

Bendat, J. S., and Piersol, A. G., (1971), *Random Data: Analysis And Measurement Procedures*, New York, Wiley-Interscience.

Benson, H., (1975), *The Relaxation Response*, New York, Morrow.

Bertsekas, D. P., (1982), *Constrained Optimization And Lagrange Multiplier Methods*, New York, Academic Press.

Bird, R. B., Stewart, W. E., and Lightfoot, E. N., (1960), *Transport Phenomena*, New York, Wiley-Interscience.

Bligh, J., (1973), *Temperature Regulation In Mammals And Other Vertebrates*, New York, American Elsevier Publishing Company, Inc.

Bligh, J., and Moore, R., (1972), *Essays On Temperature Regulation*, New York, American Elsevier Publishing Company, Inc.

Bonner, D. M., and Mills, S. E., (1964), *Heredity*, New Jersey, Prentice-Hall, Inc.

Brady, M., Hollerbach, J. M., Johnson, T. L., Lozano-Perez, T., and Mason, M. T., eds., (1982), *Robot Motion: Planning And Control*, Cambridge, Massachusetts, The MIT Press.

Brighton, C. T., editor (1977), *Bioelectrical Effects On Bone And Cartilage; No. 124 In The Series On Clinical Orthopaedics And Related Research*, Philadelphia, Penna., Lippincott.

Bronner, F., and Kleinzeller, A., eds., (1975), *Current Topics In Membranes And Transport, Volume 7*, New York, Academic Press.

Bronzino, J. D., (1977), *Technology For Patient Care*, St. Louis, Missouri, C. V. Mosby, pp. 156-188.

Brown, J. H. U., and Gann, D. S., eds., (1973), *Engineering Principles In Physiology, Vol. 1 and 2*, New York, Academic Press.

Brown, R. F., (1985), *Biomedical Systems Analysis Via Compartmental Concept*, Cambridge, Massachusetts, Abacus Press.

Buckingham, E., (1914), "On Physically Similar Systems; Illustrations tions Of The Use Of Dimensional Equations," *Physical Review*, Vol. 4, Series 2, No. 4, pp. 345-376.

Buckingham, E., (1915), "Model Experiments And The Forms Of Empirical Equations," *Transactions Of The American Society Of Mechan-*

ical Engineers, Vol. 37, pp. 263-292 (with a discussion from pages 292 to 296).

Bull, H. B., (1964), *An Introduction To Physical Biochemistry,* Philadelphia, Penna., F. A. Davis Company.

Butler, F. E., and Fleck, J. T., (1980), "Advanced Restraint Systems Modelling," Aerospace Medical Research Laboratory Technical Report Number AFAMRL-TR-80-14, Ohio, Wright-Patterson Air Force Base (NTIS No. AD-A088 029).

Butler, F. E., Fleck, J. T., and Difranco, D. A., (1983), "Modelling Of Whole-Body Response To Windblast," Aerospace Medical Research Laboratory Technical Report Number AFAMRL-TR-83-73, Ohio, Wright-Patterson Air Force Base (NTIS No. AD-B079 184).

Cena, K., and Clark, J. A., eds., (1981), *Bioengineering, Thermal Physiology And Comfort,* New York, Elsevier Scientific Publishing Company.

Chao, E. Y., and An, K. N., (1978), "Graphical Interpretation Of The Solution To The Redundant Problem In Biomechanics," *Trans. Am. Soc. Mech. Eng., J. Biomech. Eng.,* Vol. 100, pp. 159-167.

Charm, S., Paltiel, B., and Kurland, G. S., (1968), "Heat Transfer Coefficients In Blood Flow", *Biorheology,* Vol. 5, pp. 133-145.

Childress, D. S., Holmes, D. W., and Billock, J. N., (1974), "Ideas On Myoelectric Prosthetic Systems For Upper Extremity Amputees," In: *The Control Of Upper Extremity Prostheses And Orthoses,* (Edited by P. Herberts, R. Kadefors, R. Magnusson and I. Peterson); Springfield, Illinois, Thomas, pp. 86-106.

Christensen, H. N., (1962), *Biological Transport,* New York, Benjamin.

Christensen, H. N., and Cellarius, R. A., (1972), *Introduction To Bioenergetics: Thermodynamics For The Biologist,* Philadelphia, W. B. Saunders Company.

Clynes, M., (1970), "Toward A View Of Man," In: *Biomedical Engineering Systems,* (Edited by M. Clynes and J. H. Milsum), New York, McGraw-Hill Book Company.

Clynes, M., and Milsum, J. H., (1970), *Biomedical Engineering Systems,* New York, McGraw-Hill Book Company.

Colmano, G., Gregg, J. M., Schwab, W. E., Schneck, D.J., Herbert, W. G., and Berlien, C. D., (1983), "Development Of A Rat Model For Computer Recorded Sciatic-Nerve Signals (Before Injury) To Assist Repair Using Metal (Na, K, Ca) Monolayers Coated Sutures," *Virginia Journal of Science, Medical Section,* Vol. 34, No. 3, pg. 183.

Cooney, D. O., (1976), *Biomedical Engineering Principles,* New York, Marcel Dekker.

Cowherd, W. F., II, and Cadman, T. W., (1969), "Numerical Prediction Of Transfer Functions," *Instruments And Control Systems* Vol. 42, No. 5, pp. 109-113.

Cox, B., Lomax, P., Milton, A.S., and Schonbaum, E., eds., (1980) *Thermoregulatory Mechanisms And Their Therapeutic Implications,* Proceedings Of The Fourth International Symposium On The Pharmacology Of Thermoregulation Held At Oxford, July 30 to August 3, 1979; New York, S. Karger-Basel.

Crank, J.,(1956), *The Mathematics Of Diffusion*, Oxford, Clarendon.

Crowninshield, R. D., (1978), "Use of Optimization Techniques to Predict Muscle Forces," *Adv. Bioengineering*, ASME.

Crowninshield, R. D., and Brand, R. A., (1981), "A Physiologically Based Criterion of Muscle Force Prediction in Locomotion," *J. Biomech.*, Vol. 14, pp. 793-801.

Cunningham, B. A., (1977), "The Structure And Function Of Histocompatibility Antigens," *Scientific American*, Vol. 234, No. 4, pp. 96-103.

Curran, P. F., and Schultz, S. G., (1964), "Transport Across Membranes," In: *Handbook Of Physiology, Section 6: Alimentary Canal, Volume 3*, (Edited by M. B. Visscher), Baltimore, Williams and Wilkins.

Dahl, D. J., and Hoken, J. K., (1974) "The Sodium-Potassium Adenosine-Triphosphatase," *Annual Review Of Biochemistry*, Vol. 43, pp. 327-337.

Davis, R. B., III, Schneck, D. J., and Gutstein, W. H., (1984), "Pulsatile Flow Through A Branching Tube With Collapsing Walls," *Transactions Of The American Society Of Mechanical Engineers, Journal Of Fluids Engineering*, Vol. 106, No. 4, pp. 430 - 434.

Davis, R. B., III, (1986), "Musculoskeletal Biomechanics: Fundamental Measurements And Analysis," In: *Biomedical Engineering And Instrumentation: Basic Concepts And Applications*, (Edited by Joseph D. Bronzino), Boston, Prindle, Weber and Schmidt (PWS), Chapter 6, pp. 180-222.

Davson, H., and Danielli, J. F., (1952), *The Permeability Of Natural Membranes*, Cambridge, Cambridge University Press.

DeCoulon, F., (1986), *Signal Theory And Processing*, Dedham, Massachusetts, Artech House.

Del Toro, V., and Parker, S. R., (1960), *Principles Of Control Systems Engineering*, New York, McGraw-Hill Book Company.

Dempster, W. T., (1961), "Free-Body Diagrams As An Approach To The Mechanics Of Human Posture And Motion," In: *Biomechanics Of The Musculo-Skeletal System*, (Edited by F. Gaynor Evans), Springfield, Illinois, Charles C. Thomas, Chapter 5, pp. 81-135.

Denkewalter, R. G., and Hirschmann, R., (1969), "The Synthesis Of An Enzyme," *American Scientist*, Vol. 57, No. 4, pp. 389-409.

Desai, C. S., and Siriwardane, H. J., (1984), *Constitutive Laws For Engineering Materials*, Englewood Cliffs, New Jersey, Prentice-Hall, Inc.

Deutsch, D. H., (1981), "Can The Second Law Of Thermodynamics Be Circumvented?" *American Laboratory*,Vol. 13, No. 5, pp. 54-65.

Deutsch, S., and Micheli-Tzanakou, E., (1987), *Neuroelectric Systems*, New York, New York University Press.

DiStefano, J. J., III, Stubberud, A. R., and Williams, I. J., (1967), *Schaum's Outline Of Theory And Problems Of Feedback And Control Systems*, New York, McGraw-Hill Book Company, Inc.

Edwards, B., (1979), *Drawing On The Right Side Of The Brain*, New York, J. P. Tarcher.

Eyring, H., and Eyring, E. M., (1963), *Modern Chemical Kinetics*, New York, Reinhold Publishing Corporation.

Flash, T., and Hogan, N., (1985), "The Coordination Of Arm Movements: An Experimentally Confirmed Mathematical Model," *The Journal Of Neuroscience*, Vol. 5, pp. 1688-1703.

Fleck, J. T., and Butler, F. E., (1975), "Development Of An Improved Computer Model Of The Human Body And Extremity Dynamics,"Aerospace Medical Research Laboratory Technical Report Number AMRL-TR-75-14, Ohio, Wright-Patterson Air Force Base.

Fleck, J. T., Butler, F. E., and Deleys, N. J., (1982), "Validation Of The Crash Victim Simulator," Department Of Transportation Technical Reports Number DOT-HS-806-279 to 282 (Vols. 1-4, inclusive); Washington, D. C.

Fleming, D.G., (1969), "Physiology Of Nerve And Muscle," In: *Bio-Electric Phenomena*, (By Robert Plonsey), New York, McGraw-Hill Book Company, Inc., Chapter 1, pp. 1-22.

Friedman, M. H., (1986), *Principles And Models Of Biological Transport*, New York, Springer-Verlag.

Frost, A. A., and Pearson, R. G., (1961), *Kinetics And Mechanisms* New York, John Wiley And Sons.

Gainer, J. L., and Metzner, A. B., eds., (1965), *Transport Phenomena: Proceedings Of A Joint Meeting Of The American Institute Of Chemical Engineers And The Institution Of Chemical Engineers*, New York, American Institute of Chemical Engineers.

Giese, A., (1968), *Cell Physiology*, Third Edition, Philadelphia, Saunders.

Glasstone, S., Laidler, K. J., and Eyring, H., (1941), *The Theory Of Rate Processes*, New York, McGraw-Hill Book Company, Inc.

Glynn, I. M., and Karlish, S. J. D., (1975), "The Sodium Pump," *Annual Review Of Physiology*, Vol. 37, pp. 13-55.

Goldstick, T. K., (1973), "Oxygen Transport," In: *Engineering Principles In Physiology*, (Edited by J. H. U. Brown And D. S. Gann), New York, Academic Press, Volume 2, Chapter 22, pp. 257-282.

Golla, D.F., Garg, S.C., and Hughes, P.C., (1981), "Linear State-Feedback Control Of Manipulators," *Mech. Machine Theory*, Vol. 16, pp. 93-103.

Green, D. E., and Young, J. H., (1971), "Energy Transduction In Membrane Systems," *American Scientist*, Vol 59, No. 1, pp. 92-100.

Grodins, F. S., (1963), *Control Theory And Biological Systems*, New York, Columbia University Press.

Gueron, M., (1978), "Enhanced Selectivity Of Enzymes By Kinetic Proof-Reading," *American Scientist*, Vol. 66, No. 2, pp. 202-208.

Guyton, A. C., (1981), *Textbook Of Medical Physiology*, Sixth Edition, Philadelphia, Penna., W. B. Saunders Company.

Hall, J. L., and Baker, D. A., (1977), *Cell Membranes And Ion Transport*, London, England, Longman Group Ltd., pp. 85-111.

Hamilton, W. R., (1969), *Elements Of Quaternions*, New York, Chelsea Publishing Company.

Hardy, J. D., Gagge, A. P., and Stolwijk, J. A. J., eds., (1970), *Physiological And Behavioral Temperature Regulation*, Springfield, Illinois, Charles C. Thomas Publishers.

Harper, H. A., Rodwell, V. W., and Mayes, P. A., (1977), *Review Of Physiological Chemistry*, Sixteenth Edition, Los Altos, California, Lange Medical Publications.

Harris, E. J., (1960), *Transport And Accumulation In Biological Systems*, Second Edition, New York, Academic Press.

Hatze, H., (1981), *Myocybernetic Control Models Of Skeletal Muscle: Characteristics And Applications*, Muckleneuk, Pretoria, The University Of South Africa Press.

Hill, T. L., (1960), *Introduction To Statistical Thermodynamics*, Reading, Massachusetts, Addison-Wesley Publishing Co., Inc.

Hoffman, J., editor, (1964), *The Cellular Functions Of Membrane Transport*, Englewood Cliffs, New Jersey, Prentice-Hall.

Hokin, L. E., and Kokin, M. R., (1963), "Biological Transport," *Ann. Rev. of Biochemistry*, Vol. 32, pp. 553-578.

Hollerbach, J. M., (1982), "Dynamics," In: *Robot Motion: Planning And Control*, (Edited by M. Brady, J. M. Hollerbach, T. L. Johnson, T. Lozano-Perez, and M. T. Mason), Cambridge, Massachusetts, The MIT Press, Chapter 2, pp. 51-88.

Huennekens, F. M., and Whiteley, H. R., (1960), "Phosphoric Acid Anhydrides And Other Energy-Rich Compounds," In: *Comparative Biochemistry*, (Edited By M. Florkin and H.S. Mason), Vol. 1, pp. 107-180.

Huffaker, C. B., editor, (1974), *Biological Control*, New York, Plenum Press.

Israelachvili, J. N., (1985), *Intermolecular And Surface Forces*, San Diego, Calif., Academic Press/Harcourt Brace Jovanovich.

Jacob, S. W., Francone, C. A., and Lossow, W. J., (1982), *Structure And Function In Man*, Fifth Edition, Philadelphia, W. B. Saunders Company.

Jeffreys, H., (1965), *Cartesian Tensors*, London, Cambridge University Press.

Jones, C. W., (1976), *Biological Energy Conservation*, New York, Chapman and Hall Publishers.

Jones, M. N., (1979), *Biochemical Thermodynamics*, New York, Elsevier Scientific Publishing Company.

Jones, R. P., and Godfrey, K. R., (1981), "Problems In The Compartmental Modelling Of Biomedical Systems," *IEEE Conference On Control And Its Applications*, Warwick, Conference Bulletin Number 194, pp. 43-47.

Jones, R. W., (1969), "Biological Control Mechanisms," In: *Biological Engineering*, (Edited By H. P. Schwan), New York, McGraw-Hill Book Company, Inc.

Jost, W., (1960), *Diffusion In Solids, Liquids, Gases*, Second Edition, New York, Academic Press.

Kalmus, H., Editor, (1966), *Regulation And Control In Living Systems*, New York, John Wiley And Sons.

Kang, Q. S., Updike, D. P., and Salathe, E. P., (1990), *J. Biomech. Engrg.* (In Press).

Kaplan, J. H., (1985), "Ion Movements Through The Sodium Pump," *Annual Review Of Physiology*, Vol. 47, pp. 435-444.

Kaplan, W., (1952), *Advanced Calculus*, Reading, Massachusetts, Addison-Wesley, pp. 128-129.

Kays, W. M., (1966), *Convective Heat And Mass Transfer*, New York, McGraw-Hill Book Company, Inc.

Kerkut, G. A., (1971), *The Electrogenic Sodium Pump*, Bristol, Scientechnica Ltd., pp. 1-57.

Klein, M. J., (1970), "Maxwell, His Demon, And The Second Law Of Thermodynamics," *American Scientist*, Vol. 58, No. 1, pp. 84-97.

Kleinzeller, A., and Kotyk, A., eds., (1961), *Membrane Transport And Metabolism*, Prague, The Czechoslovakian Academy Of Science, New York, Academic Press.

Kline, J., Editor, (1976), *Biological Foundations Of Biomedical Engineering*, Boston, Mass., Little, Brown and Company.

Kline, J., Editor, (1987), *Handbook Of Biomedical Engineering*, San Diego, Calif., Academic Press/Harcourt Brace Jovanovich.

Klotz, I. M., (1957), *Energetics In Biochemical Reactions*, New York, Academic Press.

Klotz, I. M., (1967), *Energy Changes In Biochemical Reactions*, New York, Academic Press.

Kuffler, S., Nicholls, J., and Martin, A., (1984), *From Neuron To Brain*, Second Edition, New York, Sinauer Associates, Inc., pp. 187-205.

LaBarbera, M., and Vogel, S., (1982), "The Design Of Fluid Transport Systems In Organisms," *American Scientist*, Vol. 70, No. 1, pp. 54-60.

Lacquanti, F., and Soechting, J. F., (1982), "Coordination Of Arm And Wrist Motion During A Reaching Task," *The Journal Of Neuroscience*, Vol. 2, pp. 399-408.

Laidler, K. J., (1965), *Chemical Kinetics*, New York, McGraw-Hill Book Company, Inc.

Lamb, H.,(1898), "On The Velocity Of Sound In A Tube, As Affected By The Elasticity Of The Walls," *Manchester Memoirs*, Vol. 42, pg. 1.

Langley, L. L., (1971), *Review Of Physiology*, New York, McGraw-Hill Book Company, Inc.

Leake, C. D., (1962), "The Historical Development Of Cardiovascular Physiology," In: *Handbook Of Physiology, Section 2: Circulation, Volume 1*, (Edited By W. F. Hamilton and P. Dow), Washington, The American Physiological Society; Baltimore, The Williams and Wilkins Company, pp. 11-22.

Lee, C. P., Editor, (1987), *Structure, Biogenesis, And Assembly Of Energy Transducing Enzyme Systems*, San Diego, California, Academic Press/Harcourt Brace Jovanovich Publishers.

Lehninger, A. L., (1965), *Bioenergetics*, Menlo Park, California, W. A. Benjamin, Incorporated.

Lenchene, C., (1988), *The Na+, K+ Pump, Part B: Cellular Aspects (Physiologic Role Of The Na-K Pump)*, New York, Alan R. Liss, Inc., pp. 171-194.

Lightfoot, E. N., (1974), *Transport Phenomena And Living Systems: Biomedical Aspects Of Momentum And Mass Transport*, New York, Wiley-Interscience, John Wiley And Sons, Inc.

Llaurado, J. G., (1973), "Compartmental Approaches To Water And Electrolyte Distribution," In: *Engineering Principles In Physiology, Volume II*, (Edited By J. H. U. Brown and D. S. Gann), New York, Academic Press, Chapter 25, pp. 347-373.

Lykov, A. V., and Mikhaylov, Y. A., (1961), *Theory Of Energy And Mass Transfer*, Englewood Cliffs, New Jersey, Prentice-Hall.

Maeda, H., Kodama, S., and Kajiya, F., (1976), "System-Theoretic Considerations For Compartmental Analysis -- Transfer Functions Of Mammillary Systems," *Systems, Computers, Controls*, Vol. 7, No. 3, pp. 27-33.

May, C. D., (1974), "Food Allergy," In: *Infant Nutrition*, (Edited By S. J. Fomo), Second Edition, Philadelphia, Penna., W. B. Saunders Company.

McElroy, W. D., (1964), *Cell Physiology And Biochemistry*, New Jersey, Prentice-Hall, Inc.

McElroy, W. D., and Glass, B., eds., (1954), *The Mechanism Of Enzyme Action*, Baltimore, Md., The Johns Hopkins Press.

Mende, T. J., (1976a), "Enzymes And Pathways For ATP Formation," In: *Biological Foundations Of Biomedical Engineering*, (Edited By J. Kline), Boston, Little, Brown and Company.

Mende, T. J., (1976b), "Energetics," In: *Biological Foundations Of Biomedical Engineering*, (Edited By J. Kline), Boston, Little, Brown and Company.

Mikulecky, D. C., (1983), "Network Thermodynamics: A Candidate For A Common Language For Theoretical And Experimental Biology," *American Journal Of Physiology*, Vol. 245, (Regulatory Integrative Comp. Physiol. 14), pp. R1-R9.

Mikulecky, D. C., (1984), "Network Thermodynamics: A Simulation And Modeling Method Based On The Extension Of Thermodynamic Thinking Into The Realm Of Highly Organized Systems," *Mathematical Biosciences*, Vol. 71, pp. 1-23.

Mikulecky, D. C., and Thomas, S.R., (1979), "Some Network Thermodynamic Models Of Coupled, Dynamic Physiological Systems," *Journal Of The Franklin Institute*, Vol. 308, No. 3, pp. 308 - 326.

Milhorn, H. T., Jr., (1966), *The Application Of Control Theory To Physiological Systems*, Philadelphia, Penna., W. B. Saunders.

Milsum, J. H., (1966), *Biological Control Systems Analysis*, New York, McGraw-Hill Book Company, Inc.

Milsum, J. H., (1970), "Biological Systems Analysis And Control Theory," In: *Biomedical Engineering Systems*, (Edited By M. Clynes and J.H. Milsum), New York, McGraw-Hill Book Company, Inc., Chapter 6, pp. 212-271.

Mitvalsky, V., (1965), "Heat Transfer In Laminar Flow Of Human Blood Through Tube And Annulus," *Nature*, Vol. 206, pg. 307.

Moore-Ede, M. C., Sulzman, F. M., and Fuller, C. A., (1982), *The Clocks That Time Us: Physiology Of The Circadian Timing System*, Cambridge, Mass., The Harvard University Press.

Morrison, R.T., and Boyd, R.N., (1983), *Organic Chemistry*, Fourth Edition, Boston, Allyn and Bacon, Inc.

Morse, P. M., and Feshbach, H., (1953), *Methods Of Theoretical Physics*, Part I, New York, McGraw-Hill Book Company, Inc.

Mullins, L. J., and Brinley, F.J., (1967), *J. General Physiology*, Vol. 51, pp. 2333-2355.

Mylrea, K. C., and Abbrecht, P. H., (1973), "Use Of Systems Techniques In The Analysis Of The Control Of Erythropoiesis," In: *Advances In Chemistry, Number 118: Chemical Engineering In Medicine*, New York, The American Chemical Society.

National Aeronautics and Space Administration, (1981), *Biomedical Research In Space*, NASA CR-3487, Falls Church, Virginia, BioTechnology, Inc.

Neilands, J. B., and Stumpf, P. K., (1958), *Outlines Of Enzyme Chemistry*, Second Edition, New York, Wiley-Interscience.

Nightingale, J. M., (1956a), "Automatic Control Systems," *Machine Design*, May 17, 1956, pp. 35-42.

Nightingale, J. M., (1956b), "A Basic Outline Of Servo Mathematics," *Machine Design*, June 28, 1956, pp. 74-81.

Nightingale, J.M., (1956c), "Criteria For Evaluating Servo System Performance: Part I, Frequency Response And Stability, Nyquist Criterion, Algebraic Criteria, Response Criteria," *Machine Design*, July 26, 1956, pp. 79-84.

Nightingale, J. M., (1956d), "Criteria For Evaluating Servo System Performance: Part II, Steady-State Errors, Nyquist Plot, Transient Response, Relative Damping," *Machine Design*, August 9, 1956, pp. 106-110.

Nightingale, J. M., (1956e), "Analyzing Servo Systems," *Machine Design*, November 1, 1956, pp. 87-91.

Nightingale, J. M., (1959), "Servo Feedback Systems," *Machine Design*, August 20, 1959, pp. 155-159.

Nubar, Y., and Contini, R., (1961), "A Minimal Principle In Biomechanics," *Bull. Math. Biophysics*, Vol. 23, pp. 377-391.

Obergefell, L. A., Kaleps, I., Gardner, T. R., and Fleck, J. T., (1988a), "Articulated Total Body Model Enhancements," Armstrong Aerospace Medical Research Laboratory Technical Reports Number AAMRL-TR-88-007 thru 009, Volumes 1-3, Ohio, Wright-Patterson Air Force Base.

Obergefell, L. A., Avula, X., and Kaleps, I., (1988b), "The Uses Of The Articulated Total Body Model As A Robot Dynamics Simulation Tool," *Proceedings Of The 1988 SOAR Conference*, Dayton, Ohio, Wright State University.

Obert, E. F., (1948), *Thermodynamics*, New York, McGraw-Hill Book Company, Inc.

Parlin, R. B., and Eyring, H., (1954), "Membrane Permeability And Electrical Potential," In: *Ion Transport Across Membranes*, New York, Academic Press.

Pitzer, K. S., and Brewer, L., (1961), *Thermodynamics*, Second Edition, New York, McGraw-Hill Book Company, Inc.

Plonsey, R., (1969), *Bioelectric Phenomena*, New York, McGraw-Hill Book Company, Inc.

Pontryagin, L. S., Boltyanskii, V. G., Gamkrelidze, R. V., and Mishchenko, E.F., (1962), *The Mathematical Theory Of Optimal Processes*, (Translated From The Russian By K. N. Trirogoff, Edited By L. W. Neustadt); New York, Wiley-Interscience.

Raibert, M. H., (1977), "Analytical Equations vs. Table Look-Up For Manipulation: A Unifying Concept," *Proc. IEEE Conf. On Decision And Control (Held In New Orleans, La., December, 1977)*, New York, The IEEE, pp. 576-579.

Reiner, J.N., (1968), *The Organism As An Adaptive Control System*, Englewood Cliffs, New Jersey, Prentice-Hall, Inc.

Rice, O. K., (1967), *Statistical Mechanics, Thermodynamics and Kinetics*, San Francisco, Calif., The W. H. Freeman Company.

Roederer, J., (1975), *Introduction To The Physics And Psychophysics Of Music*, Second Edition; New York, Springer-Verlag.

Roederer, J.,(1982), "Physical And Neuropsychological Foundations Of Music," In: *Music, Mind And Brain*, (Edited By M. Clynes), New York, Plenum Press.

Roesel, C.E., (1978), *Immunology: A Self-Instructional Approach*, New York, McGraw-Hill Book Company, Inc.

Rohsenow, W. M., and Choi, H.Y., (1961), *Heat, Mass, And Momentum Transfer*, Englewood Cliffs, New Jersey, Prentice-Hall.

Ruch, T. C., and Patton, H. D., eds., (1965), *Physiology And BioPhysics*, 19th Edition, Philadelphia, Pa., W. B. Saunders Co.

Sage, A.P., and White, C.C.,III, (1977), *Optimum Systems Control*, Second Edition, Englewood Cliffs, New Jersey, Prentice-Hall.

Sandberg, I.W., (1978a), "On The Mathematical Foundations Of Compartmental Analysis In Biology, Medicine, And Ecology," *IEEE Transactions On Circuits And Systems*, Vol. CAS-25, No. 5, pp. 273-279.

Sandberg, I.W., (1978b), "A Note On The Properties Of Compartmental Systems," *IEEE Transactions On Circuits And Systems*, Vol. CAS-25, No. 6, pp. 379-380.

Sayers, B., McA., (1970), "Inferring Significance From Biological Signals," In: *Biomedical Engineering Systems*, (Edited By M. Clynes and J.H. Milsum),New York, McGraw-Hill Book Co., Inc.

Scartelli, J., (1990), "A Rationale For Sub-Cortical Involvement In Musical Response," (Personal Communication), *In Press*.

Schally, A.V., Kastin, A.J., and Arimura, A., (1977), "Hypothalamic Hormones: The Link Between Brain And Body," *American Scientist*, Vol. 65, No. 6, pp. 712-719.

Schepartz, B., (1980), *Dimensional Analysis In The Biomedical Sciences*, Springfield, Illinois, Charles C. Thomas.

Schneck, D. J., and Ostrach, S., (1975), "Pulsatile Blood Flow In A Channel Of Small Exponential Divergence: I. The Linear Approximation For Low Mean Reynolds Number," *Trans. Am. Soc. Mech. Eng., J. Fluids Engrg.*, Vol. 97, Ser. 1, No. 3, pp. 353-360.

Schneck, D. J., and Walburn, F. J., (1976), "Pulsatile Blood Flow In A Channel Of Small Exponential Divergence: II. Steady Streaming Due To The Interaction Of Viscous Effects With Convected Inertia," *Trans. Am. Soc. Mech. Eng., J. Fluids Engrg.*, Vol. 98, Ser. 1, No. 4, pp. 707-714.

Schneck, D. J., (1977), "Pulsatile Blood Flow In A Channel Of Small Exponential Divergence: III. Unsteady Flow Separation" *Trans. Am. Soc. Mech. Eng., J. Fluids Engrg.*, Vol. 99, Ser. 1, No. 2, pp. 333-338.

Schneck, D. J., (1978), "Aerodynamic Forces Exerted On An Articulated Human Body Subjected To Windblast," *Aviation, Space And Environmental Medicine*, Vol. 49, No. 1, Section II, pp. 183-190.

Schneck, D. J., (1980), "Studies Of Limb-Dislodging Forces Acting On An Ejection Seat Occupant," *Aviation, Space And Environmental Medicine*, Vol. 51, No. 3, pp. 256-264.

Schneck, D. J., (1981), "Deductive Physiologic Analysis In The Presence Of 'Will' As An Undefined Variable," *Mathematical Modelling*, Vol. 2, No. 3, pp. 191-199.

Schneck, D.J., (1983a), "Principles Of Mass Transport Across Biological Membranes," Virginia Polytechnic Institute and State University Technical Report No. VPI-E-83-07, Blacksburg, Va.

Schneck, D. J., and Davis, R. B., III, (1983b), "Ischaemia In The Heart Due To Atherosclerotic Mechanisms, Flow Anomalies And Vascular Spasm," In: *Heart Perfusion, Energetics, And Ischemia*, (Edited By L. Dintenfass, D. G. Julian, and G. V. F. Seaman), New York, Plenum Publishing Corp., pp. 223-274.

Schneck, D. J., (1983c), "Viewpoint On Heart Disease And Public Health," *ibid.*, pp. 665-667.

Schneck, D. J., (1983d), "Principles Of Bioenergetics And Metabolism In Physiologic Systems," Virginia Polytechnic Institute And State University Technical Report No. VPI-E-83-42, Blacksburg, Virginia.

Schneck, D.J., (1983e), "A Non-Dimensional Analysis Of Cardiovascular Response To Cold Stress; Part I: Identification Of The Physical Parameters That Govern The Thermoregulatory Function Of The Cardiovascular System," Naval Medical Research Institute Technical Report No. NMRI-83-51 (NTIS AD-A-138-710/9); Bethesda, Maryland.

Schneck, D. J., (1984a), "Some Effects Of Environmental Stress (Specifically, Cold Stress) On Cardiovascular Function," In: *Proceedings Of The First Carolina Biomedical Engineering Conference*, (Edited By R. L. Hellwig), Research Triangle Park, North Carolina, The Biotechnology Center, pp. 87-94.

Schneck, D. J., (1984b), *Mechanics Of Muscle*, Blacksburg, Va., Kinko's Professor's Publishing Group; (Second Edition, *In Press*, New York, New York University Press).

Schneck, D. J., (1984c), "Blood," Virginia Polytechnic Institute And State University Technical Report No. VPI-E-84-33, Blacksburg, Virginia.

Schneck, D.J., (1984d), "Historical Development Of Cardiovascular Physiology, "Virginia Polytechnic Institute And State University Technical Report No. VPI-E-84-37, Blacksburg, Virginia.

Schneck, D. J., (1985), "On The Basis Of Some Common Assumptions In Fluid Mechanics; Part I: Assumptions Concerning The Fluid Itself," Virginia Polytechnic Institute And State University Technical Report No. VPI-E-85-29, Blacksburg, Virginia.

Schneck, D. J., (1987), "Feedback Control And The Concept Of Homeostasis," *Math. Modelling*, Vol. 9, No. 12, pp. 889-900.

Schneck, D.J., (1988a), "On The Development Of A Rheological Constitutive Equation For Whole Blood," Virginia Polytechnic Institute And State University Technical Report No. VPI-E-88-14, Blacksburg, Virginia.

Schneck, D.J., and Starowicz, S.A., (1988b), "Non-Dimensional Analysis In Physiologic Research; A Case In Point: Cardiovascular Thermoregulation," In: *Developments In Theoretical And Applied Mechanics, Volume XIV*, (Edited By S.S.Y. Wang, R.M. Hackett, S.L. DeLeeuw, and A.M. Smith), University, Mississippi, The University of Mississippi Press, pp. 237-248.

Schneck, D. J., (1989), "Engineering Principles Of Physiologic Function," In: *Proceedings Of The 1989 SEM Spring Conference On Experimental Mechanics*, Bethel, Connecticut, SEM Publications, pp. 181-187.

Schneck, D.J., (1989), "The Transport Of Mass, Energy And Momentum In Physiologic Systems," *Math. Comput. and Modelling*, Vol. 12, No. 12, pp. 1473-1485.

Schneck, J. K., and Schneck, D. J., (1989), "A Physiological Basis For Recognizing The Importance Of The Fine Arts In Public Education," *VMEA Notes (Official Journal Of The Virginia Music Educators Association)*; Vol. XLI, No. 3, pp. 16-19.

Sears, F. W., and Zemansky, M. W., (1955), *University Physics*, Reading, Mass., Addison-Wesley Publishing Company, Inc.

Shepherd, J. T., and Vanhoutte, P. M., (1979), *The Human Cardiovascular System: Facts And Concepts*, New York, Raven Press.

Sherwood, T. K., Pigford, R. L., and Wilke, C. R., (1975), *Mass Transfer*, New York, McGraw-Hill Book Company, Inc.

Shilling, C. W., and Story, P., (1981), "Man In The Cold Environment: A Bibliography With Informative Abstracts," Bethesda, Maryland, Undersea Medical Society, Inc., Publication Number 50(CE-2)5-15-81.

Shires, T., Williams, J., and Brown, F.,(1959), "Changing Concept Of Salt Water And Surgery," *Texas State Journal Of Medicine*, Vol. 55, pp. 753-756.

Shires, T., Brown, F., Canizaro, P., and Somerville, N., (1960), "Distributional Changes In Extracellular Fluid During Acute Hemorrhagic Shock," *Surgical Forum*, Vol. 11, pp. 115-117.

Shires, T., Williams, J., and Brown, F., (1961), "Acute Changes In Extracellular Fluids Associated With Major Surgical Procedures," *Ann. Surg.*, Vol. 154, pp. 803-810.

Shires, T., Carrico, J., and Coln, D., (1964), "The Role Of The Extracellular Fluid In Shock," *International Anesthesiology Clinics*, Vol. 2, pp. 435-454.

Silver, F. H., (1987), *Biological Materials: Structure, Mechanical Properties, And Modeling Of Soft Tissues*, New York, The New York University Press.

Smith, C. A., and Corripio, A. B., (1985), *Principles And Practices Of Automatic Process Controls*, New York, John Wiley and Sons, Inc.

Sommerfeld, A., (1956), *Thermodynamics And Statistical Mechanics*, New York, Academic Press, Inc.

Sorensen, H. A., (1961), *Principles Of Thermodynamics*, New York, Holt, Rinehart and Winston.

Spalding, D. B., and Patankar, S.V., (1967), *Heat And Mass Transfer In Boundary Layers*, London, Morgan-Grampian.

Spells, K. E., (1960), "The Thermal Conductivities Of Some Biological Fluids," *Phys. Med. Biol.*, Vol. 5, pg. 5.

Springer, S.P., and Deutsch, G., (1981), *Left Brain, Right Brain*, San Francisco, Calif., W. H. Freeman and Company.

Squire, J., (1981), *The Structural Basis Of Muscular Contraction*, New York, Plenum Press.

Starowicz, S. A., and Schneck, D. J., (1986), "A Non-Dimensional Analysis Of Cardiovascular Response To Cold Stress; Part II: Development Of The Non-Dimensional Parameters That Govern The Thermoregulatory Function Of The Cardiovascular System," Naval Medical Research Institute Technical Report No. NMRI-86-80, (NTIS # PB 88-149/GAR), Bethesda, Maryland.

Stein, W. D., (1967), *The Movement Of Molecules Across Cell Membranes*, New York, Academic Press.

Stein, W. D., and Lieb, W. R., (1986), *Transport And Diffusion Across Cell Membranes*, San Diego, Calif., Academic Press/Harcourt Brace Jovanovich Publishers.

Stevens, C., (1979), *The Neuron, The Brain: A Scientific American Book*, Salt Lake City, Utah, W. H. Freeman and Company.

Swanson, C. P., (1964), *The Cell*, Englewood Cliffs, New Jersey, Prentice-Hall, Inc.

Taylor, A., and Prochazka, A., eds., (1981), *Muscle Receptors And Movement*, London, Macmillan.

Taylor, R. H., (1979), "Planning And Execution Of Straight Line Manipulator Trajectories," *IBM Journal Of Research And Development*, Vol. 23, pp. 424-436.

Teorell, T., (1953), *Progress In Biophysics And Biophysical Chemistry*, Vol. 3, pg. 305.

Thomas, C.L., Editor, (1981), *Taber's Cyclopedic Medical Dictionary*, 14'th Edition, Philadelphia, Pa., F. A. Davis Company.

Trimmer, J. D., (1950), *Response Of Physical Systems*, New York, John Wiley and Sons, Inc.

Tuwiner, S. B., (1962), *Diffusion And Membrane Technology*, American Chemical Society Monograph No. 156; New York, Reinhold.

Van Dijk, J. M. N., (1978), "Simulation Of Human Arm Movements Controlled By Peripheral Feedback," *Biological Cybernetics*, Vol. 29, pp. 175-185.

Victor, S. A., and Shah, V. L., (1975), "Heat Transfer To Blood Flowing In A Tube," *Biorheology*, Vol. 12, pp. 361-368.

Walburn, F. J., and Schneck, D. J., (1976), "A Constitutive Equation For Whole Human Blood," *Biorheology*, Vol. 13, No. 3, pp. 201-210.

Walburn, F. J., and Schneck, D.J., (1978), "An Experimental Technique For Quantifying Unsteady Flow Separation In Diverging Circular Channels," In: *Proceedings Of The First Mid-Atlantic Conference On Bio-Fluid Mechanics*, (Edited By D. J. Schneck), Blacksburg, Virginia, The Virginia Polytechnic Institute And State University Press, pp. 161-170.

Walburn, F. J., and Schneck, D. J., (1980), "An Experimental Investigation Of Pulsatile Laminar Flow Separation In Exponentially Diverging Tubes," In: *Biofluid Mechanics * 2: Proceedings Of The Second Mid-Atlantic Conference On Bio-Fluid Mechanics*, (Edited By D. J. Schneck), New York, Plenum Press, pp. 433-456.

Wall, F. T., (1965), *Chemical Thermodynamics*, Second Edition, San Francisco, California, W. H. Freeman and Company.

Weast, R. C., Editor, (1969), *CRC Handbook Of Chemistry And Physics*, 50'th Edition, Cleveland, Ohio, The Chemical Rubber Company.

Wessling, F. C., and Blackshear, P. L., (1973), "The Thermal Properties Of Human Blood During The Freezing Process," *J. Of Heat Transfer*, Vol. 95, pp. 246-249.

White, A., Handler, P., and Smith, E. L., (1964), *Principles Of Biochemistry*, New York, McGraw-Hill Book Company, Inc.

Wing, J., and Touchstone, R. M., (1963), "A Bibliography Of The Effects Of Temperature On Human Performance," Aerospace Medical Research Laboratories Technical Report No. AMRL-TDR-63-13; Ohio, Wright-Patterson Air Force Base.

Winter, D. A., (1979), *Biomechanics Of Human Movement*, New York, John Wiley and Sons.

Winters, J. M., and Stark, L., (1985), "Simulation Of Fundamental Movements; I: 'Systems' Analysis," In: *Proceedings Of The Seventh Annual IEEE Conference Of The Engineering In Medicine And Biology Society*, New York, The IEEE, pp. 39-50.

Womersley, J. R., (1957), "An Elastic Tube Theory Of Pulse Transmission And Oscillatory Flow In Mammalian Arteries," Wright Air Development Center Technical Report No. WADC-TR-56-614, Ohio, Wright-Patterson Air Force Base.

Yamamoto, W. S., and Brobeck, J. R., eds., (1965), *Physiological Controls And Regulation*, Philadelphia, Penna., The W. B. Saunders Company, Inc.

Yao, B. P., (1976), "Investigations Concerning The Principle Of Minimal Total Muscular Force," *J. Biomech.*, Vol. 9, pp. 413-416.

Yeagle, P., (1987), *The Membranes Of Cells*, San Diego, California, Academic Press/Harcourt Brace Jovanovich, Publishers.

Zachar, J., (1971), *Electrogenesis And Contractility In Skeletal Muscle Cells*, Publishing House Of The Slovak Academy Of Sciences.

Zuniga, E. N., and Simons, D. G., (1969), "Non-Linear Relationship Between Averaged Electromyogram Potential And Muscle Tension In Normal Subjects," *Arch. Phys. Med.*, Volume 50, pp. 613-620.

Index

About The Author

Dr. Daniel J. Schneck is a Professor Of Engineering Science and Mechanics at Virginia Polytechnic Institute and State University, where he has been since 1973, and where he directs the Program In Biomedical Engineering administered by the Department of Engineering Science and Mechanics. Trained in both engineering and medicine, Dr. Schneck's research interests are in the areas of cardiovascular and musculoskeletal mechanics. In addition to the current textbook, he is the author of a book entitled, *Mechanics Of Muscle* (the second edition of which is scheduled to be published in this series); and he has co-authored and/or edited four other texts -- one of which, *Student Manual For A Clinical Internship In Biomedical Engineering,* received in 1984 the "Outstanding Contribution To Engineering Education" award presented by the Instructional Unit of the Southeastern Section of the American Society For Engineering Education.

Dr. Schneck has served as chairman of the Biomedical Engineering Division of ASEE, and is currently serving as president of the Biomedical Engineering Society. His interest in education is further evidenced by his numerous teaching awards -- among them, the prestigious W. E. Wine Award (the highest University-wide recognition of teaching excellence awarded at VPI), and the "Outstanding Educator Of The Year Award," from the Engineering Science and Mechanics Society. Although he is dedicated to his profession and to his students, Dan's wife, Judi, insists that he went into Biomedical Engineering to support his passion for music. Dr. Schneck is also an accomplished violinist!